Melissa Craft
CIO, popular author, and Network+ expert

Network+ Certification

Take charge of the Network+ exam—
faster, smarter, *better*!

PUBLISHED BY
Microsoft Press
A Division of Microsoft Corporation
One Microsoft Way
Redmond, Washington 98052-6399

Library of Congress Cataloging-in-Publication Data
Craft, Melissa.
Faster Smarter Network+ Certification / Melissa Craft
p. cm.
Includes index.
ISBN 0-7356-1932-8
1. Computer networks. 2. Telecommunications engineers--Certification. I. Title.

TK5105.5.C72 2003
004.6--dc21 2003044151

Printed and bound in the United States of America.

1 2 3 4 5 6 7 8 9 QWE 8 7 6 5 4 3

Distributed in Canada by H.B. Fenn and Company Ltd.

A CIP catalogue record for this book is available from the British Library.

Microsoft Press books are available through booksellers and distributors worldwide. For further information about international editions, contact your local Microsoft Corporation office or contact Microsoft Press International directly at fax (425) 936-7329. Visit our Web site at www.microsoft.com/mspress. Send comments to *mspinput@microsoft.com.*

Acquisitions Editor: Hilary Long
Project Editor: Kristen Weatherby
Technical Editor: Christopher Pierce

Body Part No. X09-39103

Table of Contents

Part 4: Network Support

This book is dedicated to my mother, because she taught me how to write. And it's dedicated to you, my readers, for whom it was written. Good luck on passing the Network+ exam!

Acknowledgments

I would like to acknowledge the technical editor, Christopher Pierce, and the copyeditor, Teri Kieffer, for their astute observations, as well as for gently correcting my occasional late-night ramblings. I would like to thank Kristen Weatherby, Hilary Long, and the rest of the folks at Microsoft Press for the opportunity this book represents. I'd also like to thank my husband, Danny Meyer, for putting up with many a late-night writing stint. (Plus, thanks to my puppies, Apollo and Pooka, who kept me company regardless of the hour.)

Introduction

Getting started is an apt description for not only the beginning of this book, but also for your work toward achieving Network+ certification. CompTIA's Network+ is one of the premier certifications that professional network engineers worldwide select to advance their technical expertise and launch their careers in the Information Technology (IT) industry. Networking is one of the hottest topics in IT, and to move to the top of the IT field, you must understand networking at the detail level.

This book is intended to start you on the path to Network+ certification What's interesting about CompTIA's Network+ certification is that it teaches basic networking concepts that are not tied to any particular vendor's products. These concepts apply for any type of server or workstation that communicates across a network, and they're valid in any environment. This is one of the main reasons that people select Network+ certification—the concepts you must know to earn it will be useful for as long as networks exist.

If you run a network that uses Microsoft Windows 2000 or any of the other network operating systems (NOSs) out there, you know that certain specific concepts apply only to that particular NOS. The Network+ certification does not ignore this fact. In Network+, you are required to know the basics of Windows networks, Novell NetWare, Macintosh, UNIX, and Linux, in addition to how to network Windows workstations.

This book is organized to help you study for the Network+ examination. The majority of the material offers detailed information about networking theory and models, building from fundamentals to more complex subjects.

About CompTIA

CompTIA stands for the Computing Technology Industry Association, which was formed in 1982.. The association works with systems integrators and value added resellers. For more than two decades, CompTIA has helped computing professionals achieve certifications that validate their knowledge and also help them launch their careers and earn higher wages.

In providing a vendor-neutral information technology certification series, CompTIA offers many certifications. Just a few are listed below. More are being added as CompTIA grows with the technology industry:

- A+
- Network+
- I-Net+
- CDIA+
- Server+
- E-Biz+
- IT Project+
- Jobs+
- Certified Technical Trainer+
- Linux+
- Security+
- Home Technology Integrator+

This extensive array of certifications enables a computing professional to pursue several different skill sets of information technology. Holding such certifications is undeniable proof that you possess the qualifications, the knowledge, and the skills that are necessary to be successful in the industry.

CompTIA is an association dedicated to advancing the growth of the IT industry. With its history, global reach, and extensive membership (over 13,000 strong), CompTIA has an impact on all areas of the IT industry. It looks closely at industry standards; develops expertise among professionals in the industry; provides business solutions; and offers training, certifications, workforce development, public policy, and technology innovations.

One of the advantages of CompTIA's interest in public policy is its strong influence. When the public becomes more concerned with privacy, antitrust, security, and other technology issues, CompTIA's recommendations carry weight.

CompTIA's certifications are highly regarded throughout the industry, as evidenced by Microsoft's adoption of the A+/Network+ combination or A+/Server+ combination of certifications as acceptable replacements for an elective for the MCSA certification. These certifications encompass the theory of networking protocols through the skills of building and maintaining a server-based computing network and more.

Who Should Read This Book

Two types of people should read this book: those who seek Network+ certification and those who are maintaining a network and require networking reference materials.

This book was written with the network technician, IT manager, and systems administrator in mind, in addition to people who are breaking into the IT industry. The goal of the book is, of course, to help the reader learn the networking fundamentals as laid out in CompTIA's Network+ objectives. You aren't expected to have a complete understanding of networking before you begin using this book. However, you should understand that CompTIA established the Network+ certification to build on skill sets required for its A+ and Server+ certifications. As for understanding networking itself, you won't be expected to have this knowledge—until you finish the book.

About Network+ Certification

The Network+ certification measures technical knowledge. It is geared toward networking technicians who have had about a year's worth of experience in network administration or other type of network technology support.

One of the advantages of the Network+ certification is that it can get you started on the path toward your Microsoft Certified Systems Administrator (MCSA) certification. Microsoft currently accepts a combination of either the A+ and Server+ exams or the A+ and Network+ exams toward the MCSA elective. To achieve the MCSA, you must first pass at least one Microsoft Certified Professional (MCP) exam and obtain your MCP identification number. Then you should begin taking the A+ and Network+ (or Server+) certification exams and register with your MCP number so that CompTIA will forward your certification information to Microsoft. At that point, you can complete the remaining exams for the MCSA certification, which you can see more about at *http://www.microsoft.com/traincert/mcp/mcsa/Default.asp.*

There are more benefits to achieving certifications than simply adding letters and, in this case, a plus sign, to your name as it appears on your business card. The certification stands for the body of knowledge you have studied, an achievement that few others can claim, and a particular level of professional competence. Employers and recruiters seek out individuals with CompTIA certifications. With the proliferation of outsourcing in the IT industry, you should know that outsource clients show quite a bit more confidence in certified outsource staff members, and often require that all staff be certified. It's obvious that this certification provides an incredible advantage.

Exam Overview

In creating the Network+ certification testing program, CompTIA offers the IT industry a vendor-neutral method of ascertaining the networking knowledge of engineers, administrators, technicians, and managers in IT positions. Because the test is vendor-independent, it is useful for people working in any environment.

The Network+ certification objectives are divided into four testing domains. These domains are weighted to match the job tasks that networking technicians and engineers spend time on during their average workdays.

Twenty percent of the test focuses on media and topologies, 25 percent of it on protocols and standards, 23 percent on network implementation, and 32 percent (almost one-third) requires network support knowledge.

Knowing the material for the exam is probably the most critical part. However, when you're familiar with the way the test works, you'll find this vital to achieving a passing score. The questions in the chapters that follow will drill you on your knowledge of the technologies, mirroring the types of information you'll face on the real exam.

Successfully passing the Network+ exam will prove that you know the following:

- The OSI reference model and its protocol layers
- Network components and their features and functions
- Basic networking hardware, along with installation and configuration
- Troubleshooting skills

The 2002 version of the Network+ exam includes wireless and gigabit Ethernet technologies, which build on the basics of traditional networking concepts. There is also an increased emphasis on understanding Linux and UNIX, as well as Windows workstations and Windows servers.

You can expect the exam to present you with troubleshooting scenarios. These are intended to test you on your hands-on experience and reasoning capabilities.

Network+ Exam Objectives

The following is a list of the objectives for the CompTIA Network+ exam. To view the objectives online, you can go to *http://www.comptia.org/certification/Network/objectives.asp.*

Domain 1.0: Media and Topologies

1.1 Recognize the following logical or physical network topologies given a schematic diagram or description:

- Star/hierarchical
- Bus
- Mesh
- Ring
- Wireless

1.2 Specify the main features of 802.2 (LLC), 802.3 (Ethernet), 802.5 (token ring), 802.11b (wireless), and FDDI networking technologies, including:

- Speed
- Access
- Method
- Topology
- Media

1.3 Specify the characteristics (i.e., speed, length, topology, cable type, etc.) of the following:

- 802.3 (Ethernet) standards
- 10Base-T
- 100Base-TX
- 10Base2
- 10Base5
- 100Base-FX
- Gigabit Ethernet

1.4 Recognize the following media connectors and/or describe their uses:

- RJ-11
- RJ-45
- AUI
- BNC
- ST
- SC

1.5 Choose the appropriate media type and connectors to add a client to an existing network.
1.6 Identify the purpose, features, and functions of the following network components:

- Hubs
- Switches
- Bridges
- Routers
- Gateways
- CSU/DSU
- Network Interface Cards/ISDN adapters/system area network cards
- Wireless access points
- Modems

Domain 2.0: Protocols and Standards

2.1 Give an example/identify a MAC address.
2.2 Identify the seven layers of the OSI model and their functions.
2.3 Differentiate between the following network protocols in terms of routing, addressing schemes, interoperability, and naming conventions:

- TCP/IP
- IPX/SPX
- NetBEUI
- AppleTalk

2.4 Identify the OSI layers at which the following network components operate:

- Hubs
- Switches
- Bridges
- Routers
- Network interface cards

2.5 Define the purpose, function, and/or use of the following protocols within TCP/IP:

- IP

- TCP
- UDP
- FTP
- TFTP
- SMTP
- HTTP
- HTTPS
- POP3/IMAP4
- TELNET
- ICMP
- ARP
- NTP

2.6 Define the function of TCP/UDP ports. Identify well-known ports.
2.7 Identify the purpose of the following network services (i.e., DHCP/bootp, DNS, NAT/ICS, WINS, and SNMP).
2.8 Identify IP addresses (Ipv4, Ipv6) and their default subnet masks.
2.9 Identify the purpose of subnetting and default gateways.
2.10 Identify the differences between public versus private networks.
2.11 Identify the basic characteristics (i.e., speed, capacity, media) of the following WAN technologies:

- Packet switching vs. circuit switching
- ISDN
- FDDI
- ATM
- Frame relay
- Sonet/SDH
- T1/E1
- T3/E3
- OCx

2.12 Define the function of the following remote access protocols and services:

- RAS

- PPP
- PPTP
- ICA

2.13 Identify the following security protocols and describe their purpose and function:

- IPSec
- L2TP
- SSL
- Kerberos

Domain 3.0: Network Implementation

3.1 Identify the basic capabilities (i.e., client support, interoperability, authentication, file and print services, application support, and security) of the following server operating systems:

- UNIX/Linux
- Netware
- Windows
- Macintosh

3.2 Identify the basic capabilities of client workstations (i.e., client connectivity, local security mechanisms, and authentication).

3.3 Identify the main characteristics of VLANs.

3.4 Identify the main characteristics of network attached storage.

3.5 Identify the purpose and characteristics of fault tolerance.

3.6 Identify the purpose and characteristics of disaster recovery.

3.7 Given a remote connectivity scenario (i.e., IP, IPX, dial-up, PPPoE, authentication, physical connectivity, etc.), configure the connection.

3.8 Identify the purpose, benefits, and characteristics of using a firewall.

3.9 Identify the purpose, benefits, and characteristics of using a proxy.

3.10 Given a scenario, predict the impact of a particular security implementation on network functionality (i.e., blocking port numbers, encryption, etc.).

3.11 Given a network configuration, select the appropriate NIC and network configuration settings (DHCP, DNS, WINS, protocols, NetBIOS/host name, etc.).

Domain 4.0: Network Support

4.1 Given a troubleshooting scenario, select the appropriate TCP/IP utility from among the following:

- Tracert
- Ping
- Arp
- Netstat
- Nbtstat
- Ipconfig/Ifconfig
- Winipcfg
- Nslookup

4.2 Given a troubleshooting scenario involving a small office/home office network failure (i.e., xDSL, cable, home satellite, wireless, POTS), identify the cause of the failure.

4.3 Given a troubleshooting scenario involving a remote connectivity problem (i.e., authentication failure, protocol configuration, physical connectivity), identify the cause of the problem.

4.4 Given specific parameters, configure a client to connect to the following servers:

- UNIX/Linux
- Netware
- Windows
- Macintosh

4.5 Given a wiring task, select the appropriate tool (i.e., wire crimper, media tester/certifier, punch down tool, tone generator, optical tester, etc.).

4.6 Given a network scenario, interpret visual indicators (i.e., link lights, collision lights, etc.) to determine the nature of the problem.

4.7 Given output from a diagnostic utility (i.e., Tracert, Ping, Ipconfig, etc.), identify the utility and interpret the output.

4.8 Given a scenario, predict the impact of modifying, adding, or removing network services (i.e., DHCP, DNS, WINS, etc.) on network resources and users.

4.9 Given a network problem scenario, select an appropriate course of action based on a general troubleshooting strategy. This strategy includes the following steps:

1. Establish the symptoms.
2. Identify the affected area.
3. Establish what has changed.
4. Select the most probable cause.
5. Implement a solution.
6. Test the result.
7. Recognize the potential effects of the solution.
8. Document the solution.

4.10 Given a troubleshooting scenario involving a network with a particular physical topology (i.e., bus, star/hierarchical, mesh, ring, and wireless) and including a network diagram, identify the network area affected and the cause of the problem.

4.11 Given a network troubleshooting scenario involving a client connectivity problem (i.e., incorrect protocol/client software/authentication configuration, or insufficient rights/permission), identify the cause of the problem.

4.12 Given a network troubleshooting scenario involving a wiring/infrastructure problem, identify the cause of the problem (i.e., bad media, interference, network hardware).

Taking the Network+ Exam

Preparing for the actual exam day is stressful. You've likely burned the midnight oil studying for the test, and you've probably worked through numerous installations and configurations in a practice lab. Despite how busy you are, it is highly recommended that you spend at least 15 minutes per day reviewing material or practicing in a lab to keep the subject matter fresh in your mind.

To sign up for the exam, you can go through Thomson Prometric by calling (888) 895-6116 or accessing the Web site at *http://www.2test.com*. You'll be required to select an exam location and time. The company has added directions to its Web site so that testing centers are easy to find. The Thomson Prometric associates will be able to give you directions as well. The company accepts both credit cards and prepaid vouchers in addition to checks, as long as your check is received prior to your test date.

Tip Before taking the test, you can read the chapter summaries and the Quick Reference section at the back of the book for a fast refresher.

On your test day, you should arrive at the testing center at least 15 minutes prior to your scheduled appointment. Bring two forms of identification, at least one of them with a picture and both with signatures. As you might expect, you will not be allowed to bring in any notes, books, or other reference materials. You are also prohibited from bringing pagers, cell phones, laptops, PDAs, or other electronic equipment into the testing room. Testing centers offer the ability to store your personal belongings.

Keep in mind that some people are natural test takers, while others have more difficulty even though they know exactly the same material. Understanding how the exam works in advance might be of some help. The exam is in multiple-choice format and is given on a computer. The testing center will provide you with pencils and scratch paper. You are given the chance to mark questions that you are unsure about so that you can go back and look at them later on. For each question answered correctly, you will receive points. You should answer every question, even if you're not certain about the answer. If a question is not answered, you will receive no points for it. If you guess, you have a 25 percent chance (or better, if you have made an educated guess to eliminate any obviously wrong choices) of receiving points for the answer. As you can see, it's better to guess than to leave a question blank.

Each exam has a time limit, which is usually more than adequate for reading and responding to each question as well as going over your answers at the end. The better you know the material, the faster you will likely finish. One way you can easily avoid wasting time going back to search for questions you later believe you answered wrong is to always answer the question, even if you aren't certain about your answer. This way, if you end up taking too much time, at least you will have answered all the questions by the end of the exam. If you're unsure about a question, check the box labeled "Mark" so that you'll be able to return to it easily.

You'll encounter some very long scenario questions on the test. These are time consuming, and it's tempting to skip them with the intent to return to them later. If you do decide to skip and return, make sure you guess an answer and mark the question rather than leave it blank.

When you come across a question on the test and you're not sure about the answer, use the elimination technique. Starting at the bottom, look at each possible choice and eliminate it or keep it as a potential answer. If you can narrow your choices down to two possibilities and then take a guess, you have a 50 percent chance of being correct.

After you've finished answering every question, do a quick "click-through" to look for questions that ask you to "Choose two" or "Choose all that apply." Then make sure that you've selected all the possibilities. During this step, you'll probably find at least one question for which you missed one of the multiple answers.

Test Smart Use elimination techniques when you're unsure. Don't dwell on any single question. Mark questions that you want to return to. Don't leave questions unanswered. Run through all the questions as a final check before you finish the exam. And don't worry if you feel nervous; everyone feels the same way.

When you finally submit your answers, a score is generated. This is fairly nerve-wracking because the computer will tell you to wait while the score is being printed before revealing whether you passed the test. The testing center will provide you with a hard copy of your score, which is sealed to ensure its validity. Then the testing center will provide your scores to CompTIA, and if they're passing, the Network+ certificate will be on its way to your mailbox. If you didn't pass the first time, you are allowed to register to take the test again.

How to Use This Book

This book is intended to be both a study manual and a reference. Like the certification objectives provided by CompTIA, the book is divided into four sections:

Part I: Media and Topologies
Part II: Protocols and Standards
Part III: Network Implementation
Part IV: Network Support

Even though the sections are divided in accordance with Network+ objectives, some concepts recur throughout the book. The reason for this is simple: each chapter builds on the fundamentals from earlier chapters, and reminders of those fundamentals ensure that you fully learn the more complex concepts.

Once you've finished with the Network+ exam, this book can help you with your networking job tasks. Because CompTIA places emphasis on network implementation and troubleshooting support, the final two sections of the book should prove to be invaluable long after the test is over.

Your Study Plan

Studying for any exam can be grueling. Given that Network+ certification is geared toward networking professionals with some experience under their belts, it's quite likely that you'll be studying for the exam while holding down a full-time job. To make it easier, this book has been organized to both match the Network+ objectives as well as build up from the basics to more intricate concepts.

At the end of each chapter, you will find a chapter summary that briefly covers the chapter's key points. In addition, each chapter contains exam tips throughout and a set of several questions that will help you study.

To use this book in the most efficient manner, go through it three times, but in three different ways:

1. The first time, lightly skim through each chapter, starting from the beginning, and make certain to read the headings and the exam notes, look at the diagrams, and read anything that catches your eye. This will familiarize you with the material and prepare you for the heavy studying you'll do later.
2. The second time, read each chapter in detail. Try to spend at least 15 minutes every day reading, or more if you have time. Even in small amounts, reading daily will keep the concepts fresh in your mind.
3. The third time, answer the questions at the end of each chapter. For any question that you have trouble answering, go back through the chapter and look up the answer.

In addition to studying, develop a way to test various theories and methods in the "real world." This will be the most effective way to practice installation, configuration, and troubleshooting, all of which appear in later chapters. Ideally, you should have a lab available to practice in. Many students gather old equipment and set up practice labs in their homes. Others are able to use labs that have been provided by computing equipment vendors, such as Cisco, Microsoft, or Novell. Vendors change their methods all the time and don't necessarily have labs available in every location, but it doesn't hurt to ask.

A day or two before your exam, read through the Check Yourself (Before You Test Yourself) section at the end of this book. This process will help you understand all the concepts tested for on the exam as well as concentrate your efforts on any weak areas. By the time you take the exam, you'll know what it will take to pass it.

Support

Every effort has been made to ensure the accuracy of this book. Microsoft Press provides corrections for books at the following address:

http://mspress.microsoft.com/support/

If you have comments, questions, or ideas regarding this book, please send them to Microsoft Press via e-mail at:

mspinput@microsoft.com

or via postal mail at:

Microsoft Press
Attn: *Faster Smarter Network+ Certification Editor*
One Microsoft Way
Redmond, WA 98052-6399

Please note that product support is not offered through the above addresses.

The logo of the CompTIA Authorized Curriculum Program and the status of this or other training material as "Authorized" under the CompTIA Authorized Curriculum Program signifies that, in CompTIA's opinion, such training material covers the content of the CompTIA's related certification exam. CompTIA has not reviewed or approved the accuracy of the contents of this training material and specifically disclaims any warranties of merchantability or fitness for a particular purpose. CompTIA makes no guarantee concerning the success of persons using any such "Authorized" or other training material in order to prepare for any CompTIA certification exam.

The contents of this training material were created for the CompTIA *Network+ Certification* exam covering CompTIA certification exam objectives that were current as of *April, 2003*.

How to Become CompTIA Certified:

This training material can help you prepare for and pass a related CompTIA certification exam or exams. In order to achieve CompTIA certification, you must register for and pass a CompTIA certification exam or exams.

In order to become CompTIA certified, you must:

1. Select a certification exam provider. For more information please visit *http://www.comptia.org/certification/general_information/test_locations.asp*
2. Register for and schedule a time to take the CompTIA certification exam(s) at a convenient location.
3. Read and sign the Candidate Agreement, which will be presented at the time of the exam(s). The text of the Candidate Agreement can be found at *http://www.comptia.org/certification/general_information/candidate_agreement.asp*
4. Take and pass the CompTIA certification exam(s).

For more information about CompTIA's certifications, such as their industry acceptance, benefits, or program news, please visit *http://www.comptia.org/certification/default.asp*

CompTIA is a non-profit information technology (IT) trade association. CompTIA's certifications are designed by subject matter experts from across the IT industry. Each CompTIA certification is vendor-neutral, covers multiple technologies, and requires demonstration of skills and knowledge widely sought after by the IT industry.

To contact CompTIA with any questions or comments:

Please call + 1 630 268 1818

questions@comptia.org

Part 1

Building from Media and Topologies

CompTIA has organized Network+ certification objectives into four test domains, each of which focuses on a grouping of job tasks that networking technicians and engineers must understand to perform their jobs. Domain 1.0 is "Media and Topologies." If you are wondering what media and topologies refer to, it all starts in the wiring closet. Media is the cabling, and topologies are the shapes that cabling and data transmissions take. Twenty percent (one-fifth) of the exam will contain questions that test your understanding of the following concepts of media and topologies. You can find the Domain 1.0 objectives on CompTIA's Web site at *http://www.comptia.org/certification/network/network_objectives-domain1.asp*.

Chapter 1

Understanding Networking Media

Today's businesses usually have their networks already cabled and running. It would be rare for you to be required to install an entire cabling plant, unless you're a cabling installation vendor. But don't ignore network media as an important job task. Keep in mind that the IT industry is on the verge of changing over to new technologies, such as Gigabit Ethernet, which requires drastic changes to the existing wiring plant. You can also bet that there will be new ways of transmitting data in the future—innovations based on the technologies in use today. You might be involved in the installation, design, troubleshooting, or even initial evaluation and decision-making when your company decides to look at new networking media.

Early in my own computing career, I worked for a company that was just starting to grow. Our network changed from a mainframe environment using old-fashioned vampire clamps and shielded wiring in a ring to a NetWare network running Ethernet in a star topology. Vampire clamps were required to gain access through the thick plastic shielding of the heavy cable. They looked like staple removers, with their sharp teeth ready to bite into the wire. Many days I wore grubbies to work so that I could crawl through the space above the ceiling tiles and drop wires to new cubicles to expand the ring or to create the new star.

To do this, I had to know the maximum lengths of the cables, the correct interfaces, and, well, just about everything that this chapter is about.

Topologies

The first thing to consider about a network is its physical shape, or the design layout, which will be extremely important when you select a wiring scheme and design the wiring for a new installation. Understand that the network really has two shapes, or two types of topology; one is physical and the other is logical. The physical topology is the shape you can see, and the logical topology is the shape that the data travels in. We'll go through this difference in "Comparing Logical Topologies," later in this chapter.

When you design a new network cabling scheme, you need to consider the physical location of the network media and the components and wiring closets, not only for space considerations but also to avoid electromagnetic interference (EMI) by accidentally placing your media too close to intruding electrical appliances such as fluorescent lights. In addition, you should take into account the network's growth potential. (This is important if you want to avoid crawling around in ceilings.) Most companies require you to consider how well new networking technology will fit into your budget, how secure it can be, and sometimes whether it will provide redundancy for greater uptime. Finally, you should keep in mind how easy it is to achieve connectivity, just to help prepare for the project ahead of you.

In general, the physical topologies in use today are the bus, the star, the ring, the mesh, and wireless. Each has its own advantages and disadvantages, which I will describe later in the chapter. Larger networks often comprise hybrids of these topologies, and the topology that makes up the backbone of a wide area network (WAN) is usually of an entirely different type than the one used within each of the local area networks (LANs). This creates a hybrid that is often referred to as a tree. You will soon see that these network shapes have much to do with how the network functions.

Note The word *topology*, used in geometry, refers to the mathematics branch that examines the characteristics of shapes.

Physical Topologies

Physical topologies are all variations of two basic connectivity methods:

- Point-to-point connections
- Multipoint connections

Only two devices are involved in a point-to-point connection, with one wire (or air, in the case of wireless) sitting between them. The two devices could be two computers linked by their modems through the plain old telephone system (POTS), or they could be a computer linked to a hub using Cat5 unshielded twisted pair (UTP) cable. They could also be a server linked to an Internet router using Ethernet over fiber optic (10BaseF).

A point-to-point link is typified by two devices monopolizing the media—similar to two teenagers talking on the telephone with one another, not allowing anyone else to use the phone on either side. Because the media is not shared, there is no need to identify each device, and an address is not needed to communicate just between those devices. For example, hooking up a printer directly to your PC doesn't require an address, but hooking that same printer up to the network does. As you will see later, each device in a topology made of multiple point-to-point connections, or one that passes data through connections to other networks, needs an address.

In a multipoint connection, multiple machines share the cabling. Multipoint connections might be a group of computers strung together in a long line on an old-fashioned ThinNet (10Base2) cable, or it could be a party line of telephones, all sharing a common phone connection. In fact, even your local cable TV provider uses a multipoint system to get every person in the neighborhood hooked up. Many WAN providers link up systems using multipoint connections because it increases both bandwidth and availability of the WAN. You'll often see these types of WANs represented by a cloud in a network diagram. Just remember that a cloud means that there are mysterious groups of redundant links and paths that will get data from one side of the cloud to the other.

Note You might think that a wireless connection is multipoint because the machines all seem to share the air. But, in fact, wireless uses point-to-point connections because multiple signals do not gather up together as they would in a shared multipoint connection. Instead, each machine links up to a wireless access point (WAP) separately.

Three or more devices linked on the same wire generally constitute a multipoint connection. A multipoint link is, at its most basic, a party line or conference call; several subscribers connect to the same line and can use it simultaneously. Private conversations can be attempted but are never guaranteed.

In every multipoint connection, each device must be able to identify itself. This is where addressing at the hardware level starts. The device's address must be unique on the channel that it shares with those other devices, or else confusion reigns. Just ask any network administrator who has accidentally assigned the same logical address to two computers. It's not fun dealing with any type of addressing conflict.

Table 1-1 compares all the physical topologies, which will be discussed in the following sections.

Table 1-1 Physical Topology Comparison

Topology	Shape	Benefits	Disadvantages
Star	Central hub with multiple point-to-point connections to each separate device.	Easy to troubleshoot and configure. A single device or cable failure will not bring down the network.	Expensive because of additional cabling and central hub.
Bus	Single cable connecting several devices in multipoint connections, terminated at each end.	Uses very little cabling.	Difficult to troubleshoot. A single failure will bring the network down.
Mesh	All devices are connected to all other devices in the network.	Extremely reliable. Data has access to fastest paths and can load balance.	Uses the most cabling to implement. Has a high administrative overhead.
Ring	Closed loop of cabling connecting each device in a ring.	All devices have equal access.	Costly to implement.
Wireless	Areas that are capable of transmitting data are arranged in cells.	Devices can move throughout the network and remain connected. Extremely easy to implement.	Somewhat costly because of the newness of the technology. Troubleshooting is difficult because there is no wire.

Physical Star Topology

A star configuration is simple: Each of several devices has its own cable that connects to a central hub, or sometimes a switch, multipoint repeater, or even a Multistation Access Unit (MAU). Data passes through the hub to reach other devices on the network. Ethernet over unshielded twisted pair (UTP), whether it is 10BaseT, 100BaseT, or Gigabit, all use a star topology.

A wiring closet doesn't really look like a star. It looks more like a squid, with tentacles reaching up to the ceiling in groups and then branching out to the network's workstations. In fact, it is a star topology. Each individual wire radiates from the central hub to connect to a single device.

You will find that a star topology is most common in networks. This is mainly because of the ease of configuring and troubleshooting it. If a wire or a single port on the hub or switch goes bad, only one network node goes down, which prevents a huge impact on productivity overall (unless the entire hub or switch fails—in which case, the whole LAN goes down). However, because a star topology involves a central hub or switch as well as a lot more cabling, it costs more to implement. Figure 1-1 shows the star topology.

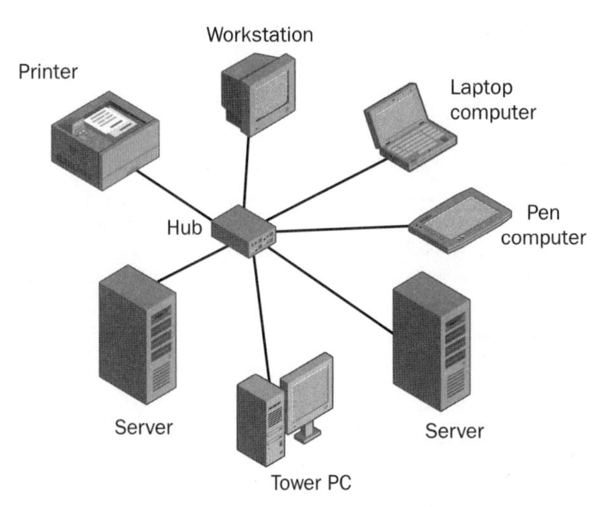

Figure 1-1 The star topology is physically a central hub with media extending to each network device.

The benefits of a star topology are:

- The ability to isolate individual devices in troubleshooting
- An intelligent central hub or switch that can help diagnose and manage the network
- Adjusting traffic levels so that computers that place heavy loads on the network are moved to separate hubs

Star topologies also have a few disadvantages:

- Lots of cabling is required.
- If a hub fails, the entire section of the network fails with it.

Physical Bus Topology

In the bus topology, every network device hooks directly into a central cable; there is no hub and no need for one. That central cable is a shared communications medium for every device that taps into it. When a device communicates across that cable, the message is seen by all the other devices, but only the true destination device will use it. The central cable is somewhat like a railroad track; any rail car can use it, and it can connect with any other rail car as needed.

Caution Don't confuse a bus topology with the system bus of the computer.

A bus topology works well as the backbone of a network because two bus topologies can be used redundantly, thus preventing the failure of the network's underpinning. Before star topologies became popular, the bus topology was the most common network type available. Most bus topologies were Ethernet ThinNet or ThickNet, discussed later in the chapter, consisting of a single cable with terminating resistors at each end. The bus topology is shown in Figure 1-2.

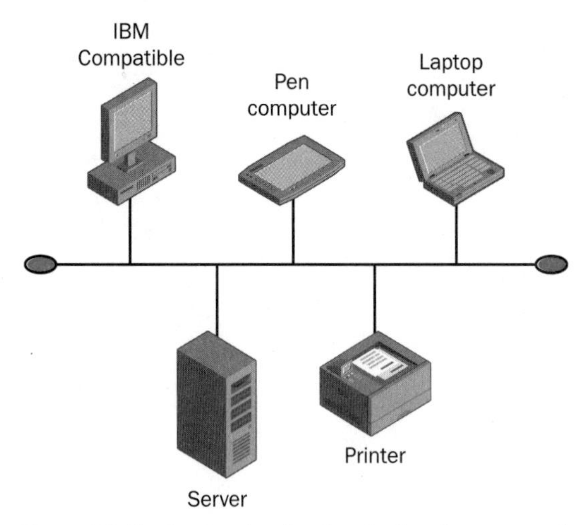

Figure 1-2 The bus topology is a length of media terminated at each end, with devices connecting into the media along its length.

Remember that the main characteristic of a bus topology is in the broadcast of data signals throughout the entire bus. In Figure 1-2 above, if Device A sends a signal to Device B, the signal is also propagated all the way to Device C and Device D, even though they will ignore the message.

There is a single advantage to a bus topology—it's cheap! You don't need a lot of cable because you string up all the computers in a straight line. Plus, you don't need a central hub.

There are three main disadvantages to a bus topology:

- You are limited with the number of devices that you can have on a single segment.
- With everything connected together in serial fashion, it's very difficult to troubleshoot. There's never an obvious point of failure.
- Any break in the cable—whether someone who's doing repairs accidentally cuts through the wire, you nail a picture on the wall and the nail goes through the wire, the wire ages and disintegrates, or (my personal favorite) a small animal decides to chew through the wire—causes the entire network segment to go down. You might think that the cable would become two separate networks, but this is not so. When the cable breaks, each of the resulting sections loses one end of termination, which is required for the bus to function.

Physical Mesh Topology

A mesh topology is unique, and a true mesh topology is practically nonexistent. Most of the networks that claim to be mesh topologies are actually partial meshes. In a true mesh, each network device must have a separate point-to-point connection to every other device in the mesh. For a three-node mesh, there are three connections; for a four-node mesh, six connections; in a five-node mesh, 10 connections; and so on. Figure 1-3 shows how cumbersome the connections can become in a mesh topology.

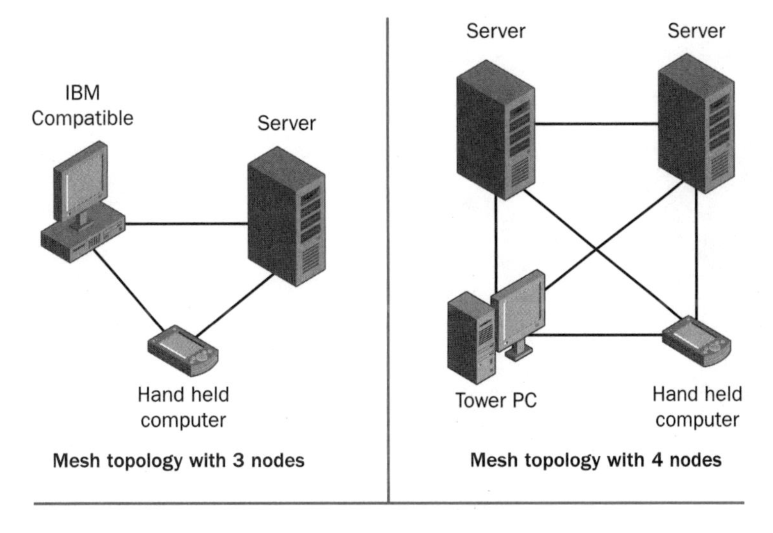

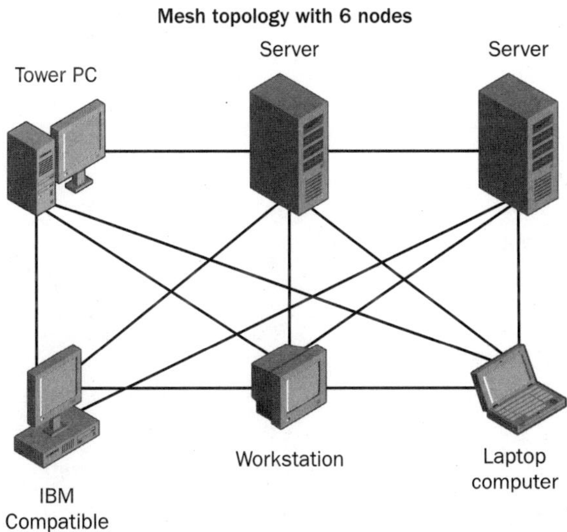

Figure 1-3 Mesh connections get cumbersome as the network size increases.

Mesh topologies are extremely useful when you want to be absolutely certain that data from one network node will reach any other network node in the mesh. If a single point-to-point connection fails, the data can travel through a different node and reach its final destination. Take, for example, the least complicated mesh topology—that of three nodes, A, B, and C, as depicted in Figure 1-4. If the connection between A and C fails, the data that A sends to C merely needs to travel through station B to arrive at the correct destination. This form of redundancy is often used in WANs, as well as corporate network backbones.

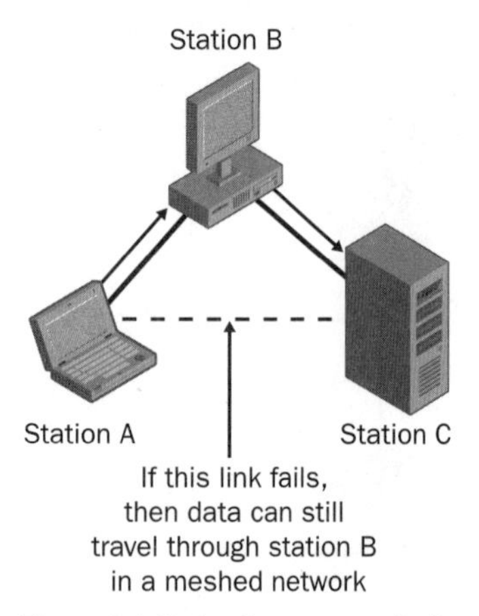
Station B
Station A
Station C
If this link fails, then data can still travel through station B in a meshed network

Figure 1-4 Redundant routes exist in a mesh network.

In a partial mesh topology, each node has at least two connections to other network devices. This means that the number of connections does not become cumbersome, but redundant routes are maintained.

The main advantage to a mesh topology is that the network has so many redundant routes that it doesn't ever shut down. A mesh network provides *availability* and *reliability*. (Think "Internet," which uses tons of redundant partial meshed routes at its backbone. The World Wide Web was named "Web" for the very fact that it holds a huge number of paths on which data can travel. It really does resemble a spider web.) In addition, this type of topology allows for (but doesn't guarantee) the ability to load balance data across multiple links. Load balancing in this manner greatly increases the network's throughput.

Management of a mesh topology, however, is not easy. An administrator would need to keep track of each network link, redundant route, network node that shares links, and so forth. The administrative overhead of a mesh topology makes it impractical for most networks.

Physical Ring Topology

The ring topology looks just like you might imagine. Each computer is connected to the next, and the final device is connected back to the first so that the entire system is a closed loop. Any single device on the ring is connected directly to two neighboring devices at all times. There are both single ring topologies and dual ring topologies. (The single ring topology is shown in Figure 1-5.)

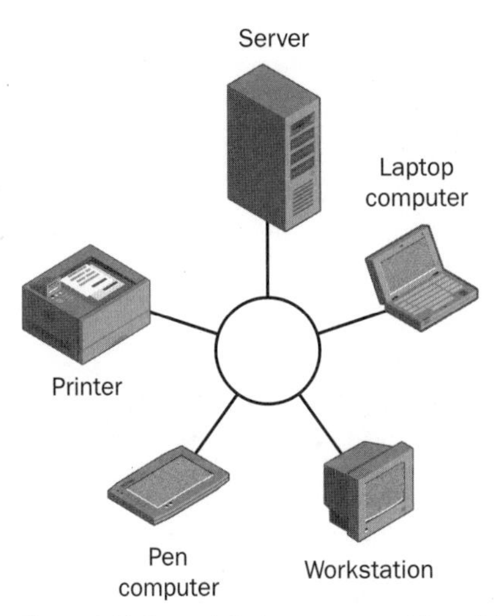

Figure 1-5 Several devices connected in serial fashion create a single ring topology, with the "last" device connected back to the "first" device in order to create a ring.

Data travels around the ring in a single direction—either clockwise or counterclockwise. In a dual ring system, data can travel in both directions to speed up throughput.

Unlike bus topologies, ring topologies require no cabling termination because the cable never really starts or ends. In most ring-based technologies, when a fault in the loop occurs, the data keeps reversing direction until it reaches the fault. When the fault is fixed, the data travels in a single direction again.

Ring topologies are fairly reliable. Since a signal is repeated at each network node, signals rarely degrade. They are also easier to troubleshoot than a bus topology (but not as easy as a star topology). One of the main disadvantages of a ring topology is that it is fairly expensive to implement because each network interface must perform the function of a signal repeater. Depending on the actual implementation of a ring topology, a break in the ring can either disable or slow down the network.

Wireless Topology

Wireless topologies seem odd at first because there are no physical wires to guide you to the actual topology shapes that they use. In fact, wireless topologies are implemented in a star, a mesh, or a cellular configuration. In the star configuration, the wireless topology is called a Basic Service Set (BSS). It consists of a wireless access point connected to a wired network, and it enables each wireless device to connect to the access point and through it to all other devices. The BSS is shown in Figure 1-6.

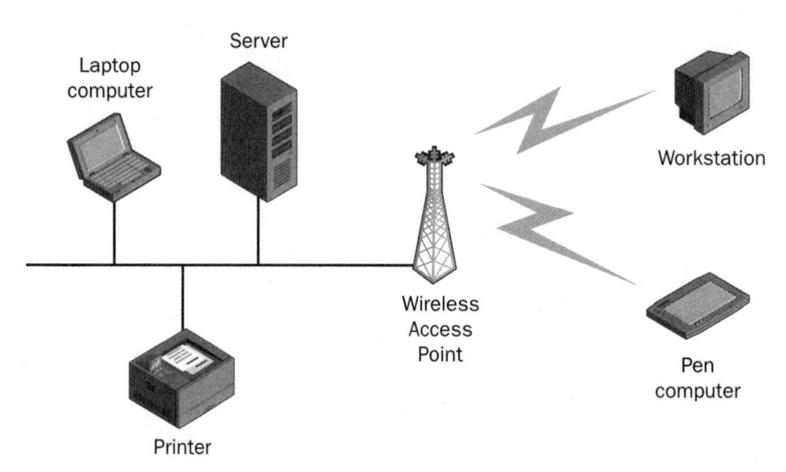

Figure 1-6 The BSS consists of wireless devices connecting to a wireless access point.

In the case of the mesh configuration, the wireless network, the Independent Basic Service Set (IBSS), enables each wireless device to connect to any other wireless device within range. The IBSS is depicted in Figure 1-7.

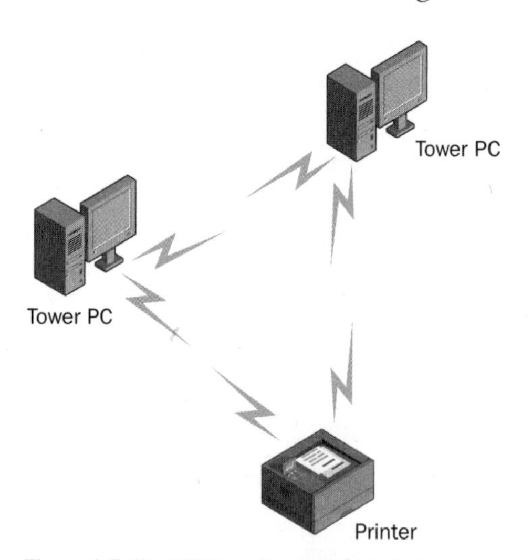

Figure 1-7 The IBSS is built as wireless devices connect to one another on an ad-hoc basis.

In the cellular topology, the wireless network, referred to as an Extended Service Set (ESS), consists of a series of overlapping wireless cells, each with its own WAP. Devices can actually move among cells and continue working seamlessly, regardless of which cell they happen to be in. It's easiest to think of this as a radio station. Imagine you're driving down a long road and you have your radio tuned to 95.5 FM. As you go along, you eventually fade out of 95.5 FM for one area, but you fade into 95.5 FM for the next area. If these two stations were playing the exact same program, you wouldn't even know that you had changed from one to another. The ESS is shown in Figure 1-8.

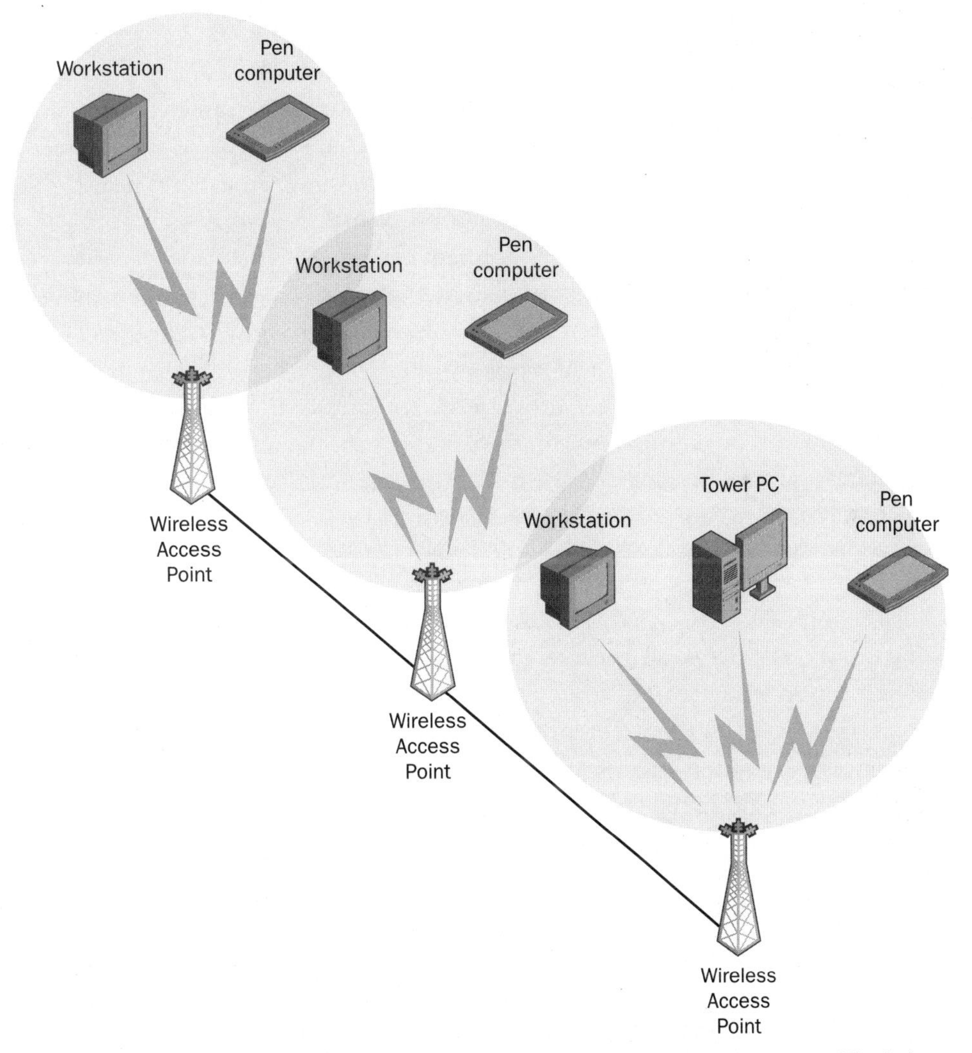

Figure 1-8 The ESS cascades wireless access points, enabling seamless access to data as a mobile wireless device moves along the network.

Wireless topologies are easy to install. Many home networks use them in a peer-to-peer formation. Peer-to-peer refers to a network where each computer on the network can access resources on every other computer. There are two issues to contend with when considering a wireless network. The first issue is cost. New technology simply costs more than old technology. Time will eventually resolve this issue. The second issue is troubleshooting network outages. Without a wire to trace back or reveal damage, the network administrator must think in terms of interference that is more or less invisible.

Comparing Logical Topologies

Movies show characters that seem to move around naturally. This is an illusion. Instead of one person moving around, the movie is actually a series of still frames that show a person in various poses. Because these frames flash by the screen so quickly, you automatically interpret the character as moving naturally. The logical procession of still frames is much different from the physical result that makes the movie appear to be a "motion picture."

This is also true of networks. You might see one thing as you examine the cabling, but the way that data moves over the network can operate in an entirely different manner. As we've discussed, the physical topology is the physical shape of the network. The logical topology, then, is the shape that the data flow takes as it travels across the network. Here's the kicker: the physical topology does not need to match the logical topology.

Test Smart On the exam, you should remember that if you can see it, it is a physical topology. Questions about the path that data travels refer to the logical topology.

The logical topology, which is the shape that the data makes when it is traveling around the network, is closely related to media access, which is the way devices access the network. While it seems that data is traveling simultaneously across the wire to all the different workstations, in truth each workstation uses the network separately. It's probably best to imagine this process like a clock hand whirling around the clock—as soon as a number (or computer) is pointed at, that computer is allowed to send data. The faster the clock hand can whirl, the more it appears that each computer is sending data at the same time. In reality, there is no central authority like a clock hand managing when the computers are allowed to transmit data. Instead, the network requires a systematic procedure to manage how devices access the media without competition.

I hesitate to claim that there are three logical topologies because the original two—ring and bus—are established. The cellular topology refers solely to wireless networks and is nearly identical to a logical bus, except that stations can move between the cells transparently. As you can tell, the names "ring" and "bus" also refer to physical topologies. They describe the shape of the data transmission rather than the physical network design.

Logical Ring Topology

As you can guess, the ring topology works by sending data from device to device in a closed loop. It's easiest to imagine the logical ring when it works on top of a physical ring, but that's not necessary. If you are ever unsure whether a topology is using a logical ring, you can check to see whether the network interface of each device uses separate transmit and receive circuits. If so, the device is functioning as a repeater and is probably connected in a logical ring.

A token ring network, which uses the Institute of Electrical and Electronics Engineers (IEEE) 802.5 specification, provides a path for data that uses a physical star topology and a logical ring topology. Most token ring IEEE 802.5 networks might seem odd because they run data in circles over physical star topologies, but this really isn't so strange. When a physical ring is broken, the entire network is down. However, when a token ring network uses a MAU as its hub of a star topology, if a cable breaks or a device is down, the MAU can intentionally skip that port and effectively create a logical ring with the remaining ports. Figure 1-9 displays a logical ring running over a physical star.

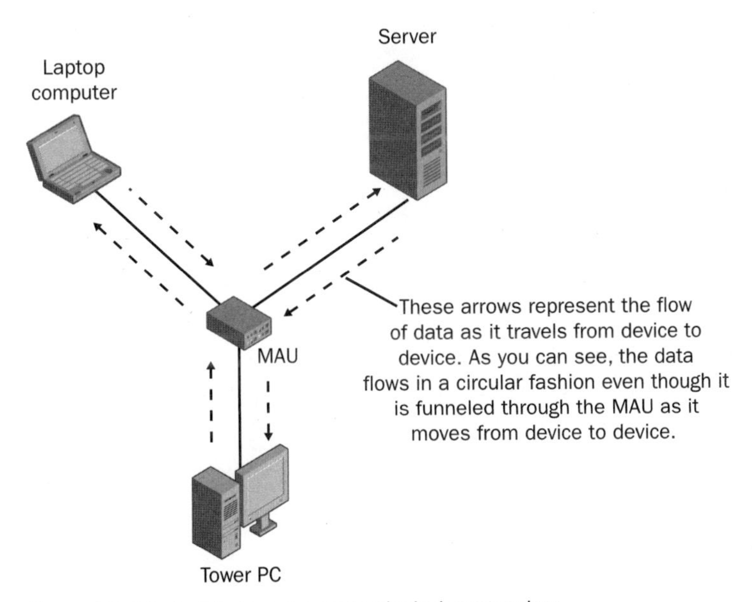

Figure 1-9 A logical ring can run over a physical star topology.

Logical Bus Topology

The logical bus topology is like a television signal. The data is transmitted throughout the entire network segment, and every device has access to the data. You can tell whether a system is a logical bus topology when every device has the ability to transmit at any time.

Cellular Topology

The cellular topology is pretty much a logical bus topology. The only difference is that each device has the ability to seamlessly move between the cells, or areas that are reachable by a particular wireless access point, without interrupting data. Cellular topologies are solely wireless.

A cellular topology has a central access point that is available to the physical region around it, which, in most implementations, resembles an upside-down ice cream cone. This is called the *cell*. From a physical perspective, there is no tangible media, just like with radio or non-cable television. Logically, the cells communicate with one other either directly or only with adjacent cells. Most of the time, you will see only cellular topologies mixed in with other types of networks. But in the future, because of the ease of configuration, you will probably see more home and small office networks running only with a cellular topology. Figure 1-10 shows a cellular topology.

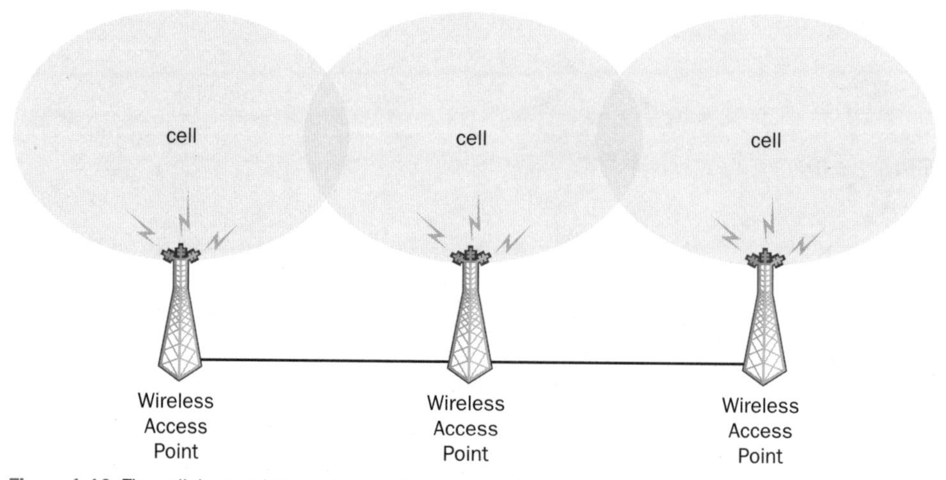

Figure 1-10 The cellular topology is used in wireless networks.

Gaining Access to the Media

Media access methods are independent of the physical and logical topologies just discussed. You'll find that there are usually just a few combinations that seem to work well, however. Media access methods are simply the rules that govern how a device can submit data to the network. Each access method will have a different effect on network traffic.

One good way to visualize media access is to compare it to going into a house. In some cases, you just walk right in and hope you don't bump into someone near the doorway. At other homes, you have to ring the doorbell and wait for someone to let you in. In yet others, a butler is always waiting there to check for your arrival. Finally, in some places, depending on who you are, the red carpet is rolled out just for you to enter before anyone else can. We'll go over each of these types of media access.

The one thing to keep in mind is that data is transmitted one message at a time. In our analogy of the door, just remember that the door is only wide enough to let one person through at a time. Once that person gets through the door, another can enter. Taking turns ensures that multiple devices can use the same media, or doorway, without having any fatal collisions.

Contention as a Method of Media Access

Contention, often called *random access*, is the media access method that acts as an open door to anyone who wants to walk in. Two types of contention methods exist for media access; they are similar, but a single difference between them changes how efficiently they operate, as I'll explain below. They are:

- Carrier Sense Multiple Access/Collision Detection (CSMA/CD)
- Carrier Sense Multiple Access/Collision Avoidance (CSMA/CA)

CSMA/CD Detects Collisions

Just as walking through someone's home without bothering to ring the doorbell, collisions can occur in CSMA/CD. Even though the device listens on the wire to find out whether other transmissions are in progress, another device can transmit at precisely the same time. Conversely, CSMA/CD's message can lag before reaching the other transmitting device, causing messages to collide. Think of it this way: in CSMA/CD, no one looks to see whether anyone else is in the doorway before walking through the door, they just make sure the door is open.

Once a device transmits, data accesses the media; if it runs into another data message, each message tells its respective sending device that a collision occurred. In the event of a collision, each device must resend the data message, contending for an available opening. The device waits a random length of time before retransmitting to maximize the chance of a successful transmission.

Note Carrier sensing is the function of listening to the channel to see whether any traffic exists. Multiple access means that many systems have access simultaneously to the same carrier. Collision detection is the process of telling the sending devices that there was a collision.

Because of the possibility of collisions, access to a contention-based network can be unpredictable. In fact, as an administrator places more devices on a network, collisions increase at a dramatic pace. While collisions are considered a part of normal operations, network managers should prevent the overuse of a network segment and consider using devices such as switches or bridges, which are described in Chapter 2, to reduce the traffic. Remember, collisions lower network performance.

Because CSMA/CD is easy to implement, it requires few resources. As such, it can be very fast if traffic is light.

The disadvantage of CSMA/CD is mainly that it doesn't cope well with collisions. As network growth increases traffic, more computers try to force larger amounts of data onto the same cable. This causes data-transfer rates to slow down and even results in *more* traffic and *more* collisions because each previously collided message must be retransmitted. CSMA/CD only detects collisions; it does not avoid them. An additional disadvantage of this method is related to the value of the data. A mission-critical server is treated the same as a mailroom workstation when each transmits data on a CSMA/CD network.

CSMA/CA Avoids Collisions

CSMA/CA is much like ringing a doorbell before entering a house. The main difference between CSMA/CD and CSMA/CA is that CSMA/CA can avoid collisions. To achieve this feat, CSMA/CA uses specific contention protocols that listen to the wire before transmitting *and* use time slices. By slicing data into time slots, CSMA/CA allows more messages to be sent seemingly simultaneously.

CSMA/CA is based on the premise that it is best to prevent collisions at the point when they are most apt to take place, for example, when the media is released from its last transmission. Using a number of time slots selected at a random basis, each client endures a forced wait period before transmitting. When beginning the transmission, the client senses the media with a request to send message. If the client senses that the media is busy, it waits another random set of time slots and then transmits again, repeating this process with longer time slots until it successfully transmits. Because of the random time slots system, the likelihood of a collision is greatly reduced.

CSMA/CA has all the same advantages and disadvantages of CSMA/CD, except for one: it requires much less administrative overhead because collisions are greatly reduced and data throughput is higher.

Token Passing

Much like a butler who grants you admission to the house at the door, token passing offers a regulated method of media access that ensures no collisions ever take place. In the token-passing scheme, a special packet of data called a *token* is passed around the network. This token authorizes network devices to transmit data. When a device receives the token and has data to transmit, it changes the control information in the token so that no other device can transmit and cause a collision. Hence, in token passing, there is no such thing as a data collision (and you will not see collision light emitting diodes (LEDs) on a token ring MAU).

Note You might find it easier to remember token passing if you think of the game of duck, duck, goose. Only the person chosen as the goose can run around the circle until another is chosen. Likewise, in token passing, only the device that is handed the token is able to send data around the network.

There is tremendous value in traffic management through the use of token passing, and it is a remarkably orderly method of network access. It offers the highest amount of data throughput in heavy traffic networks. But the most popular access method is CSMA/CD, which might have something to do with token passing's disadvantages. The token-passing method requires that every device know its neighbors so that it can receive and then forward the token. A break somewhere along the line can cause a network outage. The process of token passing requires quite a bit of computer processing resources just to decipher and forward the token each time it passes by the device, *whether or not that device needs to send or receive data*. Because of this issue, the data throughput in a small network with light traffic is much lower than it would be with CSMA/CD.

Demand Priority

Demand priority media access was developed to enable certain traffic to have priority over other traffic. In this method, each node signals the hub for transmission. If the hub is idle, the node is sent an acknowledgment packet (ACK) and the node then transmits. The hub stores the packet before forwarding it to the next destination. If the sending node is granted priority, the hub rolls out the red carpet and sends that packet with a higher priority. This eliminates rotation delays and reduces latency on the network.

One of the benefits of demand priority is a lack of collisions. Because the hub must acknowledge a transmitting device, it can easily establish that the packet will arrive and wait before acknowledging other requests. Each connection or port is given a priority setting so that an administrator can ensure that server data is transmitted at the highest priority, while less critical workstations are set at much lower priority levels. Demand polling is used in 100VGAnyLAN, which is a specification for transmitting data at a speed of 100 megabits per second over copper wiring.

Media Types

Media encompasses copper wire, fiber optic strands, and even the earth's atmosphere. Media is simply the transmission carrier for the data signal. Different types of media work at different rates of transmission and with different signaling types.

Test Smart On the Network+ exam, you will be expected to know each of these types of cabling, be able to recognize their characteristics and shapes, and know which protocols use them.

UTP, Unshielded Twisted Pair

Unshielded Twisted Pair (UTP), which you are probably most familiar with, is very flexible wiring used for the telephone and most commonly with networks. The different categories of UTP are defined by the number of pairs of wires in the cable, wire thickness, and the number of twists, which help to reduce electromagnetic interference (EMI), also known as *crosstalk*.

Most twisted pair cabling includes two or more pairs, all held within the same cable, even if only one pair is required for the usage. For example, most telephone wires have two pairs, but only one pair is required for the line to function. This makes it easy to include multiple lines within the same cable and keeps costs low. Also, different protocols require more than a single pair of wires to function, so the additional pair allows for this. When you examine a UTP cable, you will see that each wire is colored and each pair is unique. Table 1-2 describes the different grades of UTP.

Test Smart You should be familiar with Categories 3 through 6 (Cat3 through Cat6 for short) for the Network+ exam because they are data-grade cables.

Table 1-2 Categories of UTP

Grade	Max. Data Rate	Frequency	Max. Distance	Number of Pairs	Uses
Cat1	1 Mbps	1 MHz	90 meters	1 pair	Telephone and ISDN
Cat2	4 Mbps	1 MHz	90 meters	2 pairs	Token ring
Cat3	10 Mbps	16 MHz	100 meters	3 or 4 pairs	10BaseT (Can reach 100 Mbps with 100VGAnyLAN)
Cat4	16 Mbps	16 MHz	100 meters	4 pairs	Token ring
Cat5	100 Mbps 1 Gbps if using all 4 pairs	100 MHz	100 meters	4 pairs	10BaseT and 100BaseT 155 Mbps ATM Gigabit Ethernet
Cat5e	1000 Mbps	100 MHz	100 meters	4 pairs	Gigabit Ethernet
Cat6	4–10 Gbps	250 MHz	100 meters	4 pairs	Gigabit Ethernet, uses all 4 pairs

As Table 1-2 shows, the differences between Cat4 through Cat6 cables are not apparent in the wires' physical appearance. They all have the same number of wires and the same distance limitations. The true differences are in the electrical properties of the wiring, which are increasingly at higher standards.

You might also wonder what the differences in the speed and the frequency are because many of the cables seem to have the same capacity for data rate as they do for frequency. The frequency is described in megahertz, and it shows how often the carrier signal can transmit data. One hertz is completed when the signal goes from 0 to the electrical value of 1 and then back to 0.

The advantages to UTP are many. Most builders automatically install it while building, so you don't often need to install new wiring, just make certain the existing wiring meets the specifications for the protocol you intend to use. UTP is flexible, easy to install, and fairly cheap. Remember that UTP is round because the wires inside must be twisted. Flat wires are known as *telephone silver satin*.

STP, Shielded Copper Wiring

Shielded twisted pair (STP) is much like UTP, except that it has a heavy outer covering, or *shield,* that acts as a ground. STP is highly resistant to noise interference. Even though the twists in the wiring help resist noise, the lack of shielding in UTP prevents it from working in environments with high electromagnetic or radio frequency interference.

STP, which is limited to 100-meter lengths (the same as UTP), is used in token ring networks and for IBM mainframe and minicomputer environments. There is no standard for it. Since token ring networks do not require STP, it is used less and less. There are many reasons for this:

- Higher cost due to greater complexity for the cabling and connectors
- Larger size and less flexibility of the cabling
- Longer installation time

Coaxial Cables

The terms ThickNet and ThinNet refer to Ethernet over coaxial cabling, but the most common use of coax (the short form of coaxial cable) today is in standard cable TV. If you have the chance to examine a cable, you will find that it has a fairly simple design. A copper conductor lies in the center of the cable, which is surrounded by insulation. A braided or mesh outer covering surrounds the insulation. This is also a conductor. A PVC plastic jacket encases the covering.

ThickNet, or RG-8, is older and one of the first types of coaxial cable used in networks. RG-8 is strung in a physical bus topology. Its thick shielding makes it fairly immune to noise but also very rigid and difficult to work with. RG-8 requires connectors, called vampire taps, that pierce through its thick outer shielding. Both ends of the bus must be terminated with a 50-ohm resistor; without both functioning resistors, the network will fail.

ThinNet is RG-58 cable. It is far more flexible than ThickNet and much easier to work with. RG-58 cabling is also strung as a physical bus. It is capable of connecting a maximum of 30 devices on up to a 185-meter length of cable. ThinNet is constructed like ThickNet, except that the central conductor and the insulation are much thinner. British Naval Connectors (BNCs) are crimped onto the cable for connectivity, and 50-ohm resistors are required at each end of the cable. Figure 1-11 shows a coaxial cable's construction.

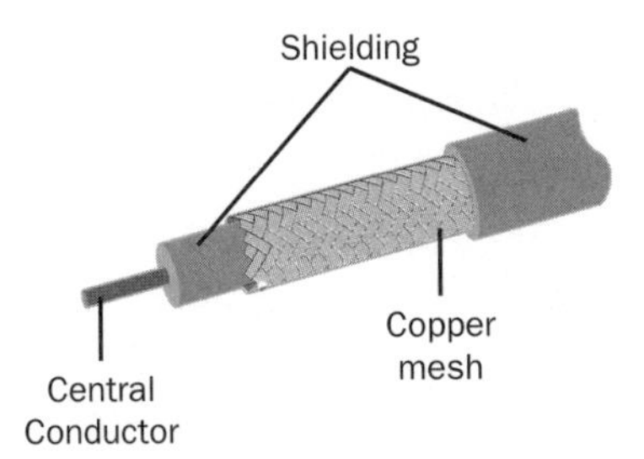

Figure 1-11 Coax cabling consists of a central conductor, surrounded by shielding, by another conductor, and finally by an outer shielding.

Note RG-58 cabling looks similar to television cable, but television cable's specification is RG-59.

Fiber Optic Cabling

Fiber optic cable is a thin strand of glass enclosed in a glass tube and shielded by a PVC plastic covering. Because the glass is so thin it does have a little bit of flexibility, but that doesn't prevent fiber cable from breaking—it is very brittle.

One huge advantage of fiber optic cabling is the extensive distance that optical signals can travel. In fact, the maximum distance of fiber optic cabling is measured in miles, rather than meters. Another benefit of fiber optics is that they are completely immune to radio frequency and electromagnetic interference.

The disadvantages of fiber optics are the special training required for a network technician to learn how to fuse the ends of the connectors to the glass core and to use the special optical equipment, which make the installation and management of such a network expensive. Figure 1-12 shows the construction of a fiber optic cable.

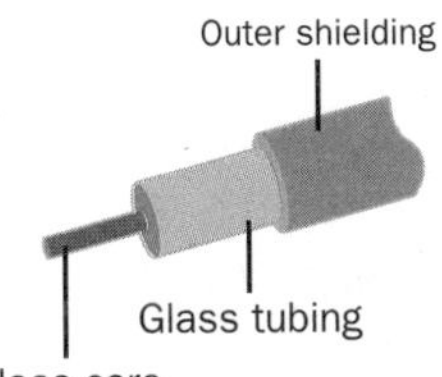

Figure 1-12 Fiber optic cable is brittle because it is made of glass fiber surrounded by a glass tube and covered in a plastic shielding.

Wireless Media

The wireless media is simply earth's atmosphere. Data transmits using the same technology as a cordless phone. For example, IEEE 802.11 specification provides for radio frequency (RF) use in the 2.4 gigahertz (GHz) range for wireless data transmissions. The data rate and the distance depend on the wireless protocol implemented. Table 1-3 compares the various types of media.

Table 1-3 Media Types

Media	Advantages	Disadvantages
UTP	Flexible, commonly used, inexpensive, easy to install.	Susceptible to EMI. Limited to 100-meter length.
STP	Resistant to EMI.	Used only in token ring and IBM networks. Limited to 100-meter length.
RG-8 (ThickNet)	Resistant to EMI.	Requires resistors and vampire taps. Rigidity makes it difficult to work with.
RG-58 (ThinNet)	Fairly flexible. Easy to network a few computers.	Requires a resistor. Limited to 185-meter length. Thinner shield makes it more susceptible to EMI than ThickNet.
Fiber optic	Immune to noise. Capable of extremely long distances.	Brittle, expensive, and requires additional training and equipment to install and manage.
Wireless	No cables to install. Inexpensive and easy to configure.	Can be a security concern. Distance is dictated by the protocol.

Media Connectors

Media connectors are dictated by the media's own specification, or in some cases by the protocol. You'll probably be familiar with most of these connectors:

- **RJ-11** Registered Jack 11 describes the common telephone connection interface, both the receptacle and the plug. Usually, it's implemented with four conductors, but it can have six. Four wires—a red/green pair and a black/yellow pair—are crimped into the plug conductor. The red/green pair is used for voice and data, and the black/yellow pair is used for low-voltage signals, such as those in phone lights.
- **RJ-45** Registered Jack 45 is the most common type of media connector for Ethernet 10BaseT, 100BaseT, and 1000BaseT networks running on UTP. It has eight conductors, which match the four pairs found in the most common grades of UTP, Cat4 through Cat6. Depending on the type of protocol, either two or all four of the pairs will be used.

Wiring an RJ-45 Connector Given the popularity of RJ-45 connectors, you should understand how to crimp a wire into one. To start, strip off the cable covering about one inch from the end. Then untwist each of the four pairs, and straighten them. Place the wires in order, and then bring them together so that they touch. (Always recheck the wiring sequence before continuing past this point.)

With the wires held tightly between your thumb and forefinger, cut all of the wires at a straight 90-degree angle at about half an inch from the end of the cable covering. Insert the wires into the connector with the clip facing down and the pins facing up. Make certain that the wires reach the end of the connector and a portion of the cable jacket is inserted into the back of the connector. Place the connector into the crimp tool, and compress the handle until it is fully closed. When you are finished, use a cable tester to ensure that the wire is working properly.

- **AUI** An Attachment Unit Interface is used for both ThickNet and ThinNet coaxial cables. It has 15 pins on one end that connect directly into the network interface card. The other end is connected to a short cable that connects to a transceiver, which connects directly to the main cable.
- **BNCs** British Naval Connectors can also be used to connect devices to a ThinNet 10Base2 Ethernet network. They look like cable TV connectors, with a central pin that connects to the primary conducting wire and is then locked into place via an outer ring that locks by turning. The connector has a "T" shape, which allows it to hook into the straight bus.
- **ST** An ST connector is a modular fiber optic connector used for duplex communications. It has a circular female connector on each side of the interface.
- **SC** An SC connector is also a modular fiber optic connector. This particular connector is provided for simplex communications, and it has rectangular female interfaces on each side of the connector.
- **RS-232** It used to be that an RS-232 (also known as EIA/TIA-232) interface was one of the most important for remote networking. This is the serial interface with either 9 or 25 pins found on every PC. Because RS-232 is a serial interface, it is used solely with analog services and modems. Most RS-232 cables run at 19.2 Kbps, but they can run data at 125 Kbps.

Networking Technologies and Characteristics

Protocols discussed in this section all refer to the topologies, media access methods, types of media, and the connectors that we have just covered. These networking technologies define the mechanical specifications and bit-stream signaling methods that are used to communicate across local area networks.

The IEEE has been involved in the development of many protocols that involve media and topologies. Table 1-4 describes the IEEE series of these protocols.

Table 1-4 The IEEE 802 Series

IEEE Specification	Description
802.2	The Logical Link Control (LLC) protocol specification used by the other 802 standards
802.3	Ethernet:10BaseT, 10Base2, 10Base5
802.3u	Fast Ethernet 100BaseT
802.4	Token bus
802.5	Token ring
802.6	Metropolitan area network based on bidirectional fiber optic bus topology
802.11a	Fast Wireless LANs up to 54 Mbps, which are just beginning to gain acceptance
802.11b	Wireless LANs up to 11 Mbps
802.12	100VGAnyLAN

Ethernet (802.3) and LLC (802.2)

There are two ways that specifications become standards. One is through standardized development, and the other is through common usage of a proprietary specification, where the usage becomes so prevalent that the specification is adopted as a standard. Ethernet is the latter. The IEEE was not the first to develop Ethernet. That honor goes to the research and development efforts of three companies in the 1970s: Digital, Intel, and Xerox, which were known collectively as DIX. Later on, the IEEE based its 802.3 standard on the DIX specification. In return, DIX updated its implementation to match the small changes made by the IEEE. Nowadays, Ethernet is used for these and several other specifications.

Ethernet 802.3 is generally implemented in conjunction with 802.2. The system uses the CSMA/CD media access method, with a logical bus topology. Physically, Ethernet can be either a star or a bus. It can use copper coaxial cabling, UTP, and fiber optics. Since Ethernet uses the broadcast system of a bus topology, each node receives every data message and examines the frame header to see whether the message is meant to be received by it. If not, the frames are discarded; if so, the frames are passed on to upper layer protocols so that the receiving application can act on them.

10BaseT

One of the most common types of Ethernet in use today is 10BaseT. This particular implementation uses four-pair UTP wiring (Cat3 or higher, but most commonly you will see Cat5) using RJ-45 connectors. Each cable is connected from each network device to a central hub in a physical star topology. Within the hub, the signals are repeated and forwarded to all other nodes on the network because it is a logical bus topology.

Older network interface cards are configured with jumpers to set addresses and interrupts. Today's network interface cards can be managed through a diagnostic program, or automatically configure themselves through plug and play technology. There is a limit of 1024 devices on an Ethernet segment, plus you can have a maximum of 1024 network segments. A UTP cable has a maximum distance of 100 meters, which is equivalent to 328 feet.

10BaseF

10BaseF is an implementation of Ethernet 802.3 over fiber optic cabling. 10BaseF offers only 10 Mbps, even though the fiber optic media has the capacity for much faster data rates. One of the implementations of 10BaseF is to connect two hubs as well as connecting hubs to workstations.

The best time to use 10BaseF is in the rewiring of a network from copper to fiber optic, when you need an intermediate protocol using the new wiring. 10BaseF is not often a permanent solution because the data rate is so low and the cabling so expensive in comparison to using UTP.

10Base2

10Base2, also called ThinNet, is one of the two Ethernet specifications that use coaxial cable. (One of the best ways to remember that 10Base2 is ThinNet, and 2 is smaller than 10Base5, which is ThickNet.) One of the most important issues to remember in an Ethernet coax wiring scheme is the 5-4-3 rule, which states that there can only be 5 segments in a series and 4 repeaters between these 5 segments, although only 3 of the segments can be populated with devices.

10Base2 uses BNC connectors and is implemented as both a physical and logical bus topology using RG-58 cabling. The minimum distance for cables between workstations must be at least a half-meter. Drop cables should not be used to connect a BNC connector to the network interface card (NIC) because this will cause signaling problems unless the NIC is terminated. 10Base2 ThinNet segments cannot be longer than 185 meters, although it is often exaggerated to 200 meters, and you can't put more than 30 devices on each populated segment. The entire cabling scheme, including all five segments, can't be longer than 925 meters.

10Base5

10Base5 is nearly identical to 10Base2, except that it uses a different type of cabling and media connector. 10Base5 is known as ThickNet because it uses the RG-8 coaxial cable. It requires an external transceiver to attach to the network interface card on each device. The transceiver is a device that translates the workstation's digital signal to a baseband cabling format. ThinNet and UTP network interface cards have built-in transceivers. Only 10Base5 ThickNet network interfaces use external transceivers.

In the 10Base5 configuration, the NIC attaches to the external transceiver using an AUI connector. The transceiver then clamps into the ThickNet cabling, which is why it is usually called a vampire tap. 10Base5 can also use BNC connectors.

For 10Base5, the following rules apply: First the 5-4-3 rule applies to ThickNet just as it did to ThinNet. In addition, the minimum cable distance between each transceiver is 2.5 meters. The maximum network segment length is 500 meters, which is where 10Base5 gets the "5" in its name. The entire set of five segments cannot exceed 2,500 meters. You can have 100 devices on a 10Base5 network segment.

100BaseFX

100BaseFX is simply Fast Ethernet over fiber. Originally, the specification was known as 100Base-X over CDDI (Copper Data Digital Interface) or FDDI (Fiber Data Digital Interface). Because the signaling is so vastly different, these two technologies were split into 100BaseFX and 100BaseTX. 100BaseFX runs over multimode fiber. There are two types of fiber in use. Multimode fiber optic cables use LEDs to transmit data and are thick enough that the light signals bounce off the walls of the fiber. The dispersion of the signal limits the length of multimode fiber. Single mode fiber optic cables use injected lasers to transmit the data along fiber optic cable with an extremely small diameter. Because the laser signal can travel straight without bouncing and dispersing, the signal can travel much farther than multimode.

100BaseT4

100BaseT4 was the specification created to upgrade 10BaseT networks over Cat3 wiring to 100 Mbps without having to replace the wiring. Using four pairs of twisted pair wiring, two of the four pairs are configured for half-duplex transmission (data can move in only one direction at a time). The other two pairs are configured as simplex transmission, which means data moves only in one direction on a pair all the time.

100BaseTX

100BaseTX, Fast Ethernet, transmits data at 100 Mbps. Leveraging the existing IEEE 802.3u standard rules, Fast Ethernet works nearly identically to 10BaseT, including that it has a physical star topology using a logical bus. 100BaseTX requires Cat5 UTP.

Gigabit Ethernet

The fastest form of Ethernet is currently Gigabit Ethernet, also known as 1000BaseT over Cat5 or higher-grade cable, using all four pairs of the cable. It uses a physical star topology with logical bus. There is also 1000BaseF, which runs over multimode fiber optic cabling. Data transmission is full-duplex, but half-duplex is also supported.

Token Ring (802.5)

Like Ethernet, the IEEE did not create the 802.5 specification until it was already developed by a vendor. IBM was the developer for token ring networking. Token ring networks use a physical star topology, but a logical ring topology. The central hub of the star is called a MAU. It uses twisted pair wiring, and you will find installations using STP, but most of them will use UTP.

The media access method for this type of networking is token passing, which you can tell by its name. Token ring also provides a method of access priority, or token seizing, in which a device can be assigned priority and is allowed to have the token and use the network more frequently. This is useful for a network server.

Token ring also provides for beaconing, a mechanism that handles network faults. In beaconing, if a device detects a network failure, it will send a beacon frame that specifies which device reported the failure, its nearest active upstream neighbor (NAUN), and the location of the failure. This process will trigger auto reconfiguration, in which the network will attempt to bypass the failed area of the ring. When a beacon frame is sent out, the MAU is said to be "beaconing." And when the MAU bypasses the failed area of the ring, it is said to have "wrapped the port." Token ring comes in two data rates, either 4 Mbps or 16 Mbps. In a token ring network, all devices must use the same speed.

Wireless (802.11b)

Wireless protocols are described under IEEE 802.11. There are several sub-specifications under development, but the one most popularly used is 802.11b. In the 802.11b specification, the network interface cards are equipped with small antennae, which transmit data to the WAP that acts as a hub, or to another wireless peer computer in a point-to-point connection. When using a WAP, 802.11b is considered to be running in *infrastructure mode*, which is either the BSS or ESS. When communicating peer to peer, 802.11b is running in *Ad-hoc mode*, which is called the IBSS.

802.11b is considered wireless Ethernet, with a data throughput of up to 11 Mbps, however this is more realistically between 2 and 4 Mbps. Distance between the wireless device and the WAP has the effect of reducing data throughput. It is similar to using a cordless phone, in that the farther you are from the telephone base, you will experience an increasingly fainter signal and higher likelihood of a dropped connection.

The most interesting configuration occurs when the ESS is used. In this configuration, several BSSs (each a wireless star with a central WAP) overlap at the edges of their "cells." These are further connected by a distributed system (DS) that enables a device to move from one cell to another. For businesses in which people move around, an ESS along with mobile devices can allow a person to access network resources, such as files and printers, even while walking along and crossing several cells.

The 802.11 specification defines wireless stations and WAP. The WAP acts as a bridge between cells, as well as between the wired and wireless network. A WAP has the properties of a radio, and also typically includes an Ethernet RJ-45 interface. The 802.11b protocol runs half-duplex. Each wireless device can send or receive, but not simultaneously. It runs at the 2.4 GHz range, which is the same as some cordless phones.

Caution When you implement a wireless network in a home or small office, make certain that all cordless phones are 900 MHz. Otherwise, you will have too much interference.

802.11b might be the current wireless protocol in use, but expect that 802.11a will surpass it. The 802.11a runs in the 5 GHz range (so you can still use your 2.4 GHz cordless phones), but best of all it can transfer data at rates up to 54 Mbps, much better than a 10BaseT network. Don't expect the 802.11a and 802.11b protocols, network interfaces, or WAPs to be compatible, either.

Fiber Optics Used in FDDI

Developed in the mid-1980s by the American National Standards Institute (ANSI), FDDI, which is pronounced "fiddey," has both a physical and a logical ring topology. Because fiber optic media can send optical signals over long distances, FDDI is commonly the backbone used to connect large campuses of businesses.

FDDI uses a dual ring topology, in which nodes are connected to each of the two rings. This topology also uses token passing for media access. One ring is considered primary and is used for all data transmission. The second ring is used for backup in case one of the brittle fiber optic cables breaks. The dual ring topology helps FDDI become self-correcting. If there is a break in the ring, a dual connected node (called a *Class A* node) can transmit back through the second ring so that it effectively becomes a single ring until the break can be fixed. Most nodes are connected only singly to the primary ring and are called *Class B* nodes. Figure 1-13 shows the FDDI ring in action as it fixes a break in the cable.

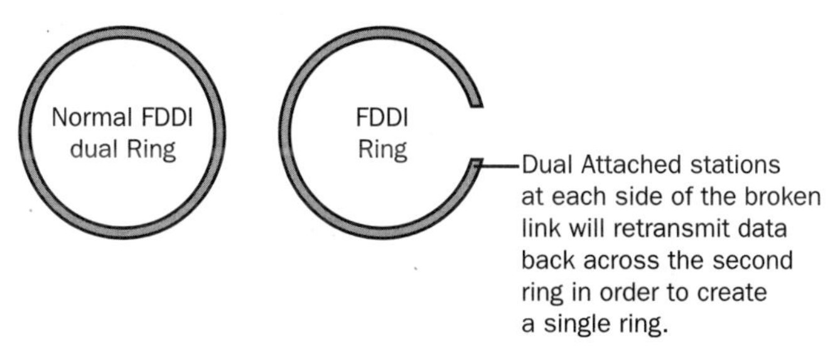

Figure 1-13 The FDDI ring is built so that it can correct a break in the cable.

The benefits of FDDI are:

- Immunity to electromagnetic and radio-frequency interference
- Long-distance capabilities
- Large capacity for data throughput
- Resistance to traditional wiretap methods

Key Points

- By understanding how networks are shaped, or their physical topologies, you will be able to better understand the installation and configuration of a network.
- By looking at how data travels around the network, or its logical topology, you will have a better understanding of how to troubleshoot networking problems.
- There are five physical topologies: The ring is a closed loop. The star uses a central hub with wires that connect to each station. The bus is a long cable terminated at each end. The mesh connects each station to each other and is often represented as a cloud. The wireless topology is a string of cells using wireless access points.
- Logically, data transmits across networks in a ring, bus, or cellular method.
- To gain access to media, networks use contention, token passing, and demand priority. Contention allows each station to transmit at any time, and collisions might result. Token passing provides each station a turn at having access to the network through the use of a specially formatted frame called a token. Demand priority offers higher throughput for certain stations by using a central hub to manage the traffic as it passes through.
- The types of media you should know are:
  - UTP, which is the most common copper cabling in use today.
  - STP, which is used in only IBM or token ring networks.
  - Coaxial cabling, RG-8, which is ThickNet, and RG-58, which is ThinNet. Both use copper wires.
  - Fiber optic cabling, a thin strand of glass surrounded by a glass tube that transmits optical signals.
  - Wireless, the earth's atmosphere.
- Media connectors hook cables up to devices' network interfaces. RJ-11 connectors are used for telephones.
  - While similar to RJ-11, RJ-45 connectors are wider, have more wires, and are the most common type of connector for UTP used in Ethernet networks.

    - AUI and BNC connectors are used with coax cables.
    - ST and SC interface to fiber optic media.
- The most common protocol series that defines physical topologies is the IEEE 802.2 and 802.3, which stands for Ethernet. It uses contention media access and a logical bus topology, usually over a physical star but sometimes over a physical bus.
    - 10BaseT is the most common 10 Mbps protocol using Cat3 or better grade UTP.
    - 10BaseF is the same, except it uses fiber optic cable.
    - 10Base2 is Ethernet over ThinNet coax, while 10Base5 is Ethernet over ThickNet coax.
    - 100BaseFX is Fast Ethernet over fiber.
    - 100BaseT4 is 100 Mbps Ethernet using lower-grade wires and four pairs.
    - 100BaseTX is Fast Ethernet over copper.
    - Gigabit Ethernet is 1000 Mbps Ethernet over copper or fiber.
- Token ring is also the IEEE 802.5 for token passing on a logical ring topology, often set up as a physical star.
- 802.11b is the IEEE's wireless protocol running at 2.4 GHz.
- FDDI is a fiber optic dual ring topology using token passing.

Chapter Review Questions

1 You are hired as a network administrator for a company of 45 people, who work in an open area of a small office and use cordless telephones. While everyone has a computer, they use "Sneakernet"; that is, they copy files to disks and carry them to other computers to share or print the file. You have been hired to develop a network for the company. Which of the following networks do you decide to implement, and why? Choose all that apply.

a) FDDI, because you need the large throughput and distance

b) Token ring, because it's rock-bottom cheap

c) Wireless, because it's easy to configure Cordless phone.

d) Ethernet 100BaseT, because it's an established system that you can easily upgrade

Either answer c or d can be correct. Answer c, wireless, is easy to configure: You may have a problem with the cordless phones in the office, so you'll need to make certain that they are running at 900 MHz. However, if you run across this type of question on the test and no mention is made that cordless phones are at 900 MHz, assume that the wireless answer is wrong. Answer d, Ethernet 100BaseT, is both established and can be easily upgraded since it runs over Cat5 or better wiring and you need only replace the network interfaces and the hubs. The drawback for this system is having to wire the network. Answer a, FDDI, is wrong; you will not need the long distance in a small office. Answer b, token ring, is wrong because it is not rock-bottom cheap.

2 Select the following protocol that uses token passing for media access:

a) FDDI

b) Wireless

c) 100BaseT

d) 10Base2

Answer a, FDDI, is correct because it uses token passing on a dual ring topology. Answer b, wireless, is incorrect because it uses a contention media access method. Answer c, 100BaseT, is incorrect because it uses contention. Answer d, 10Base2, is incorrect because it uses contention.

3 Which of the following types of media uses an RJ-45 connector?

a) RG-8

b) RG-58

c) Cat5 UTP

d) Fiber optic

Answer c, Cat5 UTP uses RJ-45 connectors. Answer a is incorrect because RG-8, or ThickNet, uses vampire taps. Answer b is incorrect because RG-58 uses either AUI or BNC connectors. Answer d is incorrect because fiber uses either ST or SC connectors.

4 You have been called in to help fix a network. Every workstation is down. What type of physical topology do you think this network is if the outage is caused by a break in the wire?

a) Ring

b) Bus

c) Mesh

d) Star

Answer b, bus, is correct. A physical bus topology must have the cable in perfect operating order for the network to communicate. If there is a break in the physical bus, the entire network fails. Answer a is incorrect; a ring topology has a method of beaconing and reversing data direction to fix itself. Answer c, mesh, is incorrect; mesh topologies offer multiple redundant routes to ensure network availability even if there is a cable break. Answer d, star, is incorrect; star networks would cause an outage only in the link to the one station connected to the broken cable.

5 Which of the following interfaces is available with either 9 or 25 pins?

a) ST

b) RJ-11

c) RJ-45

d) RS-232

Answer d, the RS-232 interface offers both a 9-pin and 25-pin connector. Answer a, ST, is incorrect because it is a fiber connector without pins. Answer b, RJ-11, is incorrect because it offers two pairs or four conductors. Answer c, RJ-45, is incorrect because it offers four pairs or eight conductors.

6 You have been called to fix a network that is down. You find that the central unit in the star is beaconing and has wrapped a port, which effectively bypasses the port, until it can be fixed. What type of network is this?

a) Gigabit Ethernet

b) Wireless 802.11b

c) Token ring

d) FDDI

Answer c, token ring, is correct. A token ring is a star configuration that, when it finds a fault in one of the wires, will begin beaconing and then wrap the port where the fault occurred to continue the ring. Answer a, Gigabit Ethernet, is incorrect because it does not have a self-repairing mechanism. Answer b, Wireless 802.11b, is incorrect because it doesn't use ports. Answer d, FDDI, is incorrect because it does not have a physical star topology to wrap ports.

7 Which of the following cables is constructed from a central copper conductor, surrounded by insulation that is surrounded by a mesh, and finally encased in the cabling cover?

a) RG-58

b) Cat3

c) Cat5

d) Fiber

Answer a, RG-58, is correct because it is a coaxial cable, which has a central conductor, insulated from a meshed conductor. Answer b, Cat3, is incorrect because it is a form of unshielded twisted pair. Answer c, Cat5, is incorrect because it is also a form of unshielded twisted pair. Answer d, Fiber, is incorrect because it does not use copper in its construction.

8 How many pairs of wires are there in a Cat5 wire using an RJ-45 connector?

a) 1

b) 2

c) 4

d) 8

Answer c, there are four pairs of wires in a Cat5 cable. Answers a, b, and d are all incorrect.

9 You have been called by a network user who has decided to set up his own workstation on the 802.3 100BaseT network. When he calls, he says that the computer will not communicate on the network. You examine his computer and find that the wire connecting the computer to the network outlet is a flat, silver wire. What is the problem?

a) There should not be a wire, because 802.3 100BaseT is wireless.

b) He used telephone wire instead of Cat5 to hook up.

c) The external transceiver and vampire tap are missing.

d) The network beaconed when he plugged in and wrapped his port.

Answer b is correct. The man used a telephone wire, called silver satin, to hook up to the network. He should have used a Cat5 wire, which is round. Answer a is incorrect because 100BaseT requires Cat5 or higher-grade wiring. Answer c is incorrect—only ThickNet using RG-8 wiring requires an external transceiver and vampire tap. Answer d is incorrect because only token ring networks beacon.

10 You are called to hook up a new network workstation. The workstation, which is running Windows XP, is 358 feet from the wiring closet. The network server is 280 feet from the wiring closet on the other side of the building, making it 638 feet away from the workstation. There are exactly 30 stations on the network segment. The network uses 10BaseT. What problems do you think you might need to overcome?

a) The network will be very slow because of the 638-foot distance.

b) The server will not be able to communicate because it will be competing with new signals on this contention-based network.

c) There are too many devices on the segment, which is allowed a maximum of 30.

d) The workstation is too far from the wiring closet.

Answer d, the workstation is too far from the wiring closet. 10BaseT is limited to 100 meters, or 328 feet of distance. Answer a is incorrect. More traffic might slow it down, but distance is not measured through a hub. Answer b is incorrect. While the server will be competing, it will still be able to function on the contention-based network. Answer c is incorrect. 10BaseT is allowed up to 1024 devices on the segment.

Chapter 2

Networking Components

By now, you should be familiar with the cabling and with the connectors to which they are attached. This chapter will discuss the equipment you plug those cables into to ensure that your data will travel from workstations to servers and printers throughout your network.

Central Hubs in Star Topologies

The word *hub* relates back to the spoked wheels that were attached to chariots and carts in ancient times and refers to the center of a wheel where the spokes meet. In more recent times, the word has come to mean the center of activity, or the meeting point where a change might take place. Today a traveler might pass through two or more airport hubs on a connecting flight across the United States. A *hub* in networking terms is the center of a network segment that acts as a place where data can arrive from one port and be forwarded out the other ports. It is aptly named, because the hub is the center of all activity on the network segment.

What Does a Repeater Have To Do with a Hub?

Hubs, which are used in star topologies, originated from repeaters, which are used in bus topologies. To understand a hub, you must first understand how a repeater works and when to use one. Hubs evolved from the original repeaters in bus topologies using ThickNet or ThinNet media. The repeater extended the length of a cable. In some cases, a network administrator connected two pieces of a cut cable or fixed a break in a cable; in others, it was necessary to extend the length of the cable beyond its signaling limitations.

Test Smart All cabling is limited in the physical length to which it can be used because signals degrade or weaken with distance; as distance grows the signal attenuates and noise interference increases. The maximum length of a cable is the point at which the signal becomes too weak to transmit data reliably.

In the case of merely fixing the break in a cable, or connecting two cables whose combined total length is less than the maximum length of that particular type of cable (185 feet for ThinNet, for example), the repeater doesn't need to make any changes to the signal. It only needs to copy the signal it receives from one port and transmit it to the other port. This is called a *passive repeater*. Figure 2-1 shows a typical passive repeater.

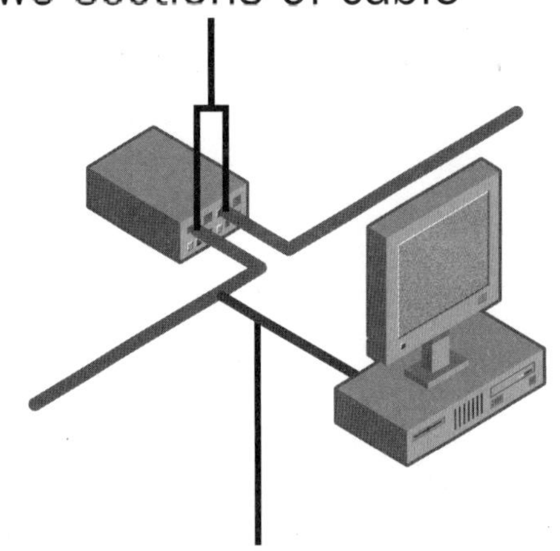
Each repeater has two ports connected to two sections of cable
Workstations are connected to drop cables, which then connect to the cable and eventually connect to the repeater

Figure 2-1 A passive repeater has two ports (sometimes two different types of connectors for each port are offered, but only two connectors can be used at any time). The ports connect two cables to repeat signals from one cable to the other, and the repeater does not have a power source.

In the case of extending the length beyond the maximum for that type of cable, the repeater does need to regenerate the signal as though it were new. This type of repeater is called an *active repeater*, and it usually requires a power source to perform this function.

Repeaters are available for both electromagnetic (usually copper cabling) and optical media, as well as wireless media. Because digital signals transmitted across electromagnetic media depend on the presence or absence of voltage, they have a tendency to dissipate quickly in comparison with both optical and analog signals. As a result, a digital network requires frequent repeating.

An optical repeater is built from a light-emitting diode (LED) or an infrared-emitting diode (IRED), along with a photo cell and an amplifier. The fiber optic repeater receives a signal and transmits it, amplified. Fiber optic repeaters are active and require a power source.

Wireless Repeaters Wireless networks have grown considerably from their humble beginnings. The oldest wireless networks that I dealt with in the mid-1990s were simple satellite dishes that pointed at one another in line of sight. The signals that traveled farther than line of sight allowed required a repeater. These systems were not reliable—a bird flying through the signal could interrupt data communications, a storm would take the link completely down, and they were very slow. In the context of older, wireless communications, a repeater consists of multiple components:

- A radio receiver that receives the incoming radio signals
- An amplifier that increases the strength of the signal
- An isolator to provide protection from a strong transmitted signal that otherwise could disable the repeater
- A transmitter to produce an outgoing signal
- Two antennae

Don't confuse wireless satellite communications with Ethernet 802.11(b) networks. Today's wireless 802.11 networks are built for transmitting inside a building or communications throughout a campus. They use a wireless access point (WAP), which more closely resembles a multiport hub in its ability to form point-to-point communications with multiple, separate wireless devices.

Hubs or Multiport Repeaters

The hub was originally called a *multiport repeater* because it forwarded data transmissions efficiently out multiple ports. It was also known as a *concentrator* because all cabling was concentrated at a central point. These terms are used interchangeably. The task of a hub is to connect two or more network nodes together in a star configuration, as shown in Figure 2-2. It receives a signal from one port and then repeats that signal out to all other ports.

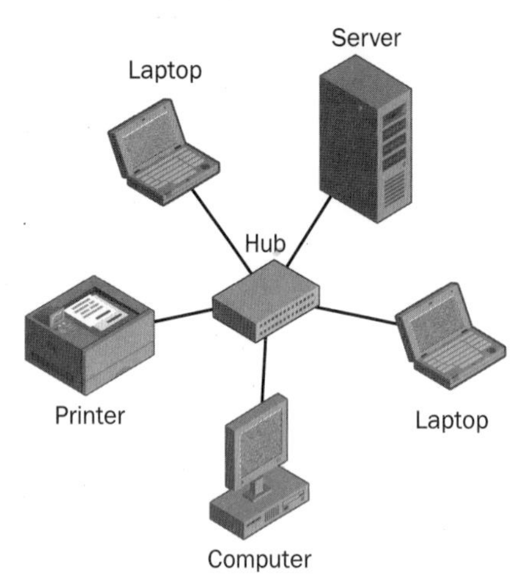

Figure 2-2 Hubs create a physical star topology.

Like passive repeaters and active repeaters, a hub can be both passive and active. The *passive hub* is merely a conduit that receives a signal and splits it out all ports. The *active hub* both retransmits and regenerates the signal out through all ports, and it requires a power source to do so. Almost all hubs used in today's networks are active. Manufacturers have added some troubleshooting tools to active hubs in the form of link lights, error lights, and collision lights.

There are times that you will run across hubs that do not have lights that immediately indicate problems. And many times, you will find that a hub just simply is not in the physical vicinity to look at—especially since wiring closets tend to be kept out of sight. In either case, you can use a network monitor to keep track of the status of links, errors, and collisions. A *network monitor* is typically an application that is capable of reading traffic as it passes on the wire. Network monitors will work with *manageable* hubs, which are considered intelligent because they can send alerts and information about their own status. Table 2-1 displays how to resolve some of the problems that can arise in a hub network.

Table 2-1 Resolving Problems on a Network Using a Hub

Problem	Possible Causes	Solution
Excessive errors	1. Bad cabling. 2. Traffic collisions.	1. Replace cables. 2. Use a switch or bridge to segment traffic.
Excessive collisions	1. Bad cabling. 2. Too much traffic.	1. Replace bad cabling. 2. Segment traffic using a switch or bridge.
Large packets	Network interfaces are transmitting incorrectly.	Replace network interface cards (NICs) and update NIC drivers.
Excessive broadcasts	Too many devices on a single broadcast segment.	Use a router to segment the broadcast traffic.

Note You can read more about how a switch, bridge, and router function in the following sections.

Hybrid Hubs

Most hubs simply include RJ-45 ports for unshielded twisted pair (UTP) cabling. Some hubs, called *hybrid hubs,* include additional ports for different types of wiring. A hybrid hub often has only one or two ports for different cable types; for example, it might have 12 ports for UTP and one for fiber optic media. The hybrid hub allows communications across two different types of cabling. A hybrid hub configuration is shown in Figure 2-3.

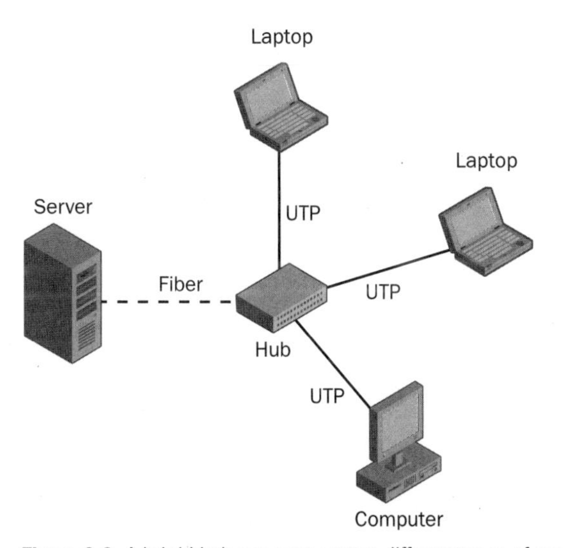

Figure 2-3 A hybrid hub can connect two different types of media into a single LAN.

Hubs in a Token Ring Network—MAUs

A *Multistation Access Unit (MAU)* is a special type of hub used for token ring networks. The word "hub" is used most often in relation to Ethernet networks, and MAU only refers to token ring networks. On the outside, the MAU looks like a hub. It connects to multiple network devices, each with a separate cable. Unlike a hub that uses a logical bus topology over a physical star, the MAU uses a logical ring topology over a physical star.

When the MAU detects a problem with a connection, the ring will beacon. Because it uses a physical star topology, the MAU can easily detect which port the problem exists on and close the port, or "wrap" it. The MAU does actively regenerate signals as it transmits data around the ring. In addition, an MAU has a ring-in and a ring-out port, which are used to connect two or more MAUs to create a larger ring, as shown in Figure 2-4.

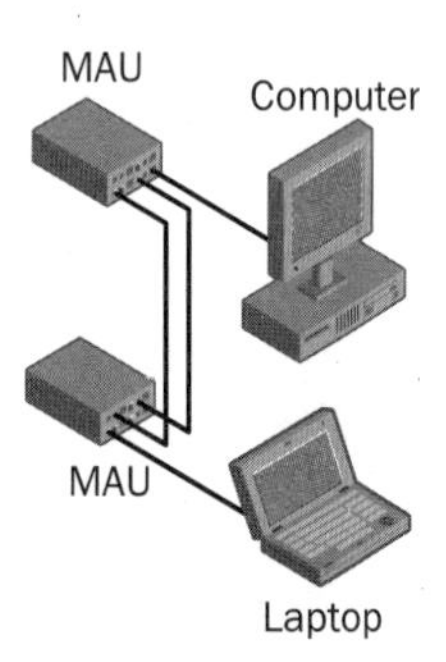

Figure 2-4 The ring-in port of one MAU is connected to the ring-out port of the second MAU and vice versa to create a larger, complete ring that all participating computers can join.

Installing a New Hub

Installing a hub is a fairly easy task that every network administrator will need to undertake at some point in time. You just need to follow some simple rules:

1. Unpack the hub and any additional hardware. Usually a hub comes with mounting hardware for a rack or rubber feet for stackable components, and sometimes with additional equipment.
2. Verify that the contents match the packing list.
3. Mount the hub either as a stackable hub, or rack-mount it.
4. Make sure the hub is located within the specified length limitations for the type of cable being used.

5. Make certain that the temperature in the location where you place the hub isn't too hot, too cold, nor too variable, to prevent equipment failure.
6. Verify that there is sufficient airflow in the hub's location to keep the equipment from overheating.
7. Ensure that the voltage in the power outlet is appropriate for the hub, and if possible, connect to a surge protector or uninterruptible power supply (UPS).

You should use straight-through cables for any port that is not marked with an X, or as MDIX. When a port is marked this way, it is intended for use with a crossover cable and is usually provided to connect multiple hubs together in a stack. It's always a good idea to check with the vendor or the information accompanying the hub to see what specially labeled ports are meant to do.

Bridges

A *network bridge* connects segments, but incongruously it breaks up the traffic by filtering data frames, which are packets of data that are transmitted across the network. Each frame contains the physical addresses of the sending node and the receiving node. When a data frame is transmitted across a network, it visits every node in the segment. Each node checks to see whether the frame is meant for it, and if so it accepts the frame; if not, the node discards it. As you can probably guess, this adds up to a lot of traffic as more devices are added to a segment.

Bridges let you determine how many frames will go to what network nodes on the segments that the bridge is connected to. The bridge learns the physical address of each device connected to those segments and stores the location in an internal lookup table. When a frame is received from either segment connected to the bridge, it looks into the table to determine whether that frame's destination address is located on the segment from which it was received or on the other segment. If the destination is on the same segment that the bridge received the frame from, the bridge discards the frame; otherwise, the bridge forwards the frame. Keep in mind that data will still be sent to every node on the originating segment—the bridge will only keep the frame from being sent to the other segment it is connected to. As shown in Figure 2-5, when device A sends data to device B, the bridge filters it out and the devices on the other segment do not see the frame. However, if device A sends data to device C, the bridge will forward that frame to the other segment.

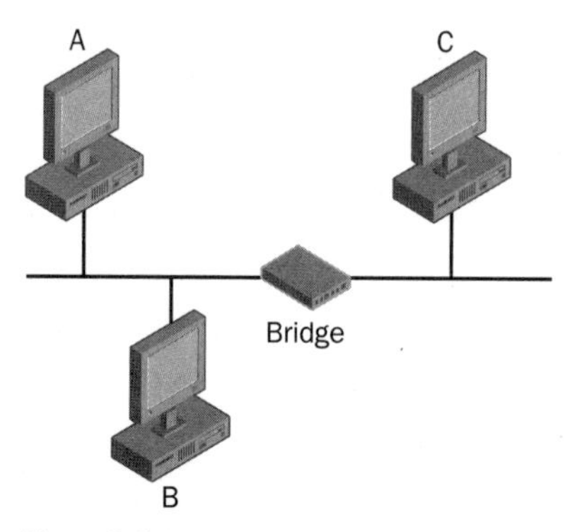

Figure 2-5 When A transmits to B, the bridge will drop the packet. When A transmits to C, the bridge will forward the packet. In this way, bridges act as filters.

Bridges can connect networks that use different types of media. For example, a bridge can connect Ethernet over fiber (100BaseFX) to Fast Ethernet over UTP (100BaseTX).

Switches

Just as a hub is considered a multiport repeater, a *switch* is basically a multiport bridge. Much of the network equipment you will encounter is similar in appearance, mainly because uniformity of size and shape allows for easy stacking or rack-mounting of equipment. A switch and a hub look nearly alike; internally, they work a lot differently.

The switch has multiple ports. Like a bridge, it learns the physical addresses of each device attached through each port, which it stores in an internal lookup table. When a frame is received from a port, the switch determines which port contains its destination physical address. If it is the same as the port from which the switch receives the frame, the frame is discarded. Otherwise, the switch creates a temporary *switched* path between the port where the frame originated and the port containing the frame's destination, and the frame is then transmitted correctly.

Because a switch contains multiple ports and can even link up with other switches, you can design a network with many separate segments logically bridged together. Another option available with a switch is the ability to dedicate switch ports to single network nodes. This creates a high-bandwidth link for the servers and workstations given such dedicated ports. Figure 2-6 shows how a switch transmits data.

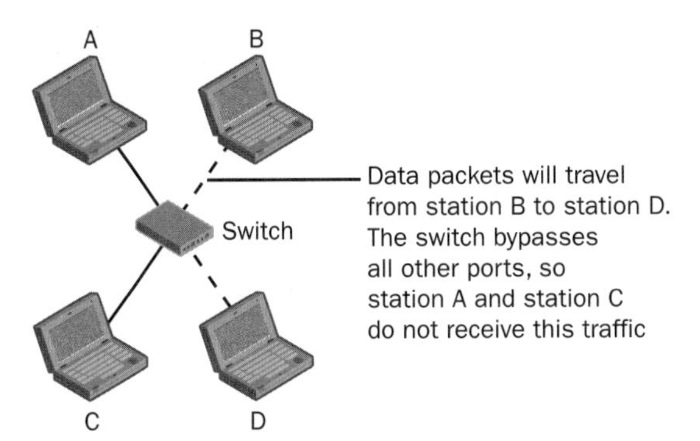

Figure 2-6 When a switch is in use, if B transmits to D, a virtual link between the two is created and no other ports are involved in the data transmission.

Routers

A *router* connects multiple LANs together. Routers are similar to bridges and switches in that they filter data. The difference is that, while bridges and switches filter data based on physical addresses, the router filters data based on the logical address, which includes the network segment address.

For routing to function, each network segment must be assigned a unique logical address. The router stores these unique addresses inside a routing table and uses them to make decisions about where to forward incoming data packets.

Routers, while resembling standard hubs in structure and size, are more closely related to a personal computer. Even though the router lacks a keyboard, monitor, and mouse, it contains the internal processors and memory components that are necessary to store data (which includes the routing table as well as large packets of data), select the fastest route, and then transmit the data.

A server or workstation with two network interfaces that passes data between two different networks is becoming common in small offices and home networks in which a single Internet connection is shared by multiple computers. Most operating systems incorporate routing functionality, including Novell NetWare, UNIX, Linux, and Windows NT/2000 servers and workstations.

Using a workstation or server as a router in a *large* internetwork, however, is not recommended. PCs are intended and designed for interactive use rather than routing needs. When a PC performs both the functions of:

- router and workstation *or*
- router and server,

it will usually function poorly, especially compared with a standalone router and separate workstations and servers.

Routers are designed to connect different types of media, different topologies, and even networks using different media access methods. A router can do this as long as the two networks share a common networking protocol, such as TCP/IP. Most routers have multiple, different types of interfaces, often connecting LANs using Ethernet over UTP to WANs using anything from frame relay to Fiber Distributed Data Interface (FDDI). Figure 2-7 shows a router connecting three networks using different topologies.

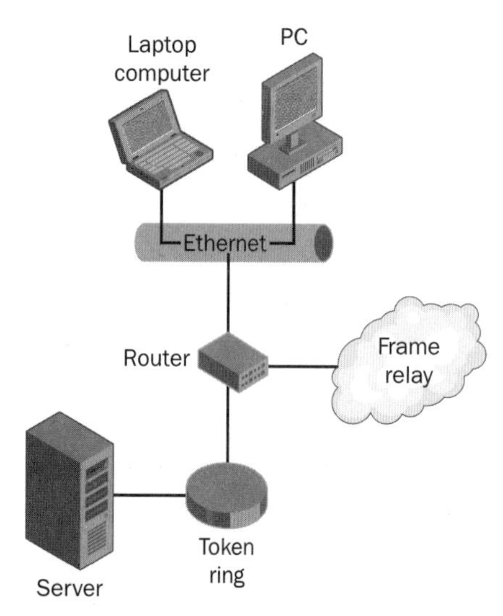

Figure 2-7 A router can connect multiple types of networks.

Notice that a router separates network broadcasts, but bridges or switches don't. Broadcasts are often used to advertise services. They can cause congestion in a large internetwork, even to the point of interrupting network traffic altogether, which is a condition sometimes caused by "broadcast storms." You can use a router to segment a network with too many broadcasts. This will reduce the amount of bandwidth broadcast traffic can use.

The router's basic functions are to determine the routes on the network and to forward packets along those routes. A router works in basically the same way, regardless of the protocol suite. Routing protocols are somewhat different in how they perform route determination or update the routing table, but the basic purpose of the router is always the same—to move data from one network to another.

Note Speaking of different protocols, a router can perform routing functions for two different protocol suites at the same time; however, it maintains a separate routing table for each protocol suite.

Most protocol suites, such as TCP/IP or IPX/SPX, include routing protocols, which are used to assist in route selection and populate the routing tables of the router. Routing protocols, which we'll discuss in more detail in later chapters, function dynamically so that an administrator does not need to manually enter routes into the routing tables whenever a router goes down or a new router is added to the network. For large internetworks, a routing protocol is a great help in network management, because it ensures ongoing and efficient network communications.

It is entirely possible for a network administrator to manually create and maintain the routing tables in a router without ever using a routing protocol. Before attempting to do this yourself, keep in mind how quickly the task increases in size. With each new router added, at least one and possibly hundreds of new routes must be added to each router's routing table.

Routing tables are the means by which a router selects the fastest or nearest path to the next "hop" on the way to a data packet's final destination. This process is done through the use of routing metrics, which are the means of determining how much distance or time a packet will require to reach the final destination.

Routing metrics are provided in different forms. A *hop* is simply a router that the packet must travel through. Metrics also measure the time that it takes to traverse a network link. This is important in the use of a backup link that is extremely slow; you would want the fastest link used before the slow link.

Ticks measure the time it takes to traverse a link. Each tick is 1/18 of a second. When the router selects a route based on tick and hop metrics, it chooses the one with the lowest number of ticks first. Then, if there are multiple routes with the same number of ticks, the router will select the route with the lowest number of hops. As you can see, this process eliminates the need for an administrator to disable slow backup links to prevent their usage. Instead, the router selects a slow link only if the fast ones are not available.

Here are the steps that a router takes:

1. Before a router looks into the routing table, however, it first examines the data that it is receiving.
2. When a device sends a packet intended for a machine on a different network segment, it places the hardware address of the router in the header so that the router will open the data packet. But instead of using the logical address of the router, the sending device puts the logical address of the destination device in the header.
3. The router will use this logical address to determine where to forward the data.

4. The router first establishes whether it knows where the destination device is located.
5. If it doesn't know the location, it will automatically forward it to the default gateway router address. For example, when you set up a network that connects to the Internet, often the default gateway router is simply the router's address that connects to the Internet, as shown in Figure 2-8. If there is no default gateway router address, the router will drop a packet if its destination isn't listed in the routing table.

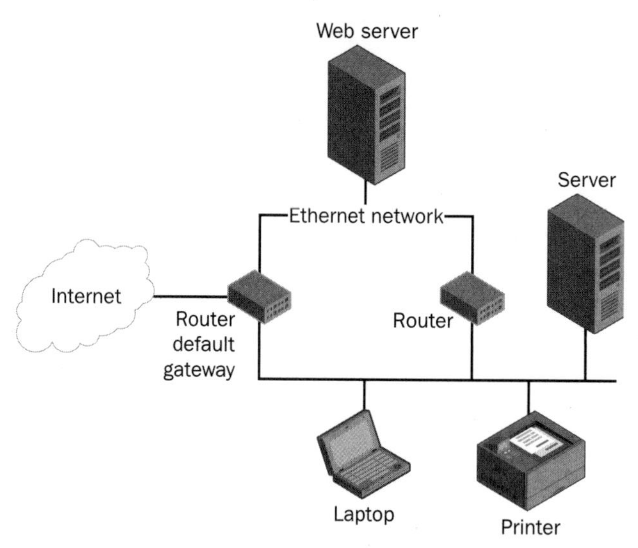

Figure 2-8 The router connected to the Internet is often considered the default gateway.

6. Once the router examines the data packet's destination address and determines that it does know how to forward the packet—that there is at least one route in its routing table to that network segment—the router selects the best possible route.
7. If other routers are on the path, the routing table will list the physical address of the next hop router in the path to the destination network segment. The router then takes that physical address and puts it in the packet so that, as it forwards the packet out, it will go to the correct router. It is important to remember that as a packet of data moves throughout an internetwork, its physical destination address changes but its logical address remains constant.
8. If the router is directly connected to the segment, it forwards the packet out to the destination device.

9 When a data packet finally reaches the router connected to the destination device's segment, the router will place the physical address of the destination device in the data packet and then transmit it.

The interesting thing to note is that in a network with multiple redundant routes that constantly change, the route is never predetermined. Prior routers might have a data packet's route listed in their routing tables, but any router at any point in the path can make an independent decision to forward the data packet differently. Each router forwards data packets on a hop-by-hop basis.

Another point about routes is that it sometimes seems that data could be passed around a network forever. If the data kept getting different information and were forwarded, hop by hop, it could simply be passed around without ever reaching a destination. In some cases, a router with old information could pass data back and forth with other routers, each believing that the route is still "alive," when, in fact, the route is no longer available. This situation is called a routing loop, and when it occurs it can cause so much congestion that the network is overwhelmed and nothing can get through. One of the safeguards against a routing loop prevents routers from using any physical address beyond a certain number of hops. In simple routing protocols, this is usually the sixteenth hop. A data packet can go through 15 routers and at the sixteenth it is dropped. Other routing protocols use a tick system, or time to live (TTL), in which the amount of time remaining is counted downward until the packet expires. This method doesn't avoid routing loops altogether, but it certainly removes the possibility of persistent immortal data packets eating up bandwidth.

Brouters

Brouters are a combination of router and bridge. This is a special type of equipment used for networks that can be either bridged or routed, based on the protocols being forwarded. Brouters are complex, fairly expensive pieces of equipment and as such are rarely used.

A Brouter transmits two types of traffic at the exact same time: bridged traffic and routed traffic. For bridged traffic, the Brouter handles the traffic the same way a bridge or switch would, forwarding data based on the physical address of the packet. This makes the bridged traffic fairly fast, but slower than if it were sent directly through a bridge because the Brouter has to determine whether the data packet should be bridged or routed.

Test Smart Routed traffic will either be received on special interfaces or identified by using a specific protocol. For example, the Brouter may be configured to bridge NetBEUI traffic and to route TCP/IP traffic. You will need to know this for the exam.

ATM Switches

You will get used to seeing the word *switch* associated with a multiport bridge, but it is often also used in routing terminology. *Asynchronous Transfer Mode (ATM) switches* are actually routers for ATM protocol traffic; they connect ATM networks together.

ATM is a protocol that provides some of the highest speeds possible through its use of *cells* instead of variable-length packets. *Cells* are small, fixed-length packets of data. In ATM, a cell is exactly 53 bytes long. Because the ATM switch knows exactly how much data to expect with each cell, it doesn't have to check the header for packet length and then check for errors. Instead, when the ATM switch receives exactly what it expected, it forwards it along at an extremely high speed.

ATM can travel at speeds of a billion bits per second, or 1 gigabit per second (Gbps), and the ATM cell can transmit voice, video, or data, making it an attractive protocol for large internetworks. Because ATM is very fast and flexible, it is often used for connecting LANs to a WAN backbone. While an ATM switch is a type of router, it is usually connected to a standard router in order to connect to other types of networks, as shown in Figure 2-9.

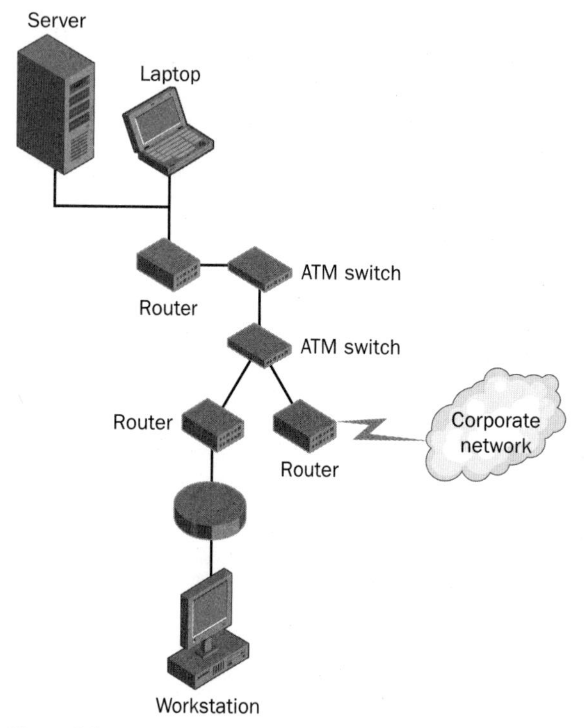

Figure 2-9 ATM switches connect to routers in order to link up to other types of network topologies.

Frame Relay Switches

Frame relay is a well-known WAN protocol that is used to transmit data packets over digital phone lines, called *digital circuits* or *leased lines*. The *frame relay digital circuit* is dedicated from one location to another and provides a specific bandwidth or throughput rate of data transmission.

A T1 line is one example of a leased line, and it is commonly used to interconnect geographically distant offices through telecommunications carriers. A T1 line is capable of 1.544 megabits per second (Mbps) of throughput. In Europe, telecommunications carriers provide E1 lines, which are similar to T1 lines but have more bandwidth. An E1 line is capable of 2.048 Mbps using 32 channels of 64 kilobits per second (Kbps). When you compare the E1 line with a T1 line, it appears nearly identical, except that the T1 line has 24 channels of 64 Kbps. The T1 line (and all other T-carrier lines) uses a process called bit-robbing for control information, which leaves it with seven bits per channel to carry data, making the effective data rate 56 Kbps for a total of 1.544 Mbps. E1 lines don't use bit-robbing, but even so, a European E1 line connected to a T1 line in North America to carry frame relay or other protocol traffic will work.

Frame relay data packets are variable in length and tend to travel in bursts. Because of its bursty nature, frame relay is better suited to data than to voice or video and is usually a WAN protocol. It's fairly common to run into a frame relay network, because the frame relay switches are relatively cheap and telephone companies usually have leased lines readily available in almost any locale.

Quite often, frame relay is used to connect multiple locations. Because the telecommunications company provides the leased line, you probably won't know how many switches or which paths in the telecom network the data will travel through to reach its final destination. Because of its unknown paths and switches, a frame relay network is often depicted as a cloud, as shown in Figure 2-10.

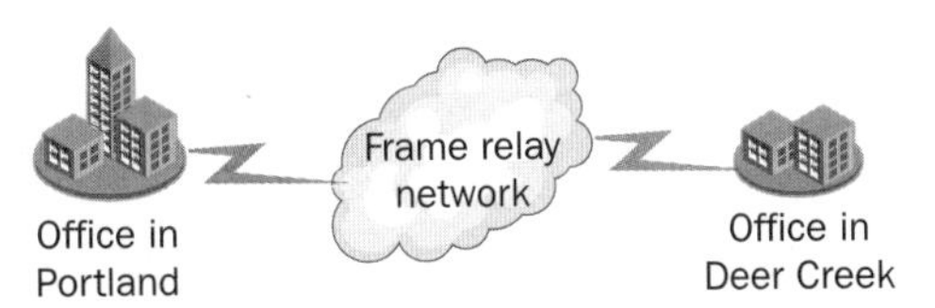

Figure 2-10 Frame relay networks are often depicted as clouds.

Setting Up a Lab You might find that reading about a network is not nearly as interesting as actually working on one. Concepts can appear abstract until you put them into practice. Hands-on experience is a plus, even if it is limited to a lab. In fact, sometimes it's even more valuable because you can intentionally break the pattern through incorrect configuration to see how the network reacts, and then when you run into the same symptoms in real life, you'll recognize them immediately.

CompTIA intended its examinations to evaluate the experience of networking technicians, administrators, and engineers, so the ability to recognize the symptoms of a problem is important throughout the test. Even if you've had years of experience, a lab can be an invaluable tool in preparing for the test. Here are some issues to consider when you set up a lab.

You'll be tested for experience with some of the more common protocols and equipment, so try to equip your lab with Ethernet over UTP, a server, two or more client workstations, and a hub. Also try to obtain an uninterruptible power supply (UPS), surge protectors, and various types of NICs and other adapters. If you can get a switch or a router, add that to the mix. If you can't get a router, install two network interfaces into one of the workstations or the server, configure routing on it, and then obtain a second hub or use a different media type on the other network. This will show you how routing works between two LANs.

Most likely, you won't have a WAN to connect to, but it's possible to create a dialup link to the Internet and route traffic from a small network to it. This is good practice. You can easily install and configure a modem on any one of the computers and use it as a router to forward traffic between the NIC and the modem out to the Internet.

Try to gather different types of cabling, whether a serial interface like RS-232 or a V.35 cable that you might use with a *Channel Service Unit/Data Service Unit (CSU/DSU)*. Then make certain that you can recognize the various cables and know what types of equipment they connect with. If you can get a different type of cabling altogether, such as ThinNet, or if you're lucky enough to get fiber, use your lab to determine what it takes to change over from one type of media to another.

On both the workstations and the server, you can get twice as much use out of them if you are able to dual-boot among different operating systems. Since most high-end workstations can run a server operating system, you can even install servers on equipment that you intended to use as clients. It's helpful to understand how a workstation will interact with different types of servers.

Finally, when you get the lab set up, the best way to learn to troubleshoot is to ask a friend to go there when you're not around and simply unplug or, if the friend is computer-savvy, change something without your knowledge. Your test will be to restore the network to its working state.

CSU/DSU

A *Channel Service Unit/Data Service Unit (CSU/DSU)* is a piece of digital equipment used only for WANs that connects to leased lines through public networks, often using frame relay. The CSU/DSU sits between the router and the leased line and prepares signals for transmission across the WAN, ensuring that data has the proper signal format and strength to transmit across the leased line.

A CSU/DSU looks like an external modem in size and shape, but it generally has more than the two ports that a modem has. One port connects to a router, one to the leased line, and one or two more are for monitoring and testing. Figure 2-11 shows a typical CSU/DSU.

Figure 2-11 A CSU/DSU provides a port to connect to a router—usually through an RS-232 interface—a port to connect to the leased line, and additional ports for monitoring and management.

The CSU/DSU acts internally as separate units: the CSU receives and transmits signals to the leased line from the LAN; it also protects the network from outside electrical interference that the other side of the unit emits. Because the CSU connects to the leased line, the telecommunications company can echo loop-back signals to test the equipment. The DSU side manages the data line control and converts the signals between the time-division multiplexed (TDM) frames on the leased line and the frames from the LAN. The DSU side regenerates signals and handles all timing errors, providing a modem-like interface between the router and the CSU side.

A typical use for CSU/DSUs is to create a link from one corporate office via frame relay through a telecommunications company's network to another office. Multiple offices often connect to the same cloud at different throughput rates (often called committed information rates [CIR] because the telecommunications company committed to that throughput) to create a WAN. This is shown in Figure 2-12.

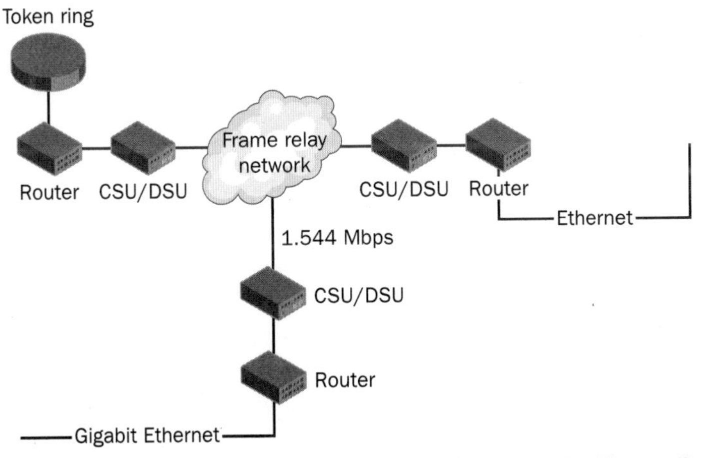

Figure 2-12 Frame relay networks can connect at different rates for different offices, yet all participate in the same frame relay "cloud."

Gateways

Gateways connect networks that use different topologies and entirely distinct network protocols, which we will learn about in more detail in the next two chapters. A gateway translates among the highest level of applications so that a computer that wouldn't normally be able to communicate with another type of computer can in fact do so.

You will see gateways in networks that use mainframes and minicomputers, where they are used for translating communications between a regular PC and the mainframe so that the PC can act as a terminal computer would. This is one of the most common methods of gateway communications, and it is shown in Figure 2-13.

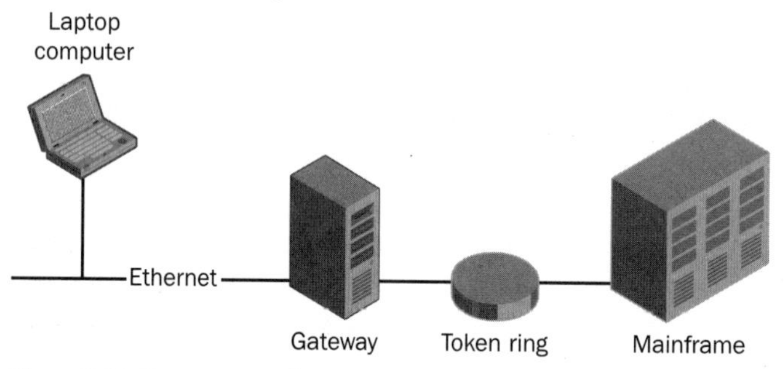

Figure 2-13 A gateway translates among entirely dissimilar networks all the way up to the application.

Gateways can connect networks that don't use TCP/IP to the Internet. For example, some products enable Novell NetWare networks to run native Internetwork Packet Exchange/Sequenced Packet Exchange (IPX/SPX) and prevent them from having to run TCP/IP natively. Not only does the gateway enable this communication, but because it handles translation services it can provide a measure of security as well. A TCP/IP attack via Telnet simply doesn't translate.

A gateway works rather simply. It receives information via one protocol on one NIC, translates it and wraps it in the other protocol, and sends it out the other NIC.

Calculating Bandwidth Needs for a WAN We've touched on a subject in this chapter that you might run into in your daily network life: the variable bandwidth that you can achieve with a telecommunications carrier. This is an important subject because the more bandwidth you have available (even unused), the more you will pay to your carrier.

Calculating your bandwidth needs depends on how much you expect to use the public network. For example, when you calculate how much bandwidth you'll need for a Web server to provide services on the Internet, you'll definitely need more than if you were only transmitting occasional e-mail messages to and from the Internet. For the latter, you could probably manage with a dialup link.

You should consider the amount of traffic that you'll be sending across the WAN. If this is the first time you've ever set up a WAN link, you'll have no data to go by. If it isn't, look at the statistics for a similar-size link elsewhere and make a more educated guess. The type of information you will need to estimate are the number of data transmissions, the average size of data transfer, and how tolerable a pause in data transmission will be.

Let's take an example of a WAN link between a sales office and a corporate office. The sales office will be sending e-mail messages, accessing a database, and uploading small amounts of data. There are 10 users, but their jobs are not tied to the computer. A dialup link is not appropriate in this case because the database application is used for point-of-sale, and the link must be available at all times that the office is open. Based on this information, you can try the lowest amount of bandwidth on a leased line, 56 Kbps, and see whether it will work well for the office. If it is not enough, your carrier will happily upgrade you. (I promise.)

In using a formula for the maximum bandwidth a link needs, multiply the number of simultaneous data transmissions by the average file size that is transferred. Because the 10 users are not tied to their computers, the simultaneous data transmissions might be only one or two every second, and that's on the high side. The average file size of an e-mail message or a database access application is probably about 4 or 5 kilobytes (K) (which you have to translate to *kilobits [Kb]* to get the correct data size for bandwidth). Given 8 bits in a byte, this comes to 32 to 40 Kb per file size. So the formula states that you will need 32 to 80 Kbps in bandwidth, but keep in mind that this is the maximum bandwidth needed. As the formula points out, 56 Kbps is probably more than adequate even during peak usage.

Wireless Access Points

A *wireless access point (WAP)* is the equipment used to interconnect with multiple wireless devices. It's similar to a cell phone tower. If you've ever been in a building or part of town where your cell phone doesn't work, you're probably too far away from a cell phone tower. As you travel across the country, your cell phone simply switches from one cell phone tower to another, and you are constantly (except where there are no towers) able to communicate. This is how wireless networking functions.

In a traditionally wired network, data is transmitted across cables made of copper or fiber optics. In the wireless network, radio waves transfer data between wireless devices. A WAP concentrates data communications in a local area, which in turn creates a cellular topology. WAPs are able to communicate with one another so that data transmissions can easily pass from one area to another as a wireless device moves between cells.

Note WAPs are often installed on walls or ceilings in key locations and ideally where data usage is concentrated, to maximize throughput. Wireless networks take advantage of the fact that radio waves can pass through environmental obstacles such as walls and ceilings.

The only other component needed in a wireless network, aside from the WAP, is a wireless adapter, though bridges are often used to connect a WAP to an existing wired network. When you see a WAP, you should pay attention to the antenna, which directs the radio waves to a specific location. Sometimes a small adjustment to an antenna can increase throughput.

A WAP usually has at least one interface that allows it to connect to a wired network. Typically, this is an RJ-45 jack for Ethernet over UTP and often can be used for either 10BaseT or 100BaseT. The WAP also contains the antenna and wireless network transmitting hardware, and usually IEEE 802.1D bridging software to enable the WAP to act as a transparent bridge. When WAPs overlap at the edges of their cellular topologies, they act as peripheral bridges to extend the network, communicating with each other to ensure transparent access to the end user. Figure 2-14 demonstrates how WAPs can bridge to a wired network.

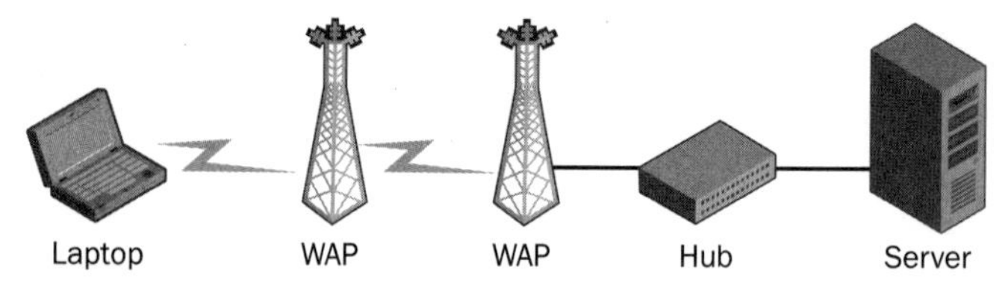

Figure 2-14 A WAP can bridge to a wired network.

Analog Signals and Modems

The word *modem* comes from two verbs: *modulate* and *demodulate*. The job of a modem is to translate the digital signals that are created on a computer to the analog signals that are understood on the plain old telephone service (POTS), which is also called the Public Switched Telephone Network (PSTN). When the modem receives the digital signal from the computer, it modulates the signal into an analog signal and transmits it across the wire. When the modem receives an analog signal, it demodulates the analog signal and translates it into a digital signal, which it then sends back to the computer.

Today's modems, which can reach speeds up to 56 Kbps, are often incorporated directly into a laptop, because they are frequently used with dialup services, as well as in PCs, which connect to the Internet. An internal modem can be an adapter card, or it can be integrated directly into the motherboard. External modems are also available and much easier to troubleshoot than internal modems because many include lights for each phase of connectivity. If a failure occurs in the connection, you need only check to see where the lights stop working and move forward from there. However, modem technology has become very stable over the past several years, so the need to troubleshoot modem connectivity is lessening. When you connect to an external modem, you will use a serial cable with an RS-232 interface with either 9 or 25 pins.

The most common configuration for a modem is a simple connection to a computer for dialing up to the Internet or to a corporate network as shown in Figure 2-15, but modems have other uses: One is a dialup link to a WAN, in which the router is connected to an external modem through a serial port and configured to dial up the telecommunications carrier network to link up. Another is a backup dialup link for a router configuration, where the router is connected via other means (such as a leased line) to a WAN, and the modem is activated only if the main link fails. You will also often find that CSU/DSUs and routers managed by the telecommunications carrier have a modem connected to an additional serial port. This modem is used not for a backup dialup link, but only for the carrier to dial in and check on the equipment's status. Finally, modems connected to fax servers are used to communicate with fax machines rather than transmit data for networking.

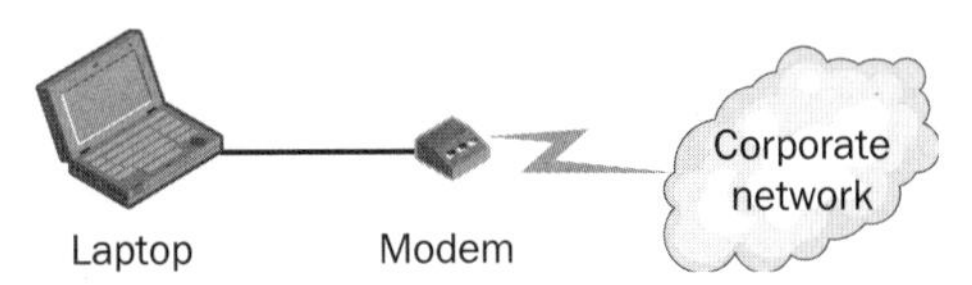

Figure 2-15 Remote users often use modems to dial into a corporate network.

NICs

As you might know, an adapter is any physical device that allows one piece of hardware to be adapted to another type of hardware. In computers, adapters are usually built into a card that is placed in a slot on the motherboard, resulting in an increase in the computer's capabilities.

A network adapter is also called a *network interface card*, or *NIC*. NICs simply allow a computer to connect to a LAN and transmit data at the speed of the type of network, which is usually greater than 4 Mbps (for a token ring network) and can achieve 1 Gbps (for a Gigabit Ethernet network). Each NIC is built specifically for a type of network protocol and topology. For example, an Ethernet 10BaseT card will not function on an Ethernet 10Base5 network because it does not have the correct interface to the media. Some cards are capable of connecting to more than one type of media or topology. For instance, you can easily find NICs that connect to either 10BaseT or 100BaseT networks. Years ago, when ThinNet was the most common network type around, I used to install cards that had both ThinNet and 10BaseT ports on them so that it would be easy to switch over the network to 10BaseT. Of course, easy is relative to the times. Today a 10/100 NIC will automatically sense the type of network it is attached to and immediately begin communications. Back then, I remember having to reconfigure each NIC's BIOS and sometimes open up the PC and alter jumpers to change over to the different media.

ISDN Adapters

You will probably hear people refer to ISDN modems for higher-speed dialup access to the network. ISDN stands for Integrated Services Digital Network, and modem is actually a misnomer in this context. Because the ISDN adapter is translating to and from digital signals, it is in reality coding and decoding, and should be called a CODEC (which is a real term, by the way).

Test Smart The exam will test you on the uses for ISDN. ISDN is one way to overcome slow dialup services in connecting small offices together, dialing up to a corporate network, or even using the Internet. An ISDN adapter with the basic rate interface can transmit data at speeds up to 128 Kbps.

Depending on where you live in the United States and how much research and development has gone into ISDN at your local telephone company, you might be able to obtain an ISDN line. There has been a lot of interest in Digital Subscriber Lines (DSL) from some telephone companies, and given the competition from high-speed cable modems, the development of ISDN has often been dropped in favor of DSL.

ISDN is a digital over telephone copper wiring technology that is capable of running over leased lines and regular telephone lines. The ISDN terminal adapter is used to establish a link by dialing up another computer and connecting. External ISDN adapters have one port to connect to the computer and another to connect to the phone line. Internal adapters, like modems, are placed in an expansion slot on a PC's motherboard.

Businesses can run ISDN over a single T1 line and have multiple ISDN lines available for users to dial into from outside the company. This is done using the ISDN Primary Rate Interface (PRI), which offers 23 B channels of 64 Kbps each and one D channel also of 64 Kbps. When an individual dials in from a home, he or she usually is provided with a Basic Rate Interface (BRI), which provides two B channels of 64 Kbps and one D channel of 16 Kbps. ISDN is a good choice for users who work remotely on a consistent basis from a single location, either a home or satellite office. For users who travel often and have to connect from hotels, ISDN isn't a good choice because it is seldom available.

System Area Network Adapters

Before we talk about System Area Network (SAN) adapters, you should know what a SAN is. SAN can also stand for Storage Area Network. A SAN provides high-speed access to data over fiber optic media or copper media without the overhead of a standard protocol suite. When it uses fiber optic media, the actual SAN can be anywhere from a few meters to several kilometers in length. The SAN over copper media does not have the capacity for distance that it does over fiber optic media.

In the SAN, several servers connected to a hub are able to share one or more databases. The SAN is separate from the LAN, as shown in Figure 2-16. Offering up to 1 Gbps throughput in traditional SAN and more recently even faster throughput, the SAN enables servers to reach data that can be located quite a distance away in order to share one or more databases. A typical SAN is switched with hubs that support four to eight nodes. Larger networks can be built from cascaded hubs.

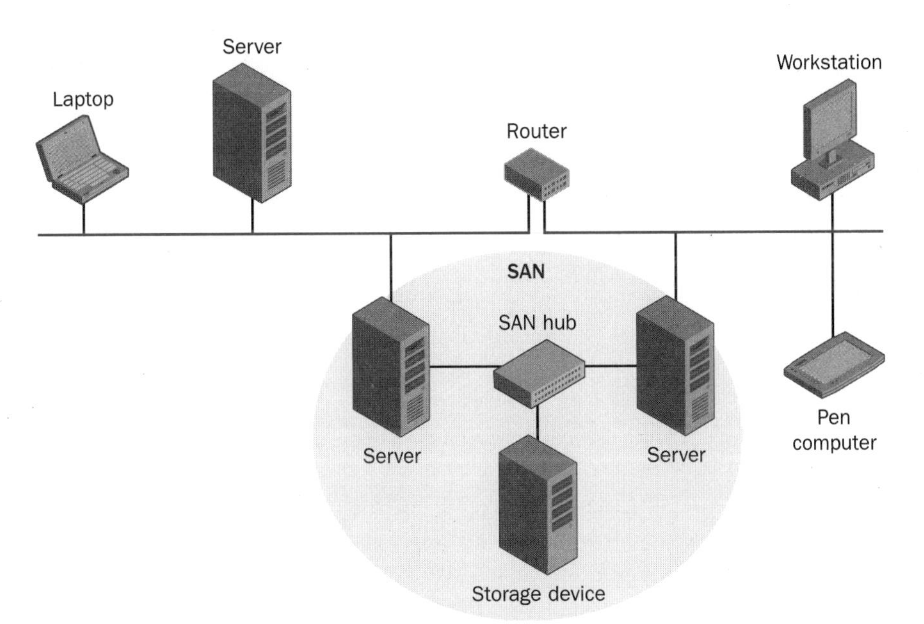

Figure 2-16 The SAN is separate from the rest of the network, so it doesn't interrupt LAN traffic, nor is it interrupted by it.

Now that you know what a SAN is, you can probably guess that the SAN adapter is the card used to connect devices to the SAN. The only devices that would connect to the SAN are the servers and storage devices that participate in sharing data. Workstations and other servers do not need to be connected to the SAN, because they will access data through the servers by transmitting across the regular LAN.

Key Points

- Networking components are the hardware devices used to help data communications across a network.
- Hubs, which form physical star topologies, are multiport repeaters. They receive a signal from one port and regenerate and transmit that signal out the remaining ports.
- A repeater is used in a physical bus topology. It lengthens the segment by regenerating the signal it receives and retransmitting it out the other port, making two cables appear as one.
- Hybrid hubs are used to connect different types of media, such as fiber optics and UTP, into a single network.
- A Multistation Access Unit (MAU) is a type of hub used in token ring networks. An MAU forms a physical star, but it functions as a logical ring topology.
- Bridges are capable of filtering out data packets by storing the physical addresses of devices in a lookup table. If a data packet is received on the same port on which its destination is located, the bridge does not forward the packet.
- Switches are multiport bridges, enabling high-speed data transmissions by combining the physical star topology of a hub with the data-filtering ability of a bridge.
- Routers forward data between different segments that can be distinct physical and logical topologies but must use the same upper-layer protocol suite such as TCP/IP.
- Brouters are a combination of a bridge and a router.
- ATM switches are special routers used for forwarding ATM cell-based traffic.
- Frame relay switches are routers that forward variable-length frame-based traffic.
- Both ATM and frame relay are typically used as backbones in a WAN.

- A CSU/DSU links a leased line to a router so that an office can communicate across a WAN.
- Gateways translate between applications on dissimilar computer networks. A gateway is often found in networks that also connect to a mainframe or minicomputer.
- Wireless access points (WAPs) are similar to cellular telephone towers in that they provide transparent data access as a wireless device moves around a network.
- The word *modem* stands for modulator/demodulator and is used to translate digital signals into analog signals before transmitting them across the plain old telephone service (POTS).
- Network interface cards (NICs) enable a device to communicate across a network.
- An ISDN adapter transmits across the Integrated Services Digital Network (ISDN) like a modem, except that it transmits digital signals across the telephone system.
- System Area Network (SAN) adapters are used for storage systems to create a network separate from the LAN that workstations and servers not participating in the SAN gain access to.

Chapter Review Questions

1 Which of the following is used with a leased line?

a) CSU/DSU

b) MAU

c) Hybrid hub

d) Repeater

Answer a, CSU/DSU, connects directly to the leased line and prepares the signal format before transmitting across the leased line. Answer b, MAU, is incorrect—it's used as the connectivity device for a token ring network, forming a physical star topology like a hub but logically transmitting data in a ring. Answer c, hybrid hub, is incorrect because it creates a LAN using two different types of media. Answer d, repeater, is incorrect because it simply repeats a signal from one side of a bus topology media to the other.

2 Which of the following is known for its ability to transmit data at high speeds?

a) MAUs

b) Bridges

c) Gateways

d) ATM switches

Answer d is correct. ATM switches are known for their ability to transmit data at high speeds mainly because they use cell switching, in which a small fixed-length packet of data can quickly be switched in and out of memory on the ATM switch. Answer a, MAUs, is incorrect because it is used with token ring networks that are not known for high speed. Answer b, bridges, is incorrect because a bridge can be used as easily in a slow network as a fast one. Answer c, gateways, is incorrect; gateways are typically slow because of the high amount of translation they must do to connect two dissimilar networks.

3 Which of the following is known to be able to transmit data across a telephone wire at speeds up to 128 Kbps?

a) Modem

b) SAN adapter

c) ISDN adapter

d) NIC

Answer c is correct. An ISDN adapter is capable of transmitting digital signals across telephone wires at speeds up to 128 Kbps. ISDN can achieve greater speeds over a leased line. Answer a, modem, is incorrect; modems generally transmit at rates up to 56 Kbps. Answer b, SAN adapter, is incorrect because a SAN adapter can reach beyond 1 Gbps. Answer d, NIC, is incorrect; a NIC will transmit data at the rate of the network, which varies based on the type of network but is generally greater than 4 Mbps.

4 Of the following, which is the best hardware device for breaking a network into two segments and filtering traffic?

a) Repeater

b) Gateway

c) Hub

d) Bridge

Answer d is correct. Bridges are used to break a network into two segments and filter traffic between them. Answer a, repeater, is incorrect because it does not filter traffic. Answer b, gateway, is incorrect because it primarily translates between two dissimilar networks. Answer c, hub, is incorrect because a hub doesn't filter traffic either.

5 Which of the following devices acts like a bridge but has multiple ports like a hub?

a) Gateway

b) Switch

c) CSU/DSU

d) SAN adapter

Answer b is correct. A switch is a multiport bridge. It looks like a hub but only transmits data to a destination port based on a lookup table similar to a bridge. Answer a, gateway, is incorrect because it doesn't have multiple ports. Answer c, CSU/DSU, is incorrect because it is used to connect a network to a leased line. Answer d, SAN adapter, is incorrect because it is used to connect a server or storage device to a system area network.

6 Which of the following will regenerate and retransmit a network signal to all ports?

a) Passive hub

b) Active hub

c) Switch

d) Bridge

Answer b is correct. Active hub is correct because it will regenerate every incoming signal and retransmit it out every port. Answer a, passive hub, is incorrect because it doesn't regenerate signals, it merely repeats them. Answer c, switch, is incorrect; a switch doesn't transmit signals through every port. Answer d, bridge, is incorrect because a bridge doesn't retransmit signals through its other port.

7 Which of the following is the WAN equipment you would find in a network that uses leased lines to connect offices?

a) Hybrid hubs

b) Brouters

c) CSU/DSU

d) WAPs

Answer c is correct. CSU/DSU is used to connect a network to a leased line. Answer a, hybrid hubs, is incorrect because they are not necessary for a leased line. Answer b, Brouters, is incorrect; Brouters aren't necessary for a network to connect to a leased line. Answer d, WAPs, is incorrect because a WAP isn't required for a network using a leased line.

8 Which of the following offers bandwidth variety from 56 Kbps to 1.544 Mbps?

a) Frame relay switch

b) Router

c) Modem

d) ATM switch

Answer a is correct. Frame relay switches offer a variety of bandwidth from 56 Kbps to 1.544 Mbps over leased lines. Answer b, router, is incorrect because a router can connect at much higher speeds, depending on the type of network it connects to. Answer c, modem, is incorrect because a modem can reach 56 Kbps. Answer d, ATM switch, is incorrect because ATM offers up to 1 Gbps in throughput.

9 Which of the following pieces of equipment is used in a physical star topology?

a) Routers

b) Bridges

c) NICs

d) Hubs

Answer d is correct. Hubs are used as the central point of a star topology. Answer a, routers, is incorrect; routers are not required for a star topology. Answer b, bridges, is incorrect because bridges usually have only two ports and can't be in the center of a physical star topology. Answer c, NICs, is both correct and incorrect–NICs are necessary for every type of topology, because they allow the devices to connect at the terminal ends of the star, but a NIC doesn't necessarily indicate a star topology.

10 You are called in to help select a new piece of equipment to expand a client's network. You know that the network must be able to pass through NetBEUI traffic, which must be bridged. It also must be able to send TCP/IP traffic, but two different subnets are required to route traffic between the old part of the network and the new addition. Which of the following is best suited to this task?

a) Switch

b) Router

c) Brouter

d) SAN adapter

Answer c is correct. Brouter is correct because you must bridge the NetBEUI traffic and route the TCP/IP traffic. A Brouter is the only piece of equipment that can perform both bridging and routing. Answer a, switch, is incorrect because it can perform only bridging services. Answer b, router, is incorrect; the router can only perform routing services. Answer d, SAN adapter, is incorrect because the SAN adapter neither bridges nor routes traffic.

Part 2

Protocols and Standards

Equipment, cabling, media, servers, printers, and workstations all make up the physical part of networking, and each requires rules for how to communicate. Protocols and standards define these rules, and the International Organization for Standardization (IOS) created the Open Systems Interconnection (OSI) protocol reference model to tie the rules together in an orderly fashion.

This section of the book describes the OSI model, along with other protocol suites, such as TCP/IP, and how they relate to the model. Knowing the OSI model and these protocol suites is a requirement of CompTIA's Network+ exam, according to Domain 2.0 objectives. You can find these objectives in their entirety at *http://www.comptia.org/certification/network/network_objectives-domain2.asp*.

Chapter 3

The OSI Model

In networking, the Open Systems Interconnection (OSI) model is one of the most useful tools available to help you understand how protocols work. But the OSI model is not a protocol in itself, and many protocols do not follow its guidelines specifically. Despite its discrepancies with various protocol stacks, the OSI model is valuable as a teaching and development tool.

In this chapter, you'll learn about the OSI protocol reference model and each of its layers in detail. They are presented by layer, from the first physical layer (also called layer 1) to the final application layer (also called layer 7). A section at the end of the chapter reviews basic concepts about the networking equipment discussed in Chapter 2 and shows where that equipment functions with reference to the OSI model.

Proposed by the International Organization for Standardization (IOS) and released in 1984, the OSI model describes a decidedly structured set of seven protocol layers interconnecting as a stack. This protocol stack model shows how the protocol layers can communicate among open systems. Keep in mind that a protocol stack of layers was not the only way that computers could communicate. Companies had already developed proprietary protocols that worked only with their own computers, which at the time were less powerful than today's PCs and as large as mainframes. Many of these proprietary protocols were just one monolith that functioned across only one type of physical media and could only provide a link between specific applications. As you can see, this entire method did nothing to enable collaboration between different types of computers. A layered model was preferable to a monolithic proprietary model because a monolithic model would have obstructed the development of applications, utilities, and networking equipment that could function in different environments with different types of computers.

Every protocol layer in the OSI model was developed to maintain the ability to connect diverse types of computers, and boundaries between layers were selected to minimize information flow across interfaces. When a layer was created in the model, it was done because differentiation was required to provide a specific function. In the following sections, you'll learn what each function of each layer is. As you can see in Figure 3-1, the model provides for seven layers: physical, data-link, network, transport, session, presentation, and application.

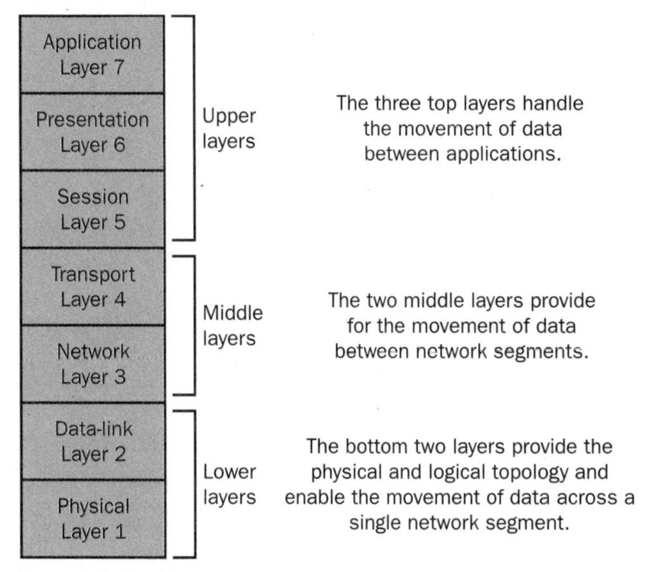

Figure 3-1 The OSI model is a layer of seven protocols. Layer 1 (physical) and layer 2 (data-link) provide for movement of data across a single network segment. Layer 3 (network) and layer 4 (transport) provide for movement of data among multiple network segments. Layer 5 (session), layer 6 (presentation), and layer 7 (application), provide for movement of data among applications.

Network products and applications can be described in part by where each fits within this layered structure. The OSI reference model is applicable to all network environments, even when the protocols in any particular suite do not fit neatly within the OSI model's layered structure. For example, the TCP/IP protocol suite, which includes programs such as Simple Mail Transfer Protocol (SMTP), is a collection of products that communicate across an internetwork. The Internet Protocol (IP) portion of the TCP/IP suite fits into the network layer; Transmission Control Protocol (TCP) fits into the transport layer; and SMTP fits somewhat loosely within the session, presentation, and application layers.

Note You'll need to know the order of the layers in the OSI model. One way is to retain it is to memorize this sentence: please do not throw sausage pizza away (physical, data link, network, transport, session, presentation, application).

Although each separate layer in the OSI model represents a protocol that provides a specific function, they can be lumped into three major types of layers based on how data is moved. The bottom two layers, physical and data-link, provide for the transport of data across a single physical segment. The middle two layers, network and transport, enable the transmission of data among multiple segments. The top three layers—session, presentation, and application—facilitate the movement of data among services or applications on different network nodes, even if the applications are on different types of computers.

Note The network and transport layers are key to routing data across an internetwork.

The entire protocol stack is structured so that data can travel from one application on a device on one network segment to another application on another device located on a different network segment. When one device transmits data to another, the data transmission must be managed specifically at both ends of the connection and in between to ensure successful communication. This process includes several aspects:

- **Interface.** Each network node must be able to access the locally attached physical media. A conversation can't take place between devices attached to two different types of physical media—such as one to copper wiring and another to a wireless network segment—unless there is a mechanism present that can translate between the two types.
- **Media access.** The manner in which devices are able to access and utilize the media. Media access can't incessantly interrupt existing transmissions; there must be a method for devices to gain access to the network segment.
- **Reliability.** Each transmission on the network must have some way to detect and correct errors to ensure that data is transmitted to the correct destination, in full and uncorrupted.
- **Signaling.** At a very basic level, every bit that is transmitted from a device must be in the same format as the device at the other end of the physical media. For example, when a positive 3 volts is considered a "1" bit on the sending device, it also must be considered a "1" on the receiving device.

One of the basic requirements for communication on a network is that both network devices must use the same protocol stack. The receiving device takes delivery of, handles, and translates the data from the sending device at a particular layer. This process is called *peer communication* because each layer corresponds to its peer on the other network node. To achieve peer communication, each layer breaks apart and encapsulates the data passed down from upper layers before it is transmitted across the network. *Encapsulation* refers to the entire data conversion procedure handled by the sending station: the sending station converts data into increasingly smaller capsules until they are mere bits, and then puts those bits back together into increasingly larger capsules at the receiving station. Encapsulation uses the following process:

1. The sending application accesses the application layer and begins to transmit a message to a receiving application.
2. The application layer forwards the data through the presentation, session, transport, network, data-link, and physical layers, in that order.
3. At each layer, the message is broken down into smaller packets; a header for that particular layer is added with control information, which are bits designated to describe how to determine whether the data is complete, uncorrupted, in the correct sequence, and so forth (and sometimes a trailer is added); and the message is then forwarded to the next layer down in the protocol stack.
4. When the message reaches the physical layer, the data is converted into signals (whether radio frequency [RF], optical, or electronic) and then transmitted across the media.
5. The data travels to the destination device, is received at the physical layer, and then is sent up to the data-link layer.
6. At each layer in the receiving network node, the headers (and trailers, if they exist) for that layer are stripped off and the data is reassembled into larger packets. In addition, if the control information includes additional instructions, those actions are performed.
7. The message is finally fully assembled at the application layer and forwarded to the destination application.

Encapsulation is shown in Figure 3-2.

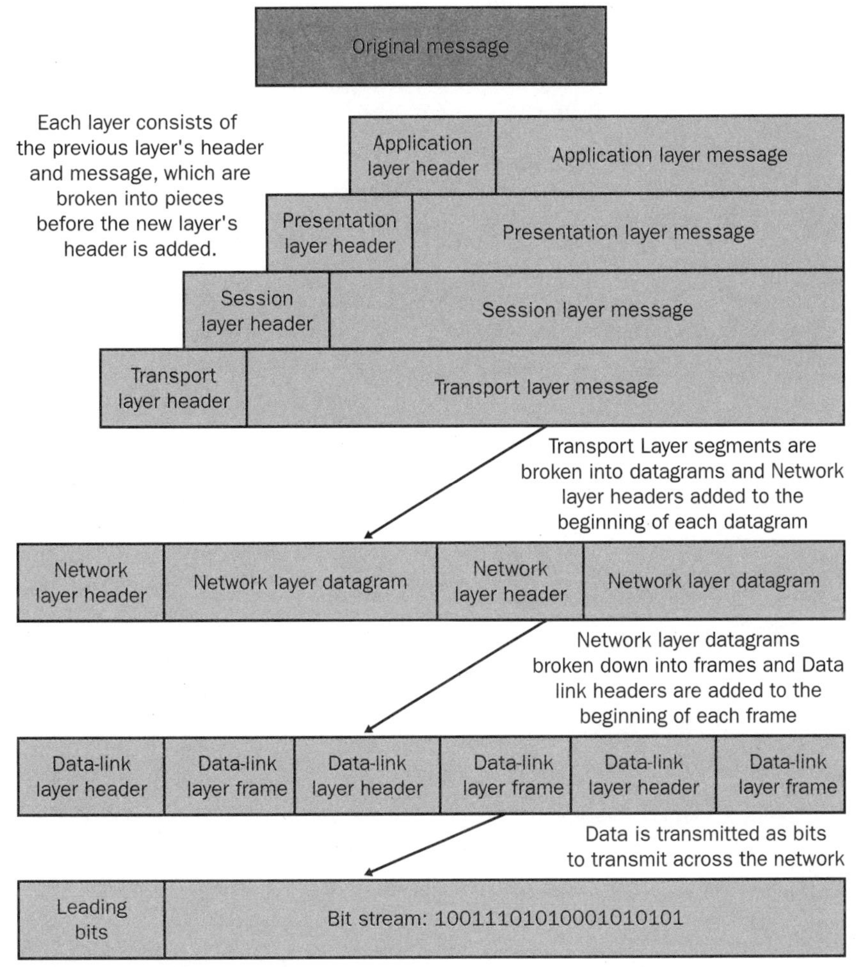

Figure 3-2 The application-layer message is added to the application-layer header. When it reaches the layer below, the header becomes part of the presentation-layer message portions, and new presentation-layer headers are added. This process continues until data reaches the physical layer, upon which the data becomes a bit stream with leading bits at the physical layer.

Test Smart In encapsulation, it's important to remember that at the sending device, each layer breaks the data down into smaller packets and adds its own header; while at the receiving device, each layer strips off the header and builds the data packets into larger packets, somewhat like a rocket that reenters earth's atmosphere. As the rocket gets closer to landing, unnecessary pieces are discarded. Each protocol layer is blind to the headers of any other protocol layer and cannot process them.

Physical Layer

The physical layer is responsible for the transmission of raw data in the form of a stream of bits across physical media. Layer 1's specifications comprise the media interface, the physical topology, and the signaling methods used across the media. Layer 1 conveys the bit stream data through electrical impulses, light, or radio signals. The bit stream travels throughout the network at the most basic electrical level. To do this, the physical layer must define the cable media (or wireless mode), the network interfaces, and all physical aspects of the network. Examples of physical-layer protocol definitions include Asynchronous Transport Mode (ATM), Gigabit Ethernet, and the RS-232 interface. Note that both ATM and Gigabit Ethernet contain specifications that cross into the Data Link layer. It is very common to find protocols that encompass more than one layer.

Note Some people refer to the OSI reference model layers by numbering them. The first layer is the physical layer. References to layer 1 signify the physical layer in the OSI model.

Understanding Signaling

Signaling is something that, once understood, forms a foundation for troubleshooting the network. Signals represent data traveling across the network media. Data transmits in binary format, either in one or zero bits. Groups of eight bits amount to a byte. (Bytes are counted by a factor of two; instead of 1000 bytes equaling a kilobyte, it is actually 1024 bytes—and 1024 kilobytes equal a megabyte.)

Digital Signals

Signals exist either in digital or analog form. Most networks use digital signals, each bit transmitted as a discrete voltage, radio frequency, or optical signal. At the physical layer, the protocol describes an encoding scheme so that the network interface of every device connecting to the network segment will understand the same digital signal to be equivalent to the same value, either one or zero.

Note While understanding signaling is important in learning about networking, for passing Network+, you won't be called upon to remember the details of encoding schemes.

- **Unipolar.** The simplest encoding scheme is unipolar. It uses either a negative voltage plus a zero voltage or a positive voltage plus a zero voltage. In all encoding schemes, voltage amounts are very small, usually around a positive or negative 3 volts, and typically no higher than 25 volts. When you compare this with the 12,000 to 35,000 volts of electrostatic discharge created when a person walks across a carpet (static electricity), 3 volts is hardly noticeable. Unipolar signals are called "unipolar" because they stay to one side of the zero-voltage pole and never use both positive and negative voltage. Because of the simplicity of this scheme, it must use a separate clocking system to synchronize signals between the sender and receiver. (A clocking system is somewhat like a metronome, the device that musicians use to make certain that they don't fall out of tempo by playing too fast or too slow. In networking, you need clocking to make sure you don't send a signal too fast for it to be received or so slowly that the receiving device shuts down the connection.) An example of a system using unipolar encoding is teletype (TTY).

- **Polar.** Polar encoding uses both a positive and a negative voltage, but never a zero voltage. Because of the wider separation between the voltages in this type of encoding, there is less vulnerability to noise; but like unipolar encoding, it requires a separate clocking signal.
- **RZ.** Return to zero (RZ) is unique in that it doesn't use the actual voltage to represent a bit value; instead, it uses the transition of a voltage to zero—meaning that a positive 3 volts switches to 0 volts, or a negative 3 volts switches to 0 volts—to represent the bit. When the voltage moves from either positive or negative voltage to neutral, it signifies either a one bit or a zero bit. This scheme avoids noise interference because noise affects voltage levels, but not the transition of a voltage level.
- **NRZ.** Like RZ, the nonreturn to zero (NRZ) encoding scheme uses voltage transitions. In the scheme, values of bits are given to transitions or the absence of transitions during a clocking time period, which is provided by a separate clocking mechanism. For example, if the voltage moves from positive to negative, it might be considered a "1" bit; while during a clocking period, if the voltage simply stays positive, it might be considered a "0" bit.
- **Biphase.** The most common types of encoding schemes that people deal with today are biphase. Biphase gives values to bits at each transition and is self-clocking. Errors are easily detected when there is an absence of a transition. Of the two types of biphase encoding, Manchester is the most common because it is used for Ethernet, and Differential Manchester is well-known because it is used for token ring.

Analog Signals

Analog signals travel in waves, like sound, and are typically carried over the plain old telephone service (POTS). Analog signals are prone to many problems because of their wavelike configuration; over long distances, they can become attenuated or stretched out so that waves are no longer distinct, and they're susceptible to noise interference. Because of the analog signal modulation, there are three different ways to determine the value of a bit: amplitude, frequency, or phase.

Amplitude is the height or loudness of the wave. Amplitude-shift keying (ASK) modulates the signal between two or more levels.

Frequency is how many waves are transmitted during a single time period. In frequency-shift keying (FSK), two different frequencies are used to give value to a bit. A faster signal might mean a "1," and a slower signal a "0."

Phase is the point at which the wave is either going up or down. Phase-shift keying (PSK) changes the phase of the signal to represent a value, either a one or a zero bit. If the phase doesn't change during a clocking period, the bit is equivalent to the opposite value.

Connecting at the Physical Layer

The connections that are supported at the physical layer are either multipoint or point-to-point. Multipoint connections are characterized by more than two network devices connecting to the same physical media. Point-to-point connections incorporate only two devices using a shared physical connection that can't be used by any other devices. Figure 3-3 shows a multipoint and a point-to-point connection.

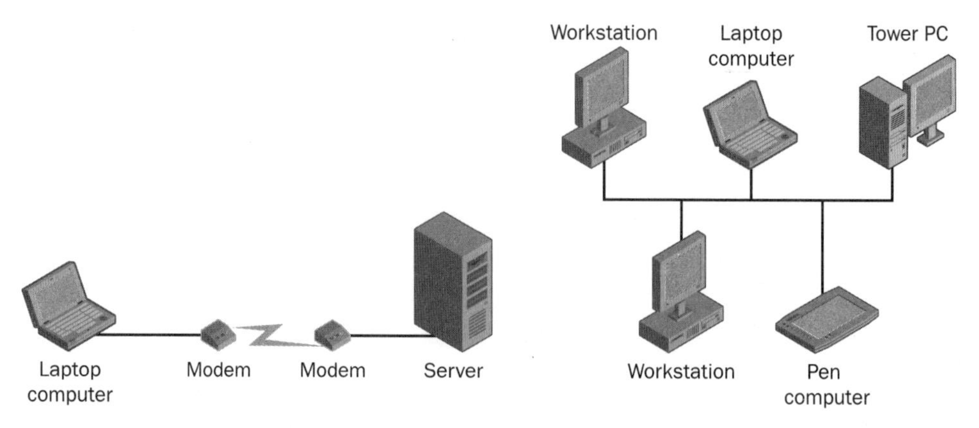

Figure 3-3 A point-to-point connection and a multipoint connection.

As discussed in Chapter 2, the physical topology is the shape of a network. Not only does the physical-layer protocol describe the signals representing bits of data, it also defines the topology that signals will travel across. Table 3-1 describes the physical topologies defined at the physical layer.

Table 3-1 Physical Topologies

Topology	Connection Type	Benefits
Star	Point-to-point	Easy to troubleshoot because there are few points of failure.
Bus	Multipoint	Easy to install and fairly cheap.
Ring	Multipoint	Never has a signal interruption.
Mesh	Point-to-point	Has high reliability because all connections are redundant.
Cellular	Point-to-point	Used for wireless only. Easy to install, difficult to secure.

Data-Link Layer

Layer 2 in the OSI reference model is called the data-link layer. It is the only layer with two sublayers: Media Access Control (MAC) and Logical Link Control (LLC). The data-link layer supplies hardware addresses, identifies errors, and manages flow control.

Within the data-link layer, the bit stream received from the physical layer is assembled into larger pieces, called *frames*. Basically, the data-link layer decodes frames into bits to send them to the physical layer at the sending device. At the receiving device, it encodes the bits into frames and reads the header before acting on the data. The control information in the frame header handles errors, flow control, and frame synchronization. Examples of protocols that work within the data-link layer include Ethernet, token ring, and ATM. You will find that protocols such as Ethernet and ATM include definitions for the physical layer as well as the data-link layer. This is very common because the two are closely related. Protocol specifications for the physical layer also often work within the data-link layer.

When the data-link layer frames the bits, it is able to insert control information into a frame header. This header provides the source and destination of the frame on that network segment.

The data-link layer also refers to the logical topology of the network. Unlike the physical shape of the network, the logical topology refers to the shape that data makes when transmitted throughout the network segment. Data takes only two logical shapes, bus and ring.

A logical bus topology describes the transmission of signals sent in such a way that the signals reach every device connected to the segment. As each device reads the information, it reads the destination address. If the address matches its own, the device accepts the transmission and begins to process it. Ethernet uses a logical bus topology, even when the physical topology is a star or bus.

Data traveling from one device to another in a serial rotation eventually returning to the first indicates a logical ring topology. This type of transmission is read by only one device at a time and never undergoes interruptions. Token ring networks use logical ring topologies even when the physical topology is a star or ring.

Media Access Control (MAC)

The first sublayer of the data-link layer is called Media Access Control (MAC). This sublayer defines how devices are able to gain access to the media. It also maintains the hardware address for the device.

The three types of MACs are:

- **Contention.** When devices access the media using the method of contention, they have immediate access to the media but only transmit when they have data. When two devices attempt to transmit data at the same time, the data collides (which results in collision lights on a hub). The more devices on a network segment, the more collisions you'll likely see. To avoid collisions, the device can check the media using a process called carrier sensing. If the device doesn't find any current data transmissions, it will send its own data. Even though the device uses this technique, collisions still take place. After a collision, both devices whose data collided will usually wait a random length of time before retransmitting to avoid a second collision.
- **Polling.** Although rarely used, polling is another MAC available, in which a central mechanism regulates network traffic by polling each device attached to the network and determining whether it is ready to transmit data. The polling mechanism acts like "Mother" in the game "Mother, May I?" If permission isn't granted, the data doesn't transmit. Polling is rarely used because it has a single, central point of failure. If something goes wrong with the polling device, the entire network goes down.
- **Token passing.** Token passing is the second most common type of MAC. It is used in both token ring and Fiber Distributed Data Interface (FDDI), which both have a ring topology. However, ARCnet, which is an older protocol that is rarely used (if at all) today, uses token passing with a bus topology. In token passing, a device gains access to the

media after it receives a special frame called a token. When the device needs to transmit, it reformats the frame, transmits the data, receives an acknowledgment, and then transmits the token to the next device.

MAC Address

The MAC address, also called the hardware address, is assigned to every network interface attached to the network segment. Each MAC address is unique to each interface. When data is received, the MAC address appears in the data-link frame header and identifies the station for which the data is destined. Network devices read the data-link header to determine whether their own MAC address is in the header; if so, the data is accepted.

Today, manufacturers assign MAC addresses to the equipment's network interface before it is delivered to consumers. Manufacturers who produce Ethernet network interface cards (NICs) assign a 6-byte address to each NIC using a designated addressing format. The manufacturer is granted a unique 3-byte address to use for the first portion of the MAC address by the Institute of Electrical and Electronics Engineers (IEEE), and then concatenates a unique 3-byte number representing that individual card for the remaining portion of the address. This method prevents a manufacturer from using the same address as another manufacturer and ensures that all MAC addresses on the network are unique. An example of an Ethernet MAC address is shown in Figure 3-4.

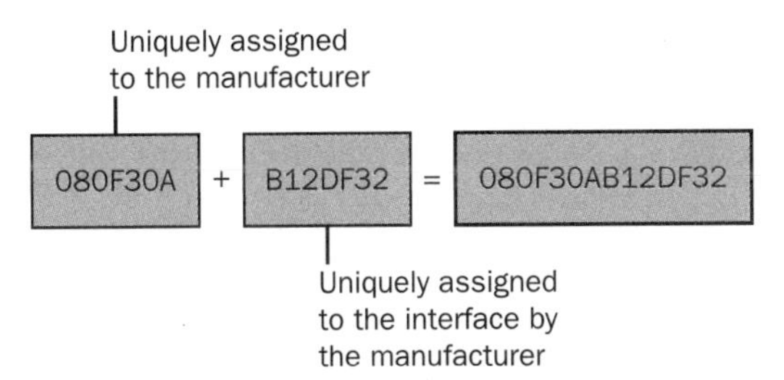

Figure 3-4 An Ethernet MAC address is a 6-byte hexadecimal number. The MAC address has a total of 12 hexadecimal digits; each digit is a number from 0 to 9 or a letter from A to F.

Caution MAC addresses are assigned by the manufacturer of the network interface. They are often called "burned-in addresses" because the manufacturer burns them into the hardware.

Logical Link Control

The second data-link sublayer is called Logical Link Control (LLC). Responsible for logical link functions of single or multiple connections, the LLC sublayer uses control packets called Protocol Data Units (PDUs). A PDU contains all the protocol information in the header necessary for the LLC to do its work. The LLC then provides services to the higher-level network layer (layer 3).

One of the services that the LLC provides to the network layer is Type 1 Operation, which is an unacknowledged, connectionless mode service. In this mode, the LLC transmits frames without knowing the destination or waiting for acknowledgment.

The other type of service that the LLC can provide the network layer is connection-mode services. In this mode, the LLC layer at both the sending and receiving stations is aware of the connection. Through control information, the LLC establishes and terminates the connection for that layer.

The LLC works with the MAC sublayer, expecting the MAC to send it data requests and status information. The LLC primarily provides flow control and frame sequencing services. These data requests are provided in a specific format that all IEEE 802.x specifications support.

Network Layer

The network layer provides the basics for internetworking, which is the transmission of data from one physical segment to a different segment. Up to this point, the physical and data-link layers handled information that stayed on the same segment. The network layer becomes more complex. Various technologies are involved in routing, including the creation of logical paths or virtual circuits to provide a conduit from a sending device to a receiving device that is independent of the physical bit stream. This process occurs at the network layer (layer 3) and involves:

- Providing an address for each network segment
- Providing a logical address for each device that is separate from the MAC address
- Routing and forwarding data
- Selecting a route if there are multiple routes to the same network segment
- Discovering the routes to other network segments
- Error handling, congestion control, and packet sequencing

Three components are required to communicate between two devices: the network path, addresses, and switching. These are shown in Figure 3-5. A network path is provided by the physical connections between the two network segments. Addresses are applied to the network segments and the devices attached to them. Switching is the process of moving data from one network segment to another, somewhat like an old-time telephone operator plugging a connection in from one telephone to another.

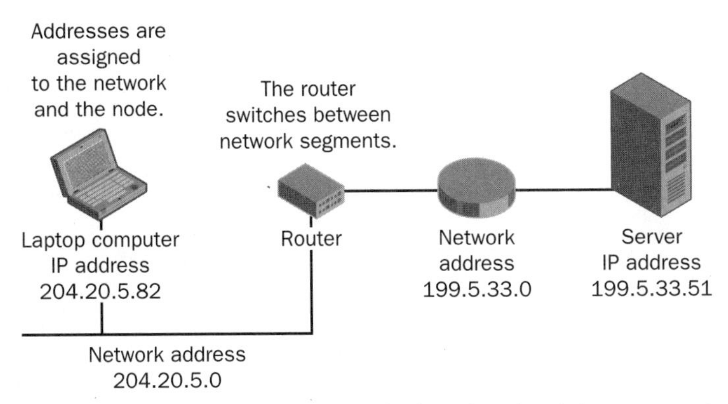

Figure 3-5 At the network layer, communication takes place between segments.

At the network layer, pieces of data, called datagrams, are reorganized into frames when they are passed to the data-link layer. Each datagram incorporates a header with data. The header includes the network layer's control information, such as the logical network address for both the sending and receiving devices, routing information, and flow control.

Addressing in layer 3 is the key to its routing capability. Network segment addresses are similar to the name of a street used in a postal address, while node addresses are much like house numbers. Each street name must be unique. To find a house on a particular street, the house number must be unique to that street. Logical network addresses are typically assigned to each segment by the network administrator. Network node addresses, however, are either assigned by the administrator or they inherit the MAC address from the data-link header and use it as a logical node address. TCP/IP uses assigned network node addresses, while the Internetwork Packet Exchange (IPX) inherits MAC addresses for the node address portion.

Understanding Switching and Routing

Switching is often another name for routing. In fact, while routing is the general term about "what" routing is, switching describes the actual "how" specific types of routing work. Switching refers to one of four different ways of moving data among different network segments on a network. These four methods are used in different types of protocols:

- Cell switching
- Circuit switching
- Message switching
- Packet switching

Of these switching types, cell switching is the newest. It is nearly identical to packet switching, except that it uses cells, which are short, fixed-length sets of data. The short length is key to the extraordinary speed that cell-switched protocols can achieve. ATM is a protocol based on cell switching. Because it uses a 53-byte cell, the ATM switches immediately know how much data in each cell is expected, which reduces latency.

Circuit switching is one of the oldest switching methods. In fact, the Public Switched Telephone Network (PSTN) uses circuit switching for phone calls. The circuit is a path set through an internetwork that is held open for data communication. If there is an idle circuit, it reduces the efficiency of the network because that idle bandwidth can't be reassigned to other communications.

Message switching improved on circuit switching by not requiring permanent paths for data transmission. In message switching, a message traverses the network separately from other messages on different paths. Each message is a series of packets. To send the message, a path opens up at the start of the message and closes at the end. As the message travels from device to device, the optimal path available can be selected based on the current network conditions, avoiding network congestion.

Packet switching, also known as independent routing, is one of the more common switching methods. Messages are broken down into packets and then sent on separate paths through the network. While packets are a much shorter length, they are variable in size. For example, a packet can be between 46 and 1500 bytes in size. Packets can also arrive at the destination device in the wrong order, in which case the message must be reassembled. Control information in the packet header includes sequence numbers to ensure that packets are assembled correctly. Packet switching uses a virtual circuit. A permanent virtual circuit (PVC) is a logical path established between two nodes, and it remains open whether or not data needs to be transmitted. A switched virtual circuit (SVC) is established at the beginning of the data transmission, dedicated for the duration, and closed once data transmission has ended. An example of a protocol using packet switching is frame relay.

Now let's examine what routing is all about. In routing, data travels from one segment to another based on router decisions about the optimal path to the next "hop." Beyond selecting a route and forwarding data, a router will also discover routes to other segments.

Routing protocols are often used for both route selection and discovery. Originally, network administrators had to enter each route statically into a routing table on each router. As the number of routers and routes grew on a network, the amount of administration required to simply manage the routes took up a tremendous amount of time. To reduce the administrative load, routing protocols were developed. With a routing protocol, route discovery and selection are dynamic and require little administrative overhead. Like static routes, dynamic routes are also stored in a routing table on each router.

As mentioned earlier, routing protocols are also used to select a route. They determine the metrics for each route—such as the cost of a route, number of hops, and so on—and store them in the routing table. The cost of the route is a number assigned to the link to give it a relative priority. When a router selects a route, it looks for the best deal and sends data along the path that is least expensive and has the lowest assigned cost. The metric of cost is determined using criteria such as how many hops, the amount of bandwidth available, and the time to transmit across the link. Hops are sometimes the only metric used for route selection. For each hop, a data transmission is expected to travel through another router. Ticks, defined as $1/18$ second, are the amount of time it takes to reach the final destination. A router that decides based on both hops and ticks will select the fastest route first (lowest number of ticks) and then, if there are still multiple routes to choose from, the closest route (lowest number of hops).

Fault tolerance is built into routers. When the least costly route becomes unavailable, the router selects the next route. But load balancing, so that data travels across two (or more) redundant routes, is not built in to most routing protocols. If load balancing is required, it must usually be configured by the network administrator.

When routing protocols perform route discovery, they use one of two methods: Distance Vector or Link State. Distance Vector routing protocols periodically broadcast the entire routing table to neighboring routers. When a router receives a neighbor's routing table, it adds any new routes to its own routing table. The problem with this method is that the router receives secondhand information. The presence of redundant routes combined with the use of secondhand information can lead to routing loops. The other problem with Distance Vector is that the entire routing table is sent in domino fashion, from router to router, so that the time it takes for the entire network to add a new route is excessively long, especially if the network is large and complex. The process of synchronizing all the routes on a network is called *convergence*. Distance Vector protocols use hops and sometimes ticks for the routing metric. Routing Information Protocol (RIP) is an example of a Distance Vector protocol.

Link State routing protocols were created to fix the problems with Distance Vector protocols. These advanced protocols use a hello process to announce a new router on the network. Then when a hello packet is sent out, neighboring routers respond. The new router collects these responses and then broadcasts all its immediately known routes to the entire internetwork. In this way, every router receives firsthand information about new routes, and the time it takes for the new routes to be added to the network is greatly reduced. Open Shortest Path First (OSPF) is an example of a Link State routing protocol in the TCP/IP protocol stack.

How the Network Layer Header Helps in Routing Information contained in the network-layer header is used to perform routing functions. Remember how data is stripped of its headers when it is received at a station? A router strips up to the network-layer header, and then after acting on the network layer's routing information, adds new headers, sends the datagram to the data-link and then physical layers, and finally forwards the data on to the next station. Not having to look at information in the upper-layer protocols lets a router perform its functions more rapidly.

The network-layer header includes the addresses of both the source device and the destination device. A router reads this information and looks in its own routing table to find the next router in the path to the destination address. Figure 3-6 shows an example of a network-layer header format for Internet Protocol (IP), the network-layer protocol in the TCP/IP protocol suite.

Version indicator	Header length indicator	Type of service	Total length	Iden-tification	Flags	Fragment offset	Time to Live (TTL)	Protocol	Header checksum	Source IP address	Destina-tion IP address

Figure 3-6 The IP header shows the types of information needed at the network layer for routing to occur.

Transport Layer

Layer 4, a.k.a. the transport layer, provides for a transparent transfer of data between the sender and receiver nodes. The control information in the transport-layer header provides end-to-end recovery and flow control. This type of control information ensures that the data is completely transferred.

One thing that's unique about the transport layer is port numbers, which are used to name the ends of logical connections. When data is received, the transport layer knows which application to forward that data to by reading the port number. In the TCP/IP protocol suite, many "well-known" ports are used for specific applications. For example, Simple Mail Transfer Protocol (SMTP) uses TCP port 25.

When data is received from upper layers, the transport-layer segments that data, applies sequence numbers to each of the segments so that they can be reassembled in the correct order, and adds the port number identifying where the data should be sent to when it is received at its destination. Then the data is sent to the network layer, through the data-link layer and the physical layer. It travels across the network, and at the destination the bit stream is reassembled into frames, and the data-link layer's frame header is stripped and assembled into datagrams for the network layer. Then the network-layer headers are stripped off. At the transport layer, the segment is assembled with the correct sequence numbers, and the data is checked for errors. If there were errors, the transport layer either initiates retransmission or notifies upper-layer protocols, being the session, presentation, and application layers. The transport-layer header is then read to discover the destination port number, and the data is passed to the correct upper-layer process.

The transport layer is the key to multiplexing services. Multiplexing services enables a node to communicate with multiple network nodes at one time. One of the jobs that an upper-layer application can do is communicate with multiple network nodes simultaneously. Plus, multiple applications can communicate on the network at the same time. These applications manage this multitasking feat by using port numbers. The use of port numbers is known as service addressing, and the transport-layer header includes both the source and destination ports. The transport-layer header also identifies the transaction so that request and response exchanges can be handled correctly. When multiple requests from upper-layer processes are sent to the transport layer, it segments each of the messages, identifies them with the port number, applies sequence numbers to the messages, and then interleaves them before sending the data to the network layer.

When data is received, the transport layer is responsible for reassembling it, separating the information that is sent to different port numbers, and then putting them in the correct order according to the sequence numbers. Then the transport layer sends this data to the correct upper-layer protocols.

The transport-layer header translates alphanumeric names to network-layer logical addresses. Because alphanumeric names are easier for people to use than memorizing jumbles of numbers, and because computers rely solely on those jumbles of numbers, a translation between the two must take place. Specific transport-layer protocols provide address/name resolution. In TCP/IP, Address Resolution Protocol (ARP) and Reverse Address Resolution Protocol (RARP) are both transport-layer protocols that translate MAC addresses to logical network-layer addresses.

Connection Orientation

Transport-layer protocols are either connection-oriented or connectionless. Connection-oriented protocols help a sending device determine whether the data it sent was received at the destination. This is also called *reliable communications*.

Reliable communications are performed through the use of acknowledgments (ACK). If an ACK packet is expected but not received in a certain time period, the data is considered undelivered. Both ACKs and negative acknowledgments (NAKs) let the sending device know that data was correctly received. TCP of the TCP/IP protocol suite is an example of a connection-oriented protocol.

Connectionless protocols do not provide reliable data transmission. The transmission might be called unreliable, but this doesn't mean that the data is not received at the destination device. Connectionless, or unreliable, means in this case that the sending device does not *expect* to receive any ACK or NAK packets. An example of a connectionless protocol is User Datagram Protocol (UDP). Connectionless protocols are faster than connection-oriented protocols because their headers are smaller and less communication needs to take place between the sender and receiver.

Session Layer

Layer 5, the session layer, establishes a connection between applications and manages that connection until it terminates it. It is aptly named because it deals with the session between two applications. When coordinating connections between applications, the session layer can create a conversation without regard to the actual path that data takes throughout the physical network.

Remote procedure calls (RPCs) function at the session layer. An RPC is an application program interface (API) that allows a remote application to execute on a computer so that it appears to the user that the application is local. This is the type of transparent communication that the session layer provides for applications like Telnet and Citrix MetaFrame ICA sessions.

Not only do session-layer protocols establish sessions, they also terminate them in an orderly manner. The session layer makes certain that resources that were being used are released for use by later sessions. Depending on the implementation of the session-layer protocol, an abrupt release of a session can sometimes be reestablished so that the session resumes where it left off.

The session layer sets the agreement on the services used and the duration of a connection. Verification of user IDs and passwords is handled in session-layer protocols. Data is transmitted in a session in three ways:

- **Simplex.** In a simplex transmission, data can travel in only one direction. One device acts as the sender, and the other is the receiver. A radio works in simplex mode.
- **Half-duplex.** Half-duplex transmissions alternate between the sender and the receiver. While each end of the session acts as both a

transmitter and receiver, the data can move only in one direction at a time. A CB radio uses half-duplex communications.

- **Full-duplex.** Full-duplex mode provides for simultaneous communications traveling in both directions. The device at each end of the session acts as both a transmitter and receiver at all times. A telephone works in full-duplex mode.

Presentation Layer

The presentation layer provides the way for data to be presented to the application. Encryption/decryption and data compression/expansion are handled at this layer. While the presentation layer is also layer 6, it is sometimes called the syntax layer.

The primary service that presentation-layer protocols provide is formatting data. When data is sent to lower layers, the presentation layer translates that data into a transfer syntax before passing it on to the session layer.

The presentation layer uses the following types of data translation:

- **Bit order.** In bit-order translation, the sending and receiving devices agree to read each byte from either the first or last bit received.
- **Byte order.** Byte-order translation syntax means that the sending device and receiving device read a string of data from either the first or last byte received. Big endian, traditionally used by Motorola processors, is where data is read from the first byte. Intel processors use little endian and read data from the last byte received.
- **Character code.** When sending data from a system that uses one type of character code to one that uses a different one, the presentation layer must agree on and translate the character code. For example, a Windows PC might use Unicode, while an IBM mainframe uses Extended Binary Coded Decimal Interchange Code (EBCDIC), and communication must be translated between these two.
- **File syntax.** The file syntax describes the meaning of bytes that are sent throughout the network.

Application Layer

Layer 7 supports end-user processes. In general, this is the only protocol layer that the end user will ever interact with. The application layer identifies the sending and receiving devices, quality of service (QoS), authentication, and privacy. All transmissions at this layer are specific to the application being used.

The most common application-layer functions are file transfer, e-mail messages, Web, chat, network printing, and additional services that connect a device to other network nodes. FTP is an example of an application-layer protocol.

Application-layer protocols connect across the internetwork in three ways:

- **Collaborative computing.** Collaborative computing provides for application-layer services that act in true client/server fashion. Both the client and the server are aware of the other's existence. The two share processing and application services and are interdependent.
- **Operating system call interception.** On the other end of the spectrum, operating system call interception methods come into play when the application being used is completely unaware of the network's existence. Using this method, the application-layer protocols intercept the application calls and then determine whether they should be redirected to a network device.
- **Remote operation.** The application layer goes into remote operation whenever a device, acting as a client, uses an application to open the console of a device acting as a server. Using this method, the user can control the server remotely.

Network Equipment and OSI Model Layers

In practice, the OSI reference model was not intended to be rigidly applied to all protocol suites; it was developed to assist network equipment manufacturers in creating devices and applications that could interoperate, avoiding proprietary and monolithic systems.

The OSI reference model allows developers to improve a piece of networking equipment or application at a specific layer (or layers, in some cases) and be assured that the product will work with products that function at other layers in the market. For example, when Gigabit Ethernet was developed, manufacturers created software drivers, NICs, hubs, and switches at the physical and data-link layers. The Gigabit Ethernet specification was easily incorporated into existing networks because there was no need for additional changes to the protocol suite, its applications, or its services that worked from the network layer up through the application layer.

NICs

Every device that connects to a network segment uses a network interface. In most cases, network interfaces are implemented as adapters, or cards, that can be added or removed from a device in a modular fashion. Sometimes a network interface is integrated into the device. NICs in workstations are either PC Cards for laptops or standard adapters, as shown in Figure 3-7.

Figure 3-7 Network interface examples include both PC Cards for laptops and PCI adapters that connect to PC motherboards.

Network interfaces perform the functions of accessing the media, transmitting and receiving bit stream data, and incorporating the MAC address. These are all functions of the physical and data-link layers.

Hubs

Hubs are devices that move data along the same physical segment. A hub creates a single physical segment by connecting multiple wires (or other media) to itself at the center. The hub is merely a connecting point for a segment and is often called a multiport repeater. Originally, a repeater copied a signal received from one port and sent it out its other port. This served to lengthen the media between two network devices. Passive repeaters are merely conduits; they forward the signals they receive in whatever form they are received, which means that they can't really lengthen the media as intended. Active repeaters not only retransmit signals, they also regenerate them. When the active repeater regenerates the signal, it can travel farther because any attenuation of the signal or noise interference is eliminated during the regeneration. In doing so, active repeaters are able to lengthen the path that a signal can take.

Because a hub is a multiport repeater, it is also available in both active and passive forms. Nearly all hubs are active. You can tell the difference between a passive and an active hub just by checking whether the hub requires power. Only active hubs require a power source to regenerate signals.

Hubs and repeaters work at the physical layer. A repeater deals only with the bit stream data and moves it to its other port. There is no need to read the data or determine addressing and error correction information in a higher layer's header.

Bridges

Like a hub, a bridge connects media segments. *Unlike* a hub, the bridge can decide what data to transfer from port to port, effectively filtering out unnecessary data transmissions. The bridge reads the destination address of data frames and decides whether the address is located on the "local" side or is "remote," where local refers to the same port from which the bridge received the data. If the data is destined for an address located on the local side, the bridge does not forward the data. If the data is destined for a device on the remote side, the bridge forwards the data.

To perform this feat, a bridge must be able to read the MAC addresses of each piece of data it receives. That requires a bridge to be able to work at the data-link layer, which includes the MAC sublayer, MAC addresses, and frames.

Switches

A switch works exactly like a bridge, except that it connects more than two segments. You might consider a switch to be a multiport bridge just as a hub is a multiport repeater. A switch contains a table that stores all MAC addresses of the devices attached to it. The table maps these addresses to the switch ports to which they are connected, and the switch uses the table to look up and forward data to the correct port. Because the data is forwarded only to one port, bandwidth is significantly increased across all ports.

Like a bridge, switches look at MAC addresses and must be able to translate data at the data-link layer.

Routers

When data needs to move from one network to another, it travels through a router. A router is either a specific type of equipment, or it is a computer configured to forward data between two or more network interfaces that are connected to different LANs. Routers store the logical addresses of the networks in a table in order to select the optimal path to forward data on. Because a router must read the network address of the data that is forwarded, it works at the network layer.

Brouters

The spelling is not a mistake. Brouters are a special type of equipment that combines the abilities of a bridge with those of a router. Depending on the ports, the configuration, and the data itself, the data can be bridged or routed. Brouters work at both the data-link layer and the network layer.

Gateways

A gateway is a special type of data-forwarding device that provides communications between completely dissimilar systems. A gateway analyzes incoming traffic and determines its destination. If any traffic is destined for the dissimilar system, the gateway translates the traffic into the correct protocol and then routes it to the other system. Gateways work at the upper three layers of the OSI reference model.

Key Points

- The OSI reference model was developed as a guideline for networking, providing for seven layers of protocols that work together to send data from an application on one network device to an application on another network device.
- Using the OSI layered model allows network manufacturers to develop new implementations for one or two layers that will be able to interoperate seamlessly with the protocols at the other layers.
- The seven layers are, in order, physical, data-link, network, transport, session, presentation, and application.
- At the physical layer, data is transmitted in bit stream format, where each bit is represented as a zero or a one.
- The bit stream travels across physical media that takes a specific form and is called a topology.
- Network interfaces, repeaters, and hubs all work at the physical layer.
- The data-link layer is the only layer with two sublayers, the MAC and LLC.
- Bridges, switches, and Brouters all work at the data-link layer.
- The network layer enables data to be sent between two different segments. This layer organizes data into datagrams and provides network-segment addresses and network-node addresses.
- Routers work at the network layer. Brouters work at the network layer as well as the data-link layer.
- The transport layer determines whether data segments are not delivered correctly to destination devices and can initiate retransmission or inform upper layers.
- The transport layer provides service addressing, through the use of port numbers. It also provides error checking and segmentation.
- The session layer establishes and manages the dialog between two nodes and ensures a proper termination.
- The presentation layer is concerned with data format, including encryption, decryption, compression, and expansion.

- The application layer provides the interface to the user. This layer provides file, print, mail, and other network services.

Chapter Review Questions

1 Which of the following works at both the data-link and network layers?

a) Router

b) Hub

c) Brouter

d) Switch

Answer c, Brouter, is correct. A Brouter functions at both the data-link and network layers. It can perform both bridging and routing functions. Answer a, router, is incorrect because a router works only at the network layer. Answer b, hub, is incorrect because a hub functions only at the physical layer in forwarding the bit stream. Answer d, switch, is incorrect because a switch functions solely at the data-link layer.

2 What type of address is looked at in a data-link layer frame?

a) The MAC address

b) The network-segment address

c) The logical-node address

d) The service address

Answer a, the MAC address, is correct. The MAC address, also known as the hardware address, is stored in the data-link layer's header for both the source and destination node. Answer b, the network-segment address, is incorrect because this address is assigned at the network layer. Answer c, the logical-node address, is incorrect because this address is assigned at the network layer. Answer d is incorrect because the service address, also known as a port, socket, or connection ID, is assigned at the transport layer.

3 Which layer transmits bit stream data?

a) Application

b) Session

c) Network

d) Physical

Answer d is correct. The physical layer transmits data in bit stream format. Answer a, application, is incorrect because the application layer provides the interface to the user. Answer b, session, is incorrect because the session layer constructs a dialog between two network devices. Answer c, network, is incorrect because the network layer enables routing between two different network segments.

4 Which OSI reference-model layer provides for dynamic route selection and discovery?

a) Physical

b) Network

c) Transport

d) Presentation

Answer b is correct. Routing protocols are found at the network layer, and they reduce administration through dynamic route selection and discovery. Answer a, physical, is incorrect because the physical layer deals with bit stream data. Answer c, transport, is incorrect because transport-layer protocols provide for reliable transmission, segmentation, and error checking. Answer d, presentation, is incorrect because presentation-layer protocols handle data format.

5 True or false? The OSI reference model allows network manufacturers to develop new network products for a specific protocol layer so that they can interoperate with other protocol layers.

a) True

b) False

True. The OSI model helps in the development of new network equipment, allowing manufacturers to innovate at a particular protocol layer and plug into other layers of a protocol suite.

6 Which of the following types of data encoding is used for analog signals?

a) Differential Manchester

b) Unipolar

c) RZ

d) ASK

Answer d, ASK, is correct. ASK, which stands for amplitude-shift keying, is an encoding format for analog signals that defines the bit based on the size of the wave. The other two types of analog encoding are phase-shift keying (PSK) and frequency-shift keying (FSK). Answer a, Differential Manchester, is incorrect because it is a digital-encoding format used for token ring. Answer b, unipolar, is incorrect because it is a digital-encoding format. Answer c, RZ, or return to zero, is incorrect because it is a digital-encoding format.

7 What type of media access method allows all devices to transmit at any time?

a) Contention

b) Polling

c) Token passing

d) Signaling

Answer a, contention, is a media access method that allows all devices to transmit at any time. Answer b, polling, is incorrect because this method requires the permission of a central device to allow data transmission. Answer c, token passing, is incorrect because a device is allowed to transmit only when it receives the token. Answer d, signaling, is incorrect because signaling is not a media access method.

8 Which of the following processes takes place at the transport layer?

a) Signaling

b) Routing

c) Token passing

d) Service addressing

Answer d, service addressing, is correct. Service addressing is the transport-layer process of assigning a port number to upper-layer processes to identify which application owns the data conversation. Answer a, signaling, is incorrect because it takes place at the physical layer. Answer b, routing, is incorrect because routing takes place at the network layer. Answer c, token passing, is incorrect because media access takes place at the data-link layer.

9 What layer handles encryption?

a) Application

b) Presentation

c) Session

d) Network

Answer b, presentation, is correct. Encryption, as well as decryption, compression, and expansion, all take place at the presentation layer. Answer a, application, is incorrect because it handles network services such as file, print, mail, and so on, which interface with the user. Answer c, session, is incorrect because it deals with establishing and managing a dialog. Answer d, network, is incorrect because it works with routing data through the internetwork.

Chapter 4

Protocol Stacks

In Chapter 3, we reviewed the OSI reference model, which defines how a protocol stack, or group of protocols, works. Protocols are software with sets of rules that define how communication takes place. You need only one protocol stack to communicate across a network. Many networks use multiple protocols to communicate. Different protocol stacks can share the same wire, but they can't communicate with one another directly without a gateway to translate between them.

This chapter will review four protocol stacks that are all commonly used in networks around the world. We'll start with the most common, Transmission Control Protocol/Internet Protocol (TCP/IP), which is used to communicate across the Internet. The remaining protocol stacks we'll review are Internetwork Packet Exchange/Sequenced Packet Exchange (IPX/SPX), NetBIOS Extended User Interface (NetBEUI), and AppleTalk. IPX/SPX is the proprietary protocol suite Novell NetWare networks use, NetBEUI is typically used by Windows and OS/2 networks, and AppleTalk is the protocol suite that was developed specifically for Apple computers.

TCP/IP

Someone invents a new way to use the Internet every day; recent innovations include videoconferencing, controlling remote computers, managing a supply chain, and even using an Internet phone. Along with these new methods, new protocols are added to the TCP/IP protocol suite. So, when you think of TCP/IP, consider that it is constantly transforming to meet people's needs. Moreover, linking to the Internet, which enables an enterprise to interact with nearly every organization and a large population of consumers, has made Internet connectivity a must for businesses. Because of the pervasiveness of the Internet, TCP/IP is often the only protocol used in a network.

The TCP/IP protocol suite was developed long before the OSI reference model. It has a model of four layers, called the DoD model since it was developed by the Department of Defense. The four layers roughly map to the OSI model, even though it has seven layers, as depicted in Figure 4-1.

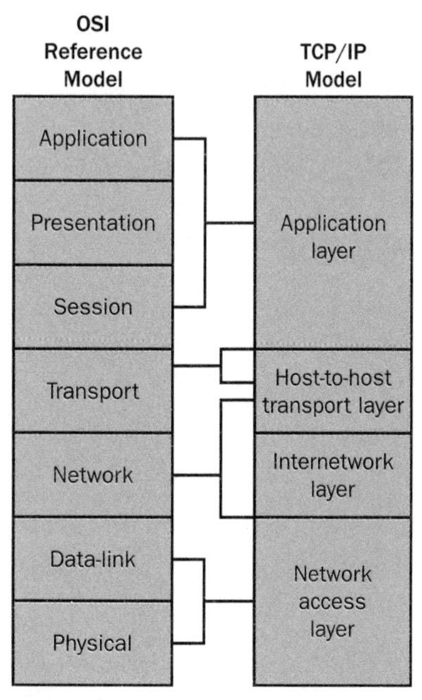

Figure 4-1 The four-layered structure of the TCP/IP model maps to the seven layers of the OSI reference model.

Because of the way TCP/IP's four protocol stack layers map to the OSI reference model's seven, individual protocols within the stack often encompass multiple OSI layer tasks. For example, the application layer of the TCP/IP model provides the application, presentation, and session layer functions offered in the OSI reference model.

There are many protocols in the TCP/IP suite. We'll examine the most important of them in Chapter 5.

Two Main Protocols, TCP and IP

The TCP/IP protocol stack begins with two main protocols, which you can probably guess are TCP and IP. *TCP*, or *Transmission Control Protocol*, handles roughly the same functions as the transport layer in the OSI model. TCP is a connection-based protocol that receives data from upper-layer services, breaks it into packets, and forwards the packets to the IP protocol. *IP* is broadly equivalent to the network layer of the OSI model (though it sits at the internetwork layer in the TCP/IP model) in that it handles the routing of packets throughout the internetwork. The IP protocol sits on top of the protocols that deliver data across the media in the network, such as token ring or Ethernet. These types of protocols are equivalent to the physical and data-link layers in the OSI reference model. Applications and services run on top of TCP or UDP (which are both located in the host-to-host transport layer of the TCP/IP model) at the TCP/IP model's application layer.

The Host-to-Host Transport Layer's TCP (and UDP)

In the TCP/IP model, the host-to-host transport layer is responsible for providing reliable transport of data between two devices, no matter what type of physical network media and systems lie in the middle. Both TCP and User Datagram Protocol (UDP) function at this layer.

Data is passed to the host-to-host transport layer as segments from the applications and services in the application layer, which is sometimes referred to as the process layer. When the segments are passed to the internetwork layer, they are broken into packets and encapsulated with an internetwork layer header. Most of these will use an IP header, but some may use headers from other protocols. For example, a request to resolve a MAC address to an IP address would use the Address Resolution Protocol (ARP) and would use ARP's own header.

One of the features common to both TCP and UDP is the use of *ports*. The port is identified by a number, which is placed in the header for both the source and destination applications so that when the data is reassembled it can be delivered to the correct one. A port number identifies which application the data is received from so that data can be reassembled and forwarded to the correct application at the receiving device.

While the word *socket* is often used interchangeably with the word *port,* a socket is actually made up of the IP address and the port number. Figure 4-2 depicts how sockets are created.

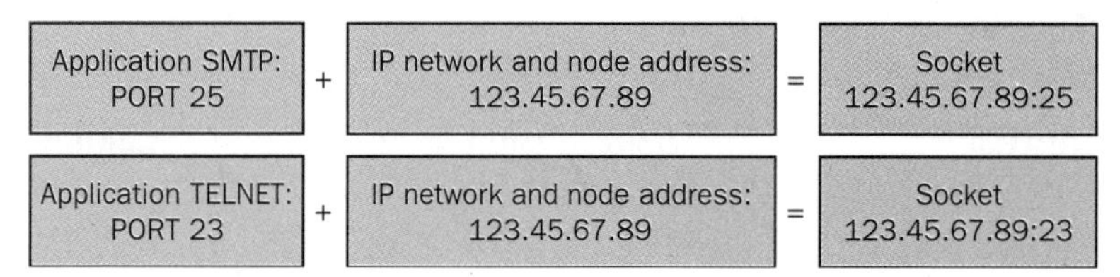

Figure 4-2 Sockets are created by adding the port number to the IP address, creating a unique identifier for each application running on a computer.

TCP, Transmission Control Protocol TCP, as a connection-based protocol, is designed to guarantee delivery of data from the sending device to the receiving device. Because TCP provides this guarantee, it requires additional control information in the header. Even more control information is used to ensure that each data packet is placed in the correct sequence order when it is received. The TCP header, shown in Figure 4-3, is the key to how TCP provides reliable, connection-oriented communications.

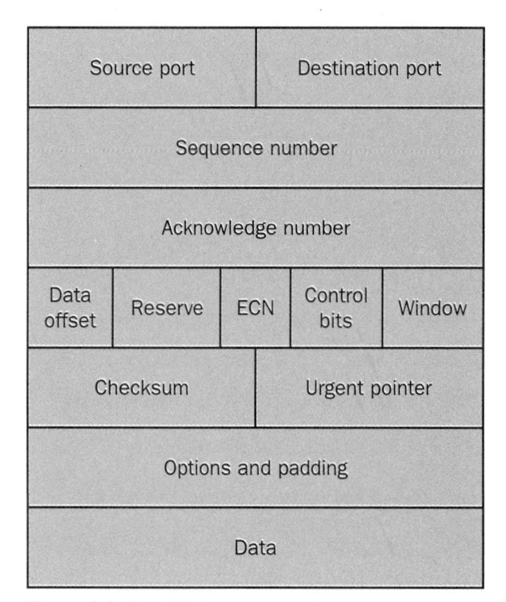

Figure 4-3 The TCP header contains a great deal of control information to help manage how data is transmitted.

TCP denotes each segment with a sequence number. When the data is received, the segments are placed in the correct order based on the sequence number and passed on to an upper-layer service. The sequence number is used not only for placing the segments in the correct order, it also identifies duplicates and missing segments. When duplicates exist, the receiving node can discard them. When segments are missing, the sending node retains copies in a buffer until it receives an ACK (acknowledgment) packet. If an ACK is expected but not received, the sending device retransmits the segment.

One of the ways TCP ensures that network bandwidth is used efficiently is through a *sliding window* mechanism. Sliding windows enable the computer to fill up the pipe at all times by sending out multiple packets simultaneously. A single ACK can be returned for an entire set of packets. The sliding window in TCP is *variable*, which means that a congested receiving device can slow down the rate of incoming packets by advertising a smaller window to the sending device. Alternatively, the receiving device can advertise a larger sliding window to make better use of available network throughput.

> **Note** Data exchange for TCP is in *full-duplex*, which means that both the sending and receiving devices can transfer data simultaneously over the same connection.

UDP, User Datagram Protocol While UDP works at the host-to-host transport layer, this protocol is not reliable. Unreliable protocols don't use ACK packets to guarantee data delivery, so why use it? The fact that a protocol is not reliable doesn't mean that it doesn't function, only that it doesn't make certain that the transfer of data is complete. Given that upper-layer protocols can have reliability mechanisms built in, reliability at the host-to-host transport layer can be redundant. So, a better question is, why use a reliable protocol at a lower layer when it isn't necessary? When you give up the reliability mechanism, you can drop a lot of the control information in the header, which makes the data segments smaller and requires a lot less reading and processing of segments at every stop in the network—and there is no need for ACK conversations between the sender and receiver. This translates into speed. UDP is not only smaller and simpler, it is much faster than TCP.

Applications use UDP for its speed, especially when reliability is not necessary, because the application itself includes those reliability mechanisms. One of the application-layer protocols that uses UDP is Trivial File Transfer Protocol (TFTP), which is known for its speed of transfer rate. Many services can use UDP instead of TCP, such as the Simple Mail Transfer Protocol (SMTP).

The UDP header, shown in Figure 4-4, consists of only four fields—the source port, destination port, length, and checksum—which are the minimum required to provide a connectionless service.

Source port	Destination port
UDP length	Checksum
Data	

Figure 4-4 The UDP header is much simpler than the TCP header, providing for much greater speed but no reliability.

- Like the TCP header, the *source* and *destination ports* identify the sending and receiving applications.
- The *length* field identifies how long the segment is, given that the data field is variable in length. If the length field and actual length of the segment are different, an error is identified.
- The *checksum* field ensures data integrity. This field is a number that represents the sum of all the bits in the data segment. The receiving device will perform a quick count of the bits and compare it with the value of the checksum. If the checksum value doesn't match the sum of the segment at the receiving node, the segment is considered in error.

Test Smart When data is sent by an application, it will either transmit it over TCP or UDP, but never both.

The Internetwork Layer and IP

Like the network layer in the OSI model, the TCP/IP model's internetwork layer deals primarily with addressing and routing data. Five protocols work at the internetwork layer, but only IP is primarily used with upper-layer services. The five protocols are listed in Table 4-1.

Table 4-1 Protocols at the Internetwork Layer

Protocol	Abbreviation	Function
Internet Protocol	IP	Handles node and network addressing, routing of data throughout the internetwork.
Internet Control Message Protocol	ICMP	Message control and error reporting.
Address Resolution Protocol	ARP	Maps an IP address to the physical (MAC) address of a device.
Internet Group Membership Protocol	IGMP	Used for multicast messages. Reports the host's multicast group membership. Multicasting allows a message to be sent to hosts that have identified themselves as being interested in receiving that content.

The IP protocol provides a connectionless delivery service, routability and address selection, and packet fragmentation and reassembly. ICMP is basically a subprotocol of IP used to send control messages, whereas ARP, RARP, and InARP are simply used to resolve IP addresses with hardware addresses.

Note The Packet InterNet Groper (PING) utility is a useful application based on ICMP. PING uses ICMP control messages to determine the status of other devices on the internetwork.

Both the OSI model and the TCP/IP model show that the layered system of protocols functions as a single unit and manages to deliver data throughout an internetwork, even though different functions are split among the layers. IP is an unreliable, connectionless protocol that relies on either TCP or the application-layer protocol to ensure that the data is received at the destination device. As with all protocol layers, the data passed to IP is split up, or *fragmented*, into smaller packets before it is sent to lower layers. At a receiving device, the IP packets are reassembled before being passed on to the host-to-host transport layer. This process is called *encapsulation*.

The IP protocol is best known for addressing network nodes and segments, which assists in the routing of data throughout the internetwork. Basically, IP provides a numerical address for a device and for the segment to which that device is attached. Through the use of the address, a packet can locate any device on the network.

How IP Addressing Works IP addressing identifies network devices without relying on an underlying physical address. An address is assigned to the network segment so that the combination of the host address and the segment address creates a unique address on the entire internetwork that allows any host to be located. When a segment is received by IP, it is broken into smaller packets, each of which is assigned an IP header with the destination and source IP addresses.

IP addresses are not specific to a piece of hardware, as a MAC address is. Instead, the IP address is assigned to network devices. The IP address for each node contains both the network segment address and the node address. It consists of 32 bits, which are most often represented as four numbers (called *octets* because they are made of eight bits), each in the range from 0 to 255, that are separated by decimal points. For example, an IP address is 193.5.201.26. If this were written in its binary form (ones and zeros), it would look like this: 11000001.00000101.11001001.00011010.

Because the IP address includes the network segment address as well as the node address, you need to understand how to determine which part is which.

The first thing you must do is determine what class of address it is. Table 4-2 describes the five IP address classes.

Table 4-2 IP Address Classes

Class	Address Range	Starting Bits
A	1.x.x.x to 126.x.x.x	0
B	128.x.x.x to 191.x.x.x	10
C	192.x.x.x to 223.x.x.x	110
D	224.x.x.x to 239.x.x.x	1110
E	240.x.x.x to 255.x.x.x	11110

You might notice in Table 4-2 that there are no addresses listing 127.x.x.x, which you might expect in the Class A address range. This is not an omission. Addresses beginning with 127 are used for loopback testing on local machines. Class A through Class C addresses are used for network nodes, but Class D and Class E addresses are reserved for multicasting, research, and development.

The first octet of an IP address tells which class of an address it belongs to, but it doesn't necessarily indicate the entire network segment address. In the following list, the part of the address that belongs to the network is denoted by an "N," while the part that belongs to the device is denoted by a "d."

- Class A: NNNNNNNN.dddddddd.dddddddd.dddddddd
- Class B: NNNNNNNN.NNNNNNNN.dddddddd.dddddddd
- Class C: NNNNNNNN.NNNNNNNN.NNNNNNNN.dddddddd

Test Smart You'll be tested on IP addressing as it relates to address classes, subnet masking, routing, and your ability to troubleshoot IP address problems. Chapter 5 discusses IP addressing in much greater detail.

Routing with IP Addressing lays the foundation for routing. Once you know addressing, you'll begin to see how the system works (or what might be broken), even if you are shown a simple diagram and IP addresses. Any computer or device that connects to multiple network segments and forwards data to other segments is a *router*. Routing is dependent on IP addressing in the TCP/IP model in the same way that the post office is dependent on street names, house names, and zip codes. The routing process follows this sequence:

1. The router receives a data packet.
2. The router looks at the destination IP address.

3. The router decides whether the destination IP address network is the same network port from which the data was received. If it is destined for that network attached to that same interface, the packet is discarded. Otherwise, the router proceeds to the next step. For example, if a packet sent to network A is received on Interface A which is attached to network A, the router drops the packet because it doesn't need to forward it. However, if a packet sent to network B is received on Interface A which is attached to network A, the router goes on to the next step in determining how to help get that packet to its destination.
4. The router looks at its routing table to determine whether it knows the path to the destination address. If it does not know, the router will send the data to its default gateway address. If it does know, the router selects a route.
5. In the case of multiple routes, the router selects the route that is fastest, and for example, it would probably select a route that involves the least number of hops through additional routers and one that used the fastest transmission speeds.
6. The router forwards the IP packet out the interface toward its destination.

DNS Names

Binary addresses composed of all ones and zeros are difficult to understand, which is one of the reasons why the dotted decimal notation is more commonly used. But even so, no type of numerical address is human-friendly; they are difficult to remember. Early on, people started mapping their affectionate nicknames for their computers to the computer's IP address. This grew into an entire system called the *Domain Naming System* (DNS).

In DNS, a fully qualified domain name (FQDN) for a computer has two parts, the host name and the DNS domain name. The host name is a word like "Pebbles." The domain name is the portion of the name that identifies the organization on the Internet. For example, microsoft.com is a domain name, and so is flintstones.bedrock.org. When you add the host name to the domain name, you get a unique name identifying that computer on the entire internetwork, as shown in Figure 4-5.

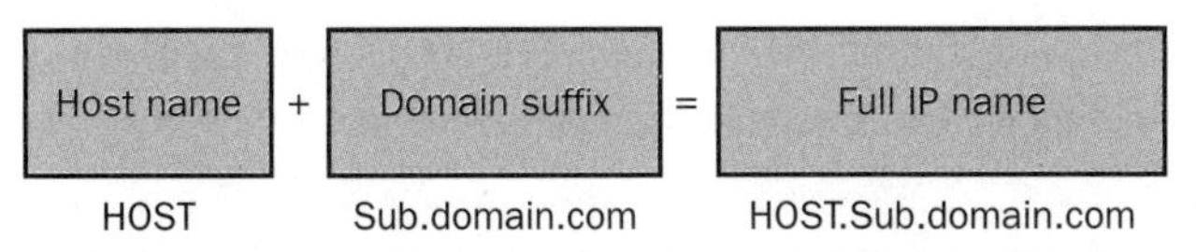

Figure 4-5 In DNS, the host name plus the domain suffix are added to each other to create the full IP name.

How to Name Network Objects Lots of objects in a network are named. Not only does each computer receive a name, but so do printers, applications, and database objects; even users receive names for use on the network. When you think of objects on the network, consider them to be the software representation of either a physical piece of hardware, a person, or a piece of software that can interact across the network. Network objects are maintained within the directory service that enables them to interact. Some directory services are simple, including little more than names, addresses and descriptions of computers, or logon IDs and passwords. Others are complex, such as Microsoft Active Directory, which is hierarchical in nature and contains references to objects ranging from applications to printers to users, as well as the assignment of rights among these objects.

Because every object provides a link to the directory service, which links to every other object in the network, objects can be a security risk. This leaves administrators with a problem: How do you give IDs to every object that people can easily use and remember, but which don't pose a security risk? It all boils down to following a few simple rules:

- Many companies use a standard naming system: the first letter of a person's first name followed by the first seven letters of the last name. Don't use the standard. Add numbers or symbols to each user's ID.
- Change the default administrator's name from "Admin" or "Administrator" to something else. Who would know that your administrator's account is YJ99mQz013, except people who should know?
- Do not use a social security number for user IDs. It will place their personal information at risk, *plus* people outside the United States don't have one.
- Enforce strict password policies that require frequent changes and the use of numbers along with letters in the password.
- Avoid symbols. In most cases, a directory service will support standard characters—0 through 9 and A through Z. If your directory service supports additional symbols, you might have problems if it interacts with an application that doesn't support those symbols.
- Create unique names for each user, printer, and computer and for other resources.
- Do not depend on the case of letters to create unique names. Joe and jOE might be considered unique in one system, but in another they would not.
- To assist users in finding shared servers and printers, include the location of the system in its name.
- Do not name workstations after their users, especially in an organization in which you move workstations around the network, or in which people can use both a workstation and a laptop.

This section has provided you with an overview of the TCP/IP protocol suite. Throughout this book, we will cover individual TCP/IP protocols and drill down into greater detail. Now, let's look at some other protocol suites that you may use in networking.

IPX/SPX

Novell first developed its Internetwork Packet Exchange/Sequenced Packet Exchange (IPX/SPX) protocol suite for use with its NetWare operating system in the 1980s. To give you a frame of reference: IPX is similar to IP in that it functions more or less at the network layer (although with some crossover into the transport layer) of the OSI reference model, and provides addressing and routing throughout the internetwork. SPX is similar to TCP in that it functions at the transport layer and provides a connection-oriented service to upper-layer applications. The similarity is in function; however, the actual implementation of the IPX/SPX suite is quite different.

Novell NetWare provides file, print, and application-sharing services, along with a directory service. All of NetWare's services have evolved over the years, but they have relied on the IPX/SPX protocol, which was for many years the dominating operating system and protocol combination used in corporate networks. The evolution of networks to incorporate the Internet has led to IPX/SPX sharing bandwidth with or being entirely put aside for TCP/IP, even in native NetWare networks.

IPX and Addressing

IPX is used by all upper-layer services to send data across an IPX/SPX network. It is a connectionless protocol, but NetWare servers are built to maintain connectivity to workstations. A connectionless protocol is problematic: IPX doesn't let the NetWare server know whether the workstations are online. So to avoid reserving resources for inactive stations, special packets called *watchdog packets* are sent to inactive stations after a set amount of idle time. If the watchdog packet is not returned with a response, the server terminates the connection.

You might wonder how IPX addresses are placed into the header so that a watchdog packet makes it to the correct workstation. The unique thing about IPX addressing is that the administrator doesn't need to assign any addresses to the nodes, only to the network segments; but nodes still receive an address. Each workstation or server on an IPX network will copy the MAC address of its network interface card (NIC) and use it as the IPX node address portion.

While IPX servers and workstations use their hardware addresses as their node addresses, administrators must assign addresses to network segments. This is done on the NetWare servers connected to those segments, not on any workstations. IPX can send data directly to upper-layer applications using service addresses, so it is not dependent on SPX to function. When IPX uses the node address in conjunction with the service address (and segment address, of course!), it can transmit data directly to an application running on a different machine.

Routing with IPX

Because IPX is connectionless, it doesn't have much overhead, which gives it a distinct advantage in speed. The header fields, shown in Figure 4-6, are often required to be set by an upper-layer protocol, such as the packet type and the destination network, destination node, and destination socket. The Transport control field is used for routing. It is set to zero at the source node and then, as the packet passes each router, it is incremented by one. When it reaches 16, the packet is considered undeliverable and is discarded.

Checksum	Length	Transport control	Packet type	Destination network	Destination node	Source network	Source node	Source socket

Figure 4-6 The IPX header includes the fields necessary to route data from one network segment to another.

IPX is always routed dynamically. The routing architecture learns network addresses automatically and maintains them. There is rarely the need to do anything special to establish routing in the NetWare environment, except in the case of unique requirements such as redundant routes.

Routing is dependent on the IPX address. IPX addresses are usually written in hexadecimal format, which means that each byte is represented by two characters in the range from A through F or numbers from 0 through 9. This consists of the 4-byte network number assigned by the network administrator, the 6-byte node number (copied from the hardware address of the NIC), and the 2-byte socket address predetermined by the application. When the network administrator assigns the 4-byte network number, it is eight characters long—for example, 4A31F80C. The IPX node address is written as 0A-56-78-9A-BB-FF.

The most confusing aspect of IPX addresses is that they can be written in multiple formats. Most of the time, when an IPX network address is written, the leading zeros are dropped as a form of shorthand; so network 0000002A becomes 2A. Notation does vary a bit in IPX addressing. Sometimes a colon is placed in the middle, so a network address of 44ABF128 becomes 44AB:F128. The zeros are never dropped from a node address, but periods or colons are often inserted every 4 hex digits (which is every 16 bits). The network address 0A-01-23-FF-45-79 is also written as 0A01:23FF:4579 or as 0A01.23FF.4579.

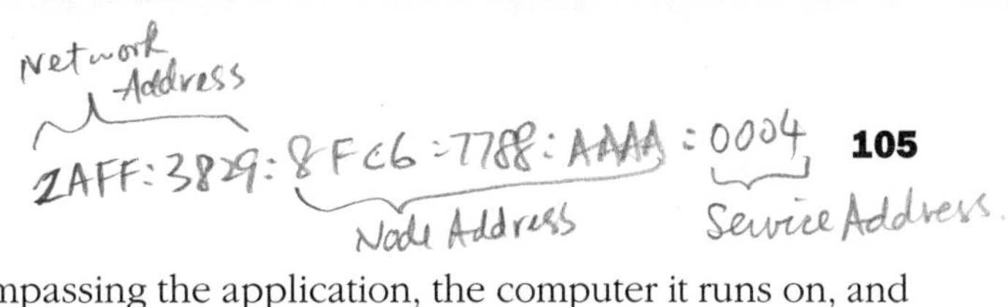

An IPX address encompassing the application, the computer it runs on, and the location in the physical network will include the network address, followed by the node address and then the service address. This means that 2AFF:3829:8FC6:7788:AAAA:0004 consists of the network address of 2AFF3829, the node address of 8F-C6-77-88-AA-AA, and the service address of 0004.

IPX routers broadcast routing information and service advertisements to all workstations on the network using Routing Information Protocol (RIP) and Service Advertising Protocol (SAP) or NetWare Link Services Protocol (NLSP). The routers share information so that any workstation on any network segment can access information about services available on other segments.

Understanding RIP and NLSP The Routing Information Protocol (RIP) used in the IPX/SPX protocol suite is a Distance Vector routing protocol for dynamic management of the IPX routing tables. Even though the names are identical, don't confuse IPX's RIP with that of the TCP/IP suite. The two are not interchangeable and do not interoperate.

RIP sends routing update messages to all neighbors of a server or router every 60 seconds. These broadcasts do not pass beyond the neighbors. Instead, the router that receives a RIP broadcast will update information in its own table and then send out a broadcast of its own to its neighbors. Like a series of falling dominoes, each router is updated with new information in turn. Because RIP uses secondhand information, bad information can result in broadcast storms that bring down the network.

NLSP is an alternative protocol that IPX routers can use for routing. NLSP is a Link State route-discovery protocol. The Link State algorithm NLSP uses enables each router to receive firsthand information about the status of other routers on the network, which prevents some of the congestion and problems inherent in RIP networks.

One additional address is used on a NetWare server—the internal network number, which logically identifies any internally running services. Because of this, NetWare servers are all automatically configured to route data between the external network and their internal network.

Figure 4-7 shows a NetWare network using IPX. If workstation A wants to access an application on server F, the following steps have to occur:

1. Workstation A discovers server F's internal address from service advertisements broadcasted to E and then to C.
2. To find out how to reach F, workstation A broadcasts a routing request through RIP.

3. C receives the routing request and returns its own IPX node address because it knows the route to F.
4. Workstation A sends out a packet with F's internal network number, 777888AB, and C's node address, AA-BB-CC-77-88-09.
5. C receives the packet, copies E's node address, CD-EF-78-A1-B2-C7, in the place of its own, and forwards it out to E.
6. E receives the packet, strips out its own node address, and copies F's node address, 01-00-2F-9A-01-27, into the header.
7. E then transmits the packet to F.

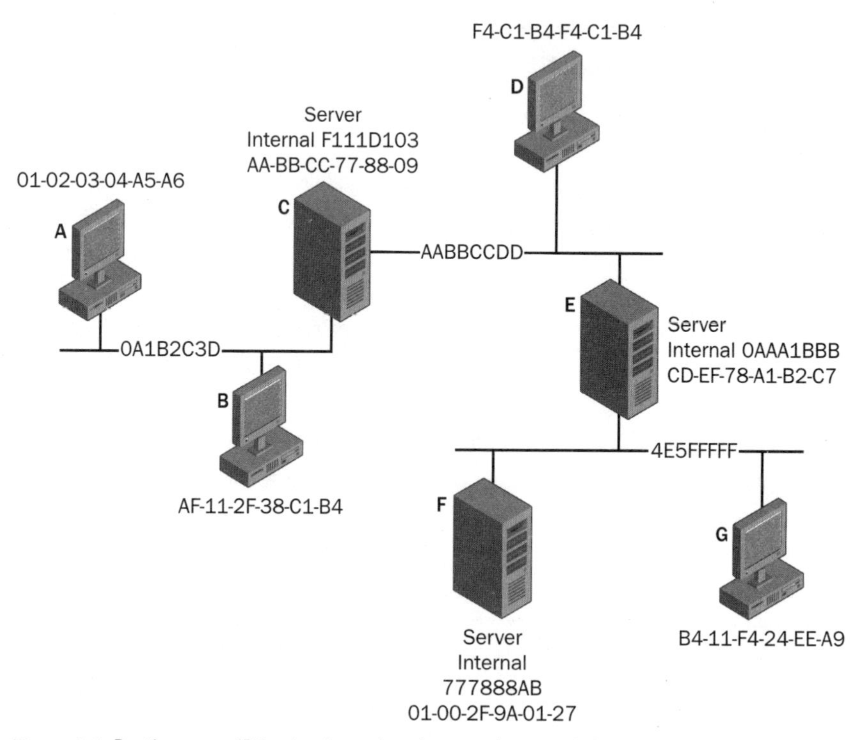

Figure 4-7 Routing on an IPX network requires the use of a server's internal network number to reach shared services.

SPX and Other Protocols

When an application requires guaranteed delivery and doesn't include the connection-oriented control information in its upper-layer protocol, it must use SPX in conjunction with IPX. SPX operates at the transport layer, providing connection-oriented transport functionality to any application that requires it.

Since SPX provides guaranteed delivery of data, why not use it with every data transmission? Good question. The main reason is the overhead that SPX places on the network. SPX must send acknowledgment packets in both directions to ensure that data delivery takes place. The sending node transmits data first, and then the destination node sends back an ACK. If the sending node doesn't receive an ACK, it requests an ACK or retransmits the data. As you can imagine, if you have an application that is also sending and receiving ACK packets, this process becomes redundant, not to mention inefficient. As part of its delivery service, SPX provides error checking, end-to-end flow control, sequencing, and error correction. These services do add more overhead, but they also work together to guarantee a reliable data transmission.

The IPX/SPX protocol suite includes several services that function at the upper layers of the OSI reference model. These services add functionality to the suite, enabling users to take full advantage of the network's capabilities.

- **NetWare shell:** The NetWare shell works on IPX/SPX workstations within the NetWare client application. It transparently looks at calls and intercepts them to see whether they're intended for the network. If the call is meant for network resources, the NetWare shell redirects the call by passing it on to the next appropriate lower-layer protocol and ultimately sending it to the correct resource. The NetWare shell provides an excellent intermediary between applications that are not network-aware and the network itself. For example, a word processor that is not network-aware must have a printing request go through the NetWare shell when the print job is directed at a network printer.
- **Service Advertising Protocol (SAP):** SAP ensures that all network nodes are aware of the services being shared on the network. SAP broadcasts are sent from a server to its neighbors every 60 seconds. Neighbors update their SAP tables with new information and forward that to their other neighbors. The service number is 2 bytes in hex format. For example, every NetWare file server sends out a SAP address of 0004 to inform stations that it is sharing files. It sends out a SAP address of 0007 to inform stations that it is sharing printers.
- **NetWare Core Protocol (NCP):** NCP is the protocol used to access file, print, and security services. Through NCP, workstations perceive remote services in the same way as local services.

Data-Link Layer Encapsulation

The process of encapsulation manages to assemble data frames from packets and then pass them to the physical media. When IPX passes a datagram to the data-link layer protocols, it encapsulates the frames with a specific header.

Ethernet has three types of encapsulation: Ethernet 802.2, Ethernet 802.3, and Ethernet II, which are known as the *frame type*. It's critical that all network devices use the same encapsulation type. If a workstation is using a different type of encapsulation than the server, the two can't communicate. You can use multiple encapsulation types on the network, but this will add more traffic to the network, so it's best to select only one encapsulation type where multiple types are available.

Test Smart Older versions of NetWare use Ethernet 802.3 as the default encapsulation, while newer versions use Ethernet 802.2. If you change out a NIC in an older server, or on any computer in a NetWare network, make certain you specify the right frame type. You might have to troubleshoot this problem on the exam (or in real life).

NetBEUI

The NetBIOS Extended User Interface (NetBEUI) protocol is the closest thing you might ever find to automatic. It was designed to be small and efficient to work well on small LANs that didn't require routing to other networks. In fact, NetBEUI (pronounced net-BOO-ee) doesn't support routing at all. Because NetBEUI doesn't support routing, multiple network segments must be bridged or switched.

NetBIOS Names

NetBEUI, as an extension of NetBIOS, uses NetBIOS names, which are always 16 bytes in length. If the name assigned is not that long, it will be padded with extra bits to reach that length. (In Windows implementations, you can type a name 15 characters in length, and Windows will add its own character for the sixteenth, which is used similarly to port numbers in TCP/IP.) NetBIOS names can include both characters and numbers, as well as a few symbols.

NetBIOS names are in a name space that is considered *flat*. This means that there is simply a list of names in one large group, with no hierarchy or subdivision within the group. While objects such as workstations, servers, printers, and users are assigned names, networks are not. Remember that a network segment must have a name or a number for routing to take place.

You can assign more than one name to a computer in a NetBEUI network. NetBIOS provides for multiple aliases (or names) for a single node. One name is considered permanent, and is usually copied from the NIC's hardware address. This particular name is usually called NETBIOS_NAME_1.

You might recognize NetBIOS names and the rules surrounding them because they are often the default in a Windows network. Legacy Windows workstations and servers are assigned NetBIOS names that are held within a flat workgroup name space. Windows 2000 networks use a hierarchical name space in the Active Directory. To ensure backward compatibility, NetBIOS names are still supported. With the Active Directory, the NetBIOS name is usually made to be identical to the host name, except without the domain name concatenated to the end.

Interoperability

One way to use NetBEUI in a routed environment is to tunnel it through a routable protocol. The most common implementation is NetBIOS over TCP/IP, which is typically used to carry the Server Message Block (SMB) protocol on Windows networks.

When using NetBIOS over TCP/IP, the key is to resolve the NetBIOS name to an IP address. Name resolution is simply a list that maps each name to an IP address. This was originally handled in an LMHOSTS file, a simple text file in which each line represents an IP address mapped to a NetBIOS name. The file can also include additional comments or parameters. The problem with LMHOSTS files is that a NetBIOS workstation must be updated with all changes or other systems become unreachable, even if they're online.

The Windows Internet Naming Service (WINS) is provided by Windows servers that map the IP addresses of each workstation to their NetBIOS name. WINS servers centralize the mapping system and enable a single point of administration, reducing administrative workload.

Protocols within NetBEUI include:

- **Name Management Protocol (NMP):** NMP broadcasts a system's new name to the network. If there is no objection, the name is registered.
- **NetBIOS User Datagram Protocol (UDP):** UDP is a connectionless, unreliable datagram-delivery protocol. It is similar to IPX, except that it doesn't provide for routing. A system can send UDP packets to another single system, or create a broadcast message using UDP.
- **NetBIOS Diagnostic and Monitoring Protocol (DMP):** DMP obtains status information about the local machine and remote systems on the network.
- **NetBIOS Session Management Protocol (SMP):** SMP manages the sessions among processes that are running on two different systems on the network. SMP runs at full-duplex, which means that both systems can send and receive data simultaneously. An SMP session will last until at least one of the systems ends it.

- **Server Message Block (SMB):** SMB is an application-layer protocol used frequently by Microsoft Windows systems, as well as by IBM OS/2. SMB implements messaging, session control, and file and printer sharing. It's roughly analogous to the NetWare Core Protocol (NCP) in the IPX/SPX protocol suite.

AppleTalk

Apple Computer developed AppleTalk to connect Apple computers together in a peer-to-peer configuration for the purpose of file and printer sharing. The protocol is simple, fairly inexpensive, and flexible for a small network. Peer-to-peer means that each station can act as a server, by sharing, and a client, by accessing other stations' resources.

AppleTalk is probably the chattiest protocol, because constant and frequent broadcasts are transmitted to ensure connectivity. This makes AppleTalk simply unsuitable for use on large internetworks. At some point, the number of broadcasts overwhelms the network, and normal network communication becomes impossible. Because Apple computers can and do use TCP/IP as readily as AppleTalk, TCP/IP is increasingly selected as the only protocol suite on an Apple network.

The AppleTalk Address

Addresses in AppleTalk networks are assigned dynamically, so there is seldom any need for an administrator to handle the assignment of network or node addresses.

When an AppleTalk system first starts up, it generates its own random network layer address, which it broadcasts out to the network to see whether that address is already in use. If the address is already in use, the device in conflict responds that the address is taken. The other device must then generate a new node address and broadcast it to see whether it is being used, repeating the process until it finds an unused address.

An AppleTalk device can have a node address between 1 and 254. This means that a maximum of 253 AppleTalk devices can reside on any single AppleTalk network address. If more addresses are required, an administrator can assign a second network address to the same physical segment. Each network address is called a *cable range*. In the case of multiple cable ranges assigned to a single segment, a router is needed. These are called seed routers, and they dynamically assign node addresses without end users being aware of the networking transactions.

Addresses are written as the network number followed by the node address and finally the socket address. They would appear in the format of NNNN.ddd (ss). In an address of 1234.12 (04), 1234 is the network number, 12 is the node, and 04 is the socket.

Protocols in the AppleTalk Suite

Apple developed a newer version of the AppleTalk protocol suite in 1989, called AppleTalk Phase 2. There are multiple protocols in the suite, each with a different function working in relation to the other protocols in a layered fashion. Figure 4-8 shows the AppleTalk protocol suite relative to the OSI reference model.

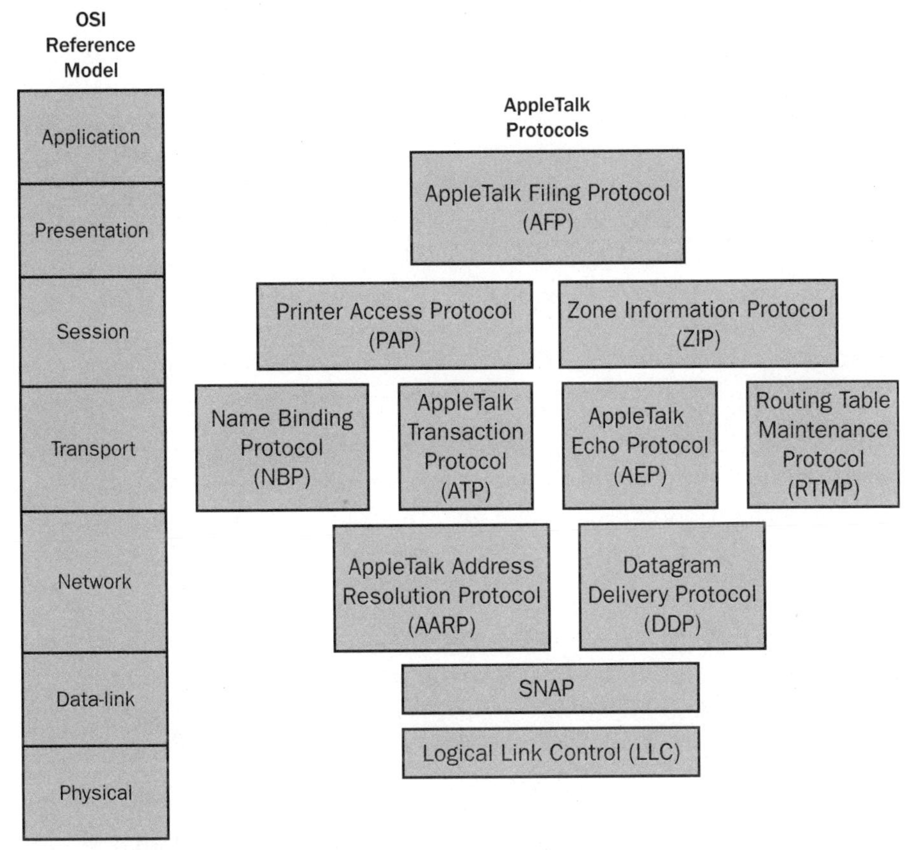

Figure 4-8 The AppleTalk protocols function at various layers in the OSI reference model.

- **AppleTalk Address Resolution Protocol (AARP):** AARP maps the AppleTalk node addresses to underlying addresses at the data-link layer, such as the physical address of an Ethernet card. AARP makes it possible for AppleTalk to run on any media.

- **AppleTalk Echo Protocol (AEP):** AEP is a simple diagnostic protocol that provides straightforward status information about other AppleTalk devices.
- **AppleTalk Filing Protocol (AFP):** AFP provides the ability for AppleTalk stations to share files. Apple files have two data structures, a data fork and a resource fork. AFP enables network devices to access either the raw data in the data fork, or to access the file through icons and drivers.
- **AppleTalk Transaction Protocol (ATP):** ATP is often used by upper-layer services to provide a connection-oriented, guaranteed delivery service through the use of acknowledgment packets and flow control. ATP uses a bitmap token that corresponds to eight segments per transaction. The bitmap position corresponds to the segment's position in the transmission. The bit on the far right represents the first data segment. A "0" in a position acknowledges that the data was received. A "1" in a position states that the data request is outstanding.
- **Datagram Delivery Protocol (DDP):** DDP provides datagram-delivery service, node addressing, and routing for upper-layer services. It functions at the network layer. Two types of frames are available in DDP, short format and long format. The long format is the only one used in AppleTalk Phase 2; it includes fields for the destination network, node, and socket.
- **Name Binding Protocol (NBP):** NBP registers names and binds them to socket addresses. Using NBP, AppleTalk computers can look up a name to find the address associated with it, and then transmit data as appropriate.
- **Printer Access Protocol (PAP):** PAP is used strictly to convey the connection status of printers and manage the transfer of printing data. An AppleTalk client uses PAP to manage its virtual connection to AppleTalk printers and the devices that are sharing them.
- **Routing Table Maintenance Protocol (RTMP):** RTMP handles routing information. Based on Distance Vector routing, AppleTalk routers broadcast their known routes to neighboring routers. The neighbors update their internal tables with new information and then broadcast their own tables to their own neighbors. RTMP includes the Split Horizon routing algorithm. Split Horizon means that RTMP for-

wards only information about directly connected networks, reducing traffic overhead.

- **Zone Information Protocol (ZIP):** In AppleTalk, each network is given a zone name. ZIP maps network numbers to zone names. Usually AppleTalk routers implement ZIP and gather the information from RTMP frames. Multiple network numbers in sequential order can belong to a single zone.

Key Points

- Internet Protocol (IP) works at the internetwork layer of the TCP/IP model, which provides functions similar to the network layer of the OSI reference model.
- IP provides addressing, routing, and a connectionless data-delivery service.
- TCP provides a connection-oriented, reliable data-delivery service.
- User Datagram Protocol (UDP) functions at the host-to-host transport layer of the TCP/IP model and provides a connectionless data-delivery service.
- Upper-layer protocols can use either TCP or UDP, but not both simultaneously.
- The Domain Naming Service (DNS) provides host and domain name management in the TCP/IP suite. Each host is given a name and placed in a hierarchical domain system. DNS maps host names to IP addresses.
- Internetwork Packet Exchange (IPX) provides addressing, routing, and a connectionless data-delivery service. It works at the network layer of the OSI reference model with some crossover into the transport layer.
- Sequenced Packet Exchange (SPX) functions at the transport layer of the OSI reference model and provides a connection-oriented, reliable data-delivery service.
- Upper-layer protocols must use IPX when transmitting data, but only those that require guaranteed delivery need to use SPX.
- NetBIOS Enhanced User Interface (NetBEUI) is a nonroutable protocol developed for use in small LANs.

- NetBEUI uses NetBIOS names that are 16 bytes in length. It does not provide for network names or addresses.
- Windows Internet Naming Service (WINS) maps NetBIOS names to IP addresses.
- Peer-to-peer networks enable file sharing by any computer on the network, so that a computer can be both a server and a client simultaneously.
- There can be up to 253 AppleTalk nodes per segment address, because the node can be numbered from 1 to 254.
- There can be multiple segment addresses (called cable ranges) assigned to a single physical segment, enabling more than 253 AppleTalk devices per physical network.

Chapter Review Questions

1 Select the protocol suite that is used to communicate across the Internet:

a) TCP/IP

b) IPX/SPX

c) NetBEUI

d) AppleTalk

Answer a, TCP/IP, is the protocol that is used to communicate across the Internet. As such, it is constantly evolving, with new protocols being added to the suite. Answer b, IPX/SPX, is incorrect because it is the protocol suite developed by Novell for NetWare networks. Answer c, NetBEUI, is incorrect because it is a nonroutable protocol and can't be transmitted natively across the Internet, which depends on routing. Answer d, AppleTalk, is incorrect because it was developed by Apple Computer for peer-to-peer sharing in Apple networks.

2 Which protocol is responsible for addresses in the format 202.5.88.253?

a) DDP

b) IPX

c) NMP

d) IP

Answer d, IP, provides for addresses in the dotted decimal notation, which is four numbers from 0 to 255, separated by decimal points. Answer a, DDP, is incorrect because it provides for AppleTalk addresses in the format of a four-digit network address, followed by a node address of three digits from 1 to 254, and then by a two-digit socket address. Answer b, IPX, is incorrect because it provides for IPX/SPX addresses in hexadecimal format, typically separated by colons. Answer c, NMP, is incorrect because it provides for name management in the NetBEUI protocol suite.

3 If an application needs to send a packet with guaranteed delivery over the Internet, but its protocol doesn't include the control information to do so, which of the following protocols should it use at the transport layer?

a) SPX

b) UDP

c) TCP

d) ATP

Answer c, TCP, is correct. TCP provides a connection-oriented, guaranteed data-delivery service in the suite that is used on the Internet. Answer a, SPX, is incorrect because it is in the protocol suite used for NetWare networks. Answer b, UDP, is incorrect because while it can communicate natively across the Internet, it doesn't provide connection-oriented data-delivery services. Answer d, ATP, is incorrect because it is used for AppleTalk networks.

4 Which of the following maps IP addresses to host names on a TCP/IP network?

a) WINS

b) DNS

c) NBP

d) SAP

Answer b, DNS, is correct because it maps IP addresses to the names used on native TCP/IP networks. Answer a, WINS, is incorrect because it maps IP addresses to NetBIOS names in a network that uses both TCP/IP and NetBEUI. Answer c, NBP, is incorrect because it is the Name Binding Protocol used in NetBEUI. Answer d, SAP, is incorrect because it is the protocol used to advertise services on an IPX network.

5 Which of the following AppleTalk protocols maintains routing information?

a) RTMP

b) RIP

c) NLSP

d) DDP

Answer a, RTMP, is the AppleTalk protocol that maintains routing information on an AppleTalk network. Answer b, RIP, is incorrect because it is not an AppleTalk protocol. There is a version of RIP in the TCP/IP suite, and another in IPX/SPX. Answer c, NLSP, is incorrect because it is not an AppleTalk protocol, and it works in the IPX/SPX suite. Answer d, DDP, is incorrect because, while it is an AppleTalk protocol, it is responsible for addressing and routing, not maintaining the routing information in a table.

6 Which of the four protocol suites allows a network node to dynamically assign itself an address, or collect one from a seed router?

a) TCP/IP

b) IPX/SPX

c) NetBEUI

d) AppleTalk

Answer d, AppleTalk, is correct. In AppleTalk, each device broadcasts its own address, randomly selected from 1 to 254, and determines whether the address is in use already on the network, or the device is dynamically assigned that address from a seed router. Answer a, TCP/IP, is incorrect. IP addresses are assigned to each network device. Answer b, IPX/SPX, is incorrect because the node address is copied from the physical address of the network interface. Answer c, NetBEUI, is incorrect because it uses 16-byte names assigned by an administrator.

7 Which address must be assigned by an administrator in a NetWare network?

a) Node address

b) Network address

c) Socket address

d) Zone address

Answer b, network address, is correct. An administrator must assign the network address to each server's network interfaces. Answer a, node address, is incorrect. The IPX address for nodes is copied from the hardware address of the NIC. Answer c, socket address, is incorrect because sockets are predetermined. Answer d, zone address, is incorrect because there is no such thing.

8 You have been called in to work on a network for a growing company that has Windows NT 4.0 servers and Windows workstations running in a single Windows NT domain and single physical segment. The network administrator has added a router to the network to connect to a new building on the company campus but can't get the traffic to pass through the router. What is most likely the problem?

a) The network is using TCP/IP, and the router doesn't know how to route it.

b) The network is running only AppleTalk, and there is too much traffic for it to pass through the router.

c) The network is running NWLink, which is its version of IPX/SPX, and the workstations on the other end aren't running that protocol suite.

d) The network is using NetBEUI, which is not routable, so no communication can take place.

Answer d is correct. The most likely problem is that the network was built using NetBEUI, which is not routable. NetBEUI used to be the default protocol for Windows networks, and small networks were often built solely using it. To fix this problem, you should switch to a routable protocol such as TCP/IP. Answer a is incorrect because TCP/IP is so prevalent that all routers (except specialized ones) know how to route it. Answer b is incorrect because AppleTalk shouldn't cause congestion on a single device like a router–it would cause it across the entire network. Answer c is possible, but it is not the most likely solution, because NWLink has never been the default on a Windows network. Plus the administrator would probably have known that the stations on the other end needed to be configured to use it.

9 Which two of the following can be used to map a NetBIOS name to an IP address?

a) WINS

b) DNS

c) NetBEUI

d) LMHOSTS

Answer a, WINS, and answer d, LMHOSTS, are both correct. WINS is the service used to map an IP address to a NetBIOS name. LMHOSTS is the file that resides on an individual machine to hold IP addresses to NetBIOS name mappings. Answer b, DNS, is incorrect because it maps IP addresses to IP names. Answer c, NetBEUI, is incorrect because it is a protocol suite.

10 You have been called in to troubleshoot a problem for a user. The user runs a Macintosh computer and wants to set up a Web browser to access the local intranet, but the computer won't access any network resources at all. No other users on the network are experiencing the same difficulty. What do you determine to be the problem?

a) NLSP is not logging the routing information for that network segment.

b) The user is using UDP, and because it is not guaranteed, the information is not reaching its destination.

c) The user's machine is configured for AppleTalk but not for TCP/IP.

d) The user's IP address conflicts with its AppleTalk address.

Answer c is correct. The problem is most likely that the Macintosh is configured to use AppleTalk but not TCP/IP, which is required to access intranet or Internet resources. Answer a is incorrect because NLSP is not required for any Internet or intranet access. Answer b is incorrect because UDP is used only when connection-oriented delivery is not needed and the user would be unaware of it. Answer d is incorrect because it is not possible for an IP address to conflict with an AppleTalk address. Different protocols can coexist on a network.

Chapter 5

Forming an IP Address

Most networks are connected to the Internet and are able to communicate across it via e-mail messages, file transfers, and other network applications. Because TCP/IP is the Internet's protocol suite, it has become the default protocol for most corporate networks as well as small home networks, so it follows that knowing how to use TCP/IP is one of the top skills required by a network engineer.

This chapter explains how to construct an IP subnet and manage IP addresses on a network. We'll consider both IPv4 and IPv6 addresses and how private and public network address translation works. Finally we'll go over TCP/IP protocol applications and protocol functions.

The Most Common Form of IP—IPv4

The version of IP used on the Internet today is IP version 4, or *IPv4*. In IPv4, each IP address is composed of two portions—one that identifies the network and the other that identifies the host. The IPv4 specification provides for address classes with subnet masks that determine which portion belongs to the network and which belongs to the host. There are five address classes. The first bits of the binary address determine which class the address belongs to. The only address classes in use are classes A, B, and C. Classes D and E are reserved for multicasting and research, respectively.

IPv4 addresses are 32 bits in length and commonly written as four decimals separated by decimal points. These four decimals are the value of a binary octet—or eight bits that are summed up and translated into standard numerals. Each of the decimals is between 0 and 255 in value. An example of an IPv4 address is 120.3.8.250. The class of the address is identified as follows:

- Class A addresses have a first octet value between 1 and 126.
- Class B addresses have a first octet value between 128 and 191. The addresses beginning with 127 are reserved for loopback testing. You can test this by pinging 127.0.0.1 on any computer.
- Class C addresses have a first octet value between 192 and 223.
- Class D addresses have a first octet value between 224 and 239.
- Class E addresses have a first octet value between 240 and 254.

IPv4 Subnetting

Understanding IP subnetting begins with grasping the limitations that older networks were faced with. In Chapter 1, we reviewed different types of network topologies and their limitations. The earliest topologies were limited in both the number of nodes per segment, as well as the length of segments.

For example, Ethernet 10Base5 had a limitation for a segment of 500 meters, with 100 devices for the first, third, and fifth segments because of the 5-4-3 rule, which in brief means you can have five segments and four repeaters, with only three populated segments. This translates to a maximum of 300 devices before the administrator had to route or bridge to another network. Given that a class A or class B network offers 16,777,214 or 65,534 nodes, respectively, the administrator is looking at a lot of wasted addresses. Conversely, the administrator was allowed only 254 hosts in a class C network and had to determine how to apply more hosts to a single network. These problems were solved by subnetting and supernetting.

There is a boundary for each class between the network portion of the address and the node portion of the address. This is determined by the class of the address. *Subnetting* is the process of shifting the boundary of the network part of an IP address into the node share. When you shift this boundary to the right, it creates more logical network addresses with fewer hosts per network address. This makes it possible to subdivide a network address so that it is a better match to the underlying physical network. We will cover how to use subnetting to break a network down into smaller pieces later in this chapter.

In *supernetting*, you shift the boundary of the host portion into the network section and create fewer networks with more hosts per network. This method combines two or more class C addresses and applies them to a large network segment with more than 254 hosts.

Subnetting is used more often than supernetting because it is more common to find an internetwork consisting of segments with fewer than 254 hosts. Network administrators tend to keep segments small to reduce traffic congestion. Table 5-1 describes the different types of class addresses, the number of network addresses, and the number of hosts per address in each class. This is important because subnetting will affect those numbers. Class D and class E addresses, which are used for multicasting and research, respectively, have been added to the table for your information, but are not applicable to subnetting because they are not applied as network IP addresses.

You will see that the bit pattern for the first octet of each address is the determining factor for the numbers included in the address. For example, in any class A address, you can add up all the bits—whether ones or zeros—in the first octet and they will be less than or equal to 127 (except addresses beginning with 127, which are only used for loopback testing) because the very first bit in a class A address is always zero.

Table 5-1 Classes of IP Addresses

Class	Bit Pattern	First Octet Range of Addresses	Number of Network Addresses	Number of Hosts Per Network	Default Mask
A	0xxxxxxx	1-126	126	16,777,214	255.0.0.0
B	10xxxxxx	128-191	16,384	65,534	255.255.0.0
C	110xxxxx	192-223	2,097,152	254	255.255.255.0
D	1110xxxx	224-239	N/A	N/A	N/A
E	1111xxxx	240-254	N/A	N/A	N/A

The default mask shown in Table 5-1 is the series of 1s and 0s that are used to tell the computer which part of the address is the network portion and which is the host portion. For example, in a class C address, 202.5.8.16, the default mask is 255.255.255.0. If you look at this in binary format, you can tell which part is meant for the network and which for the host by the locations of the ones and zeros.

11001010.00000101.00001000.00010000: class C address 202.5.8.16

11111111.11111111.11111111.00000000: default mask

The network address is 202.5.8.0, and the host address is 0.0.0.16.

If you add 3 bits to subnet the address, you will receive a mask of 11111111.11111111.11111111.11100000 or 255.255.255.224. This will create six subnets with 30 hosts per subnet. The reason that there are only six subnets is that there must always be at least one "1" bit and at least one "0" bit in each subnet value where you shifted the boundary of the network portion into the host portion. A subnet with either all ones or all zeroes can cause equipment that depends on this "classful" addressing to behave unpredictably. Given this, the network IDs of the subnets are:

11001010.00000101.00001000.001 or 202.5.8.32
11001010.00000101.00001000.010 or 202.5.8.64
11001010.00000101.00001000.011 or 202.5.8.96
11001010.00000101.00001000.100 or 202.5.8.128
11001010.00000101.00001000.101 or 202.5.8.160
11001010.00000101.00001000.110 or 202.5.8.192

The subnet mask forms a filter that passes through the network and host portions of the IP address, using a process called *bitwise ANDing*. In bitwise ANDing, 1 and 1 = 1, whereas 1 and 0 = 0, 0 and 1 = 0, and 0 and 0 = 0. Early standards prevent subnets from being composed of all ones or all zeros. Because a subnet can't consist of all ones or all zeros, the two potential subnets for 11001010.00000101.00001000.000 (202.5.8.0) and 11001010.00000101.00001000.111 (202.5.8.224) are discarded. In addition, you can't add just a single bit to any subnet mask because you'll end up with all ones or all zeros for the potential subnet mask portion, which will be discarded. All ones and all zeros for host addresses are also discarded. These host addresses are interpreted for special functions such as all ones being equivalent to broadcasts.

When you look at one of the subnets, you'll notice that 5 bits are available for host addresses. They can be 00001 (1) through 11110 (30), because host addresses consisting of all zeros or all ones are discarded. This leaves 30 individual addresses for each subnet.

Classless InterDomain Routing (CIDR)

Now that you are aware of how wasteful a classful addressing scheme is, you can appreciate why people would use Classless InterDomain Routing (CIDR) schemes. Take note that CIDR is used in most network equipment today, but there are a few legacy pieces of equipment such as old Token Ring network interface cards (NICs) that cannot use CIDR addressing. If you don't know what is on the network, then stick to the old rules. If you are using a network with no legacy equipment, then you can probably go ahead and use CIDR.

In CIDR, it is possible to use all ones and all zeros as subnet addresses. This means that you can shift the boundary between the network and host portion by a single bit and create two subnets as a result. The only unusable addresses are the host addresses that consist of all ones or all zeros. Host addresses that are composed of all ones in the host portion are broadcast addresses. Host addresses that are composed of all zeroes mean "this host" or are used in various commands for special reasons. Table 5-2 shows the masks that you may find in a Class C network using CIDR.

Table 5-2 CIDR Subnets in a Class C Network

Number of Bits Added	Mask	Available Subnets	Number of Addresses in Each Subnet
1	255.255.255.128	2	126
2	255.255.255.192	4	62
3	255.255.255.224	8	30
4	255.255.255.240	16	14
5	255.255.255.248	32	6
6	255.255.255.252	64	2

As you can see, a lot of addresses are available for use in each of these subnet masks. In addition, you have more subnets. This is a far better system than the original classful addressing.

A shortcut to identify the subnet mask is simply to insert a slash (/) followed by the number of bits in the subnet mask. For example, a default mask for a class A address would be written as 82.1.5.33/8. A class C address with a subnet mask of 255.255.255.192 is written as 198.5.1.131/26.

Translating Between Binary and Decimal Translation from binary to decimal format can be a helpful tool in determining IP addresses and subnets. In fact, network engineers who deal with routers on a daily basis can usually do the conversion in their heads. The position of a bit in a binary number represents a value in powers of two. Each bit value doubles as you move from one bit position to the next from right to left. Because you will need to know only the binary value of an octet, you really need to memorize just the values of the first eight bit positions. The values of the bits are as follows in Table 5-3.

Table 5-3 Bit Values Translated to Decimal

Bit Number	Bit 7	Bit 6	Bit 5	Bit 4	Bit 3	Bit 2	Bit 1	Bit 0
Decimal	128	64	32	16	8	4	2	1
Power of 2	2^7	2^6	2^5	2^4	2^3	2^2	2^1	2^0

To determine the decimal value of a binary octet, you simply add up the values for each bit that is represented as a one. So, 10101010 is 128 + 32 + 8 + 2 = 170.

To determine the binary octet from a decimal value, you determine which is the largest value you can subtract from the decimal and place a 1 for that bit value. Here's an example:

1. The binary octet 118 begins with a 0 because you can't subtract 128 from it.
2. However, you can subtract 64, so the next bit is a 1.
3. You then subtract 64 from 118, which leaves 54.
4. Because you can subtract 32 from 54, you place a 1 in the next bit.
5. When you subtract 32 from 54, your result is 22.
6. You can subtract 16 from 22, so you place a 1 next.
7. You can't subtract an 8 from 6, so you place a 0 in that bit value.
8. This still leaves 6, from which you subtract 4 and place a 1 in the next bit value.
9. And finally you have 2, from which you subtract 2, place a 1 in the 2 bit value, and because there is nothing left, you place a 0 in the final bit.
10. This results in 01110110.

Practice this technique by converting several dotted decimal IP addresses to binary. Then try converting random octets of ones and zeros to decimal value until you have the conversion technique down.

Test Smart Binary values you should memorize for the test include: 128, 192, 224, 240, 248, 252, 254, and 255. These are the subnet mask values you might be tested on. 128 = 10000000, 192 = 11000000, 224 = 11100000, 240 = 11110000, 248 = 11111000, 252 = 11111100, 254 = 11111110, and 255 = 11111111.

The trickiest problem in subnetting is determining the valid subnet numbers and the ranges of IP addresses within them. With a class B address of 170.8.0.0 and a subnet mask of 255.255.252.0, the first valid subnet number is the lowest number within the subnet mask: 252 = 11111100. The first valid subnet will be the last "1" value in that octet, which in this case is 4. Therefore, the first subnet number is 170.8.4.0. The remaining subnet numbers result from simply counting by fours until you reach 170.8.252.0. This means that 170.8.8.0 is the next subnet, and then 170.8.12.0, and so on. If the subnet mask were 255.255.224.0, the last octet would be 11100000 and the first subnet would be 170.8.32.0. Then you would count by 32s until you reached 170.8.224.0.

After you know what the IP network addresses are, you must then determine what the range of host addresses is for each subnet. In the network address 170.8.4.0, the host portion is 10 binary digits: 00.00000000. The range is from 00.00000001 to 11.11111110. This equates to a range of 170.8.4.1 to 170.8.7.254.

When you are subnetting, remember that the more bits you use for creating a subnet, the fewer will be available for hosts. This means that every subnet can support a dwindling number of hosts. Table 5-4 shows the effect of subnetting and the number of supported hosts for each address class.

Table 5-4 Effect of Subnetting on Each Address Class

Number of Bits Added	Class A Hosts Per Subnet	Class B Hosts Per Subnet	Class C Hosts Per Subnet
2	4,194,302	16,382	62
3	2,097,150	8,190	30
4	1,048,574	4,094	14
5	524,286	2,046	6
6	262,142	1,022	2
7	131,070	510	N/A
8	65,534	254	N/A
9	32,766	126	N/A
10	16,382	62	N/A
11	8,190	30	N/A
12	4,094	14	N/A
13	2,046	6	N/A
14	1,022	2	N/A
15	510	N/A	N/A
16	254	N/A	N/A
17	126	N/A	N/A
18	62	N/A	N/A
19	30	N/A	N/A
20	14	N/A	N/A
21	6	N/A	N/A
22	2	N/A	N/A

You can calculate the number of subnets with the formula 2^n for CIDR (or 2^n – 2 in classful addressing), where n equals the number of bits in the subnet field. You can calculate the number of hosts using the same formula, except use the number of bits in the host portion. To determine the total number of hosts you can have with a particular subnet mask, multiply the number of subnets by the number of hosts per subnet. Let's look at an example. A class C address 196.5.88.0 means 254 node addresses are available on a single network segment.

You can break this network up into 16 subnets of 14 nodes each by adding 4 bits to the default mask. The new mask would be 255.255.255.240.

Test Smart Remember that when you are determining an address range for a subnet, the first host is one bit higher than the subnet ID. If the subnet is 140.9.88.64, the first address is 140.9.88.65.

The Latest Version of IP—IPv6

Internet Protocol version 6 (IPv6) was designed by the Internet Engineering Task Force (IETF) to overcome the limitations of IPv4 addresses. As more and more people use the Internet, available IP addresses are becoming increasingly scarce. IPv6 includes improvements in other areas such as routing and network auto reconfiguration. The plan is for IPv6 addresses to coexist with IPv4 addresses as people migrate to the new addressing method.

An IPv6 address is 128 bits in length, exactly four times as long as an IPv4 address. It is divided into eight 16-bit sections, which are converted into 4-digit hexadecimal numbers separated by colons. An example of an IPv6 address is 402A:0000:2F7C:0001:08BB:FFE3:728D:095A. The IPv6 address notation can be reduced further by removing any leading zeros but maintaining at least a single digit in every piece. For example, 402A:0:2F7C:1:8BB:FFE3:728D:95A.

IPv6 has several types of addresses, each of which is assigned to an interface, not a node. The following list briefly describes them:

- *Unicast* addresses identify a single interface and are used when a single host sends a message to another single host.
- *Multicast* addresses identify multiple interfaces when a single host is sending messages to many hosts.
- *Anycast* addresses identify a single interface among many interfaces and are used when a single host sends a message to the nearest interface available.
- *Site-local* addresses are available to any private network to use internally without the need to register the address. They are used for networks that aren't connected to the Internet, or for networks that aren't directly reachable because of a firewall.
- The *loopback* address is 0:0:0:0:0:0:0:1 (also written as ::1). Loopback addresses identify a logical loopback interface, which allows the node to send packets to itself for testing purposes.

- *IPv4-compatible* addresses are used by hosts with both IPv6 and IPv4 protocols. The address is written as 0:0:0:0:0:0:123.45.67.89.
- *IPv4 mapped* addresses are used only when an IPv6 node internally maps to a node that uses only IPv4. The address is written as 0:0:0:0:0:FFFF:123.45.67.89 (or ::FFFF:123.45.67.89).
- *6to4* addresses are used when two hosts are applying both IPv6 and IPv4 but must communicate across IPv4 routers by tunneling. Such an address has a 48-bit prefix created from the initial 16 bit of 2002: combined with the IPv4 address of the node.

Converting from binary to hexadecimal is not as easy as from binary to decimal because we don't use hexadecimal in daily life. Each hexadecimal alphanumeric character represents four binary bits. Basically, you should use the same values for 0 through 9 (for example, 0011 is equal to 3), and then consider A to be the same value as a decimal 10, B as equivalent to 11, C to 12, D to 13, E to 14, and F to 15. This means that F is represented as 1111 in binary format.

Installing IPv6 on Microsoft Windows XP IPv6 is new enough that you won't find it as part of most operating systems, but you will find it in Windows XP. The process of installing IPv6 is simple.

1. Click Start and then Run.
2. Type **cmd** in the window, and press Enter to open the command prompt. Alternatively, you can select Command Prompt from the Accessories menu.
3. At the prompt, type **ipv6 install**.

The protocol will then be installed, but it won't appear in the list of installed protocols under Network Connections.

Default Gateways

In a large internetwork, it can become extremely cumbersome for a router to include the addresses of every possible network segment. The larger the routing table, the longer it takes to select a route, creating latency in the transmission of data. This becomes more obviously wasteful on a network such as that shown in Figure 5-1, in which all traffic received by the interface from a stub network is sent out the router's other interface. A stub network is similar to a cul-de-sac in a neighborhood—there is only one way in and out of it. In this case, all that is needed is a single logical route in the routing table pointing all outbound traffic to go out one interface. That is the function of a default gateway.

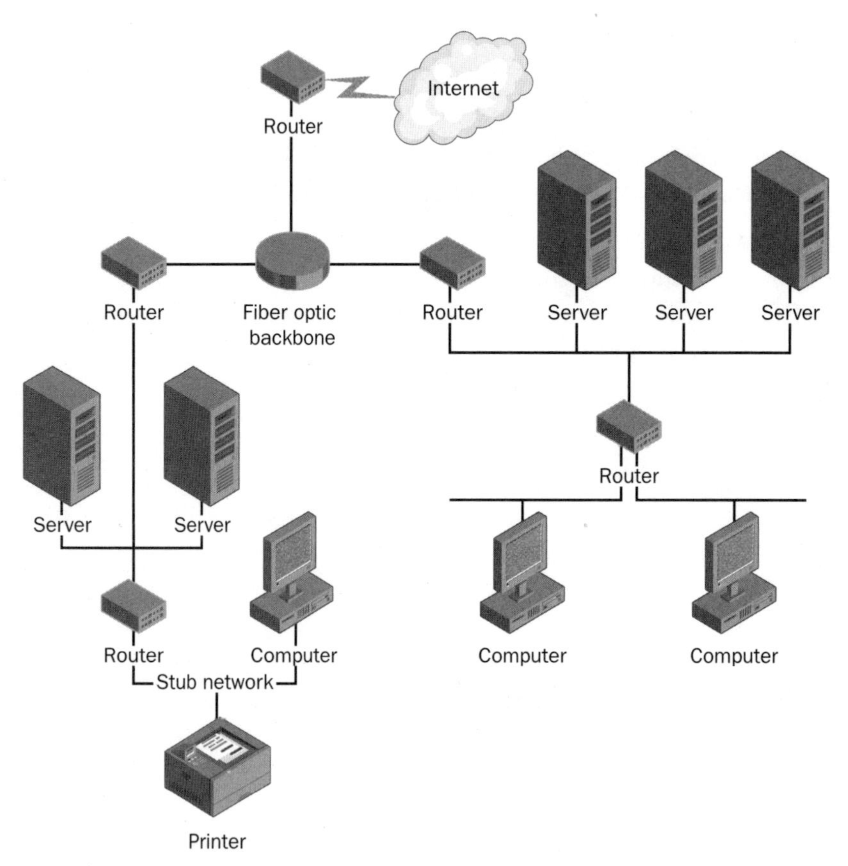

Figure 5-1 Any traffic that doesn't belong on the local network of a stub network is simply forwarded to the next router in the network. A default gateway can carry out this task.

To be precise, the default gateway is an IP address of a router that is designated for any traffic for which the route is unknown. On a computer connected to an internetwork, the default gateway is the IP address of a router on that computer's segment. On a computer connected to the Internet through an ISP, the default gateway is the Internet Service Provider's router IP address, which is used for all Internet-bound traffic. To see the default gateway on any Windows machine, you can type **ipconfig** at the command prompt, or type **winipcfg**. (Depending on which version of Windows you are running, one of these will function as long as TCP/IP is installed.)

When a network router receives data, and it has an address for a default gateway, it follows this process:

1. The router looks at the IP header to determine the destination IP address.

2 The router determines whether the destination IP address is on the same network segment the data was received from.

3 If the destination IP address is for a different network, the router looks in its routing table for a route to that address.

4 If the router does not have a route listed for that destination network, it selects the default gateway address.

The main disadvantage to default gateways is in the case of packets with bad IP addresses. When a router doesn't have a default gateway, it will discard any packet received that has a bad IP address. However, when the router has a default gateway, that packet is sent through the network until it reaches a router with the capacity to discard it, or when it reaches its infinity hop. For example, when using Routing Information Protocol (RIP), if a packet is forwarded 15 times, the sixteenth router will discard it because it has reached logical infinity. The problem in this case is that the router is consuming processing power and network bandwidth. This can become a problem when significant numbers of packets have bad IP addresses.

Figure 5-2 shows a network with several routers and two network hosts. For host A, the default gateway is the address of router B's interface to its same segment, which is 199.2.2.3. For host F, the default gateway is the address of router D's interface to its same network segment, which is 202.2.2.8. For router D, you might want to set up a default gateway route pointing to router C's 201.2.2.200 interface, because router D's other interface (202.2.2.8) is connected to a stub network.

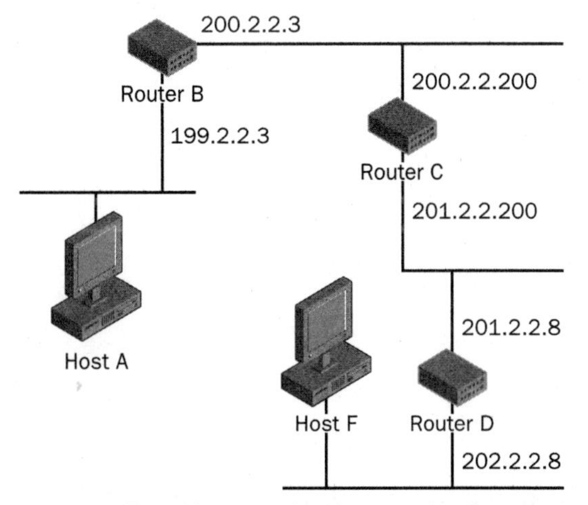

Figure 5-2 The addresses assigned to router interfaces are used as the default gateway for hosts on the same segment as each interface.

Public and Private IP Networks

A *public* IP network address is directly connected to the Internet. A *private* IP network address is used by networks that are either completely disconnected from the Internet or that use network address translation (NAT) to increase the number of IP addresses available on the internal network.

Three private IP addresses are available, one for each class of IP address. The private class A network address is 10.0.0.0/8; the private class B network address is 172.16.0.0/12; and the private class C network address is 192.168.0.0/16. Anyone using an internal IP network can make free use of these addresses. The only caveat is that they must be disconnected from the Internet or hidden behind a NAT or proxy server. Automatic Private IP Addressing (APIPA) provides for one more private class B network address of 169.254.0.0, but it is used in a special way. It is intended as a failover when there is no DHCP server available to assign an address. Any Windows 98, Windows Me, Windows 2000, and Windows XP computer will select an IP address from this range along with a subnet mask of 255.255.0.0 and then can communicate with the other computers on the same local area network. When a DHCP server comes online, the address is replaced with one assigned by the DHCP server.

NAT is used to translate IP addresses from one network and map them to a different IP address known within another network. The private IP address is used on the network that is designated as internal, and it is mapped to a public IP address used on the public Internet.

A network administrator creates a NAT table to perform public-to-private and private-to-public IP address mapping. NAT is intended for use on a router. A company with a single class C address, but far more than 254 hosts, can use the private class B address along with NAT configured with dynamic mapping to a pool of public IP addresses, which will enable all users to access the Internet.

TCP/IP Protocol Functions

The TCP/IP protocol stack has many applications available to end users and network administrators. These protocols enable a network administrator to ensure network connectivity and allow users to send electronic messages and download files. Some of the protocols are essential to common functions such as Web browsing. We'll review the most common protocols in the TCP/IP suite.

Application

Application-layer services in the TCP/IP model encompass the application, presentation, and session layers in the OSI reference model. These protocols provide the primary data for a network administrator, and are the ones users will be most likely to interface with.

Telnet

Many people use remote control applications to access computers at their workplace from outside the network. In remote control, a session appears in which the user is able to manage the files on the remote computer, although the session appears to be functioning locally. Telnet is an early version of a remote control application. Figure 5-3 shows a Telnet session login prompt.

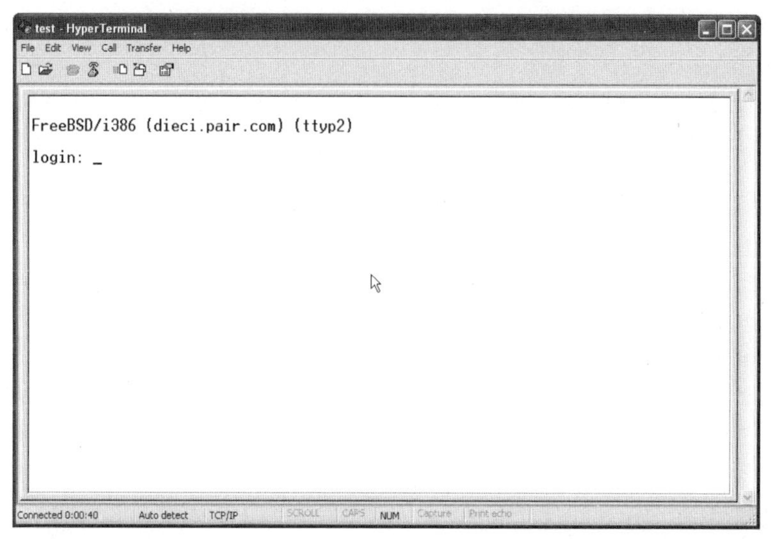

Figure 5-3 Telnet is a character-based session, even when used in graphical programs such as HyperTerminal.

Telnet is very basic; it offers solely character-based access to another computer. If you want to see a person's graphical desktop, you would need a different type of protocol, such as Remote Desktop Protocol (RDP), Independent Computing Architecture (ICA), or X Windows.

Telnet acts as a user command with an underlying Transmission Control Protocol/Internet Protocol (TCP/IP) protocol that handles the establishment, maintenance, and termination of a remote session. The difference between using Telnet and a protocol such as File Transfer Protocol (FTP), which we'll discuss next, is that Telnet logs you directly on to the remote host, and you see a window into that session on your local computer. A typical Telnet command might be as follows:

```
telnet my.librarypc.portsmouth.edu
```

Because this particular host is invalid, this command will have no result. However, if it were a valid host the remote computer would ask you to log on with a user ID and password. A correct ID and password would allow you to log on and execute Telnet commands.

You can often use Telnet to manage equipment that lacks a monitor. For example, most routers have Telnet enabled so that the administrator can log in and manage the router.

Telnet also provides a quick check to make certain that network connectivity is functioning. Because Telnet sits at the application layer, if it can connect to a remote host, you can be certain that network connectivity between the two hosts is operational, as well as all lower-layer protocols.

FTP

One of the earliest uses of the Internet, long before Web browsing came along, was transferring files between computers. The File Transfer Protocol (FTP) is used to connect to remote computers, list shared files, and either upload or download files between local and remote computers. FTP runs over TCP, which provides a connection-oriented, guaranteed data-delivery service.

FTP is a character-based command interface, although many FTP applications have graphical interfaces. FTP is still used for file transfer purposes, most commonly as a central FTP server with files available for download. Web browsers can make FTP requests to download programs from links selected on a Web page.

You should become familiar with the basic commands available in an FTP session. To begin a character-based command session on a Windows computer, follow these steps.

1. Open a Command prompt window, type **ftp** at the prompt, and press Enter.

   This will begin an FTP session on the local machine but will not initialize a connection to another machine.
2. Without a connection to another machine, you will not be able to do anything. To connect, type **open *nameofmachine.domain.com*** or **open *123.45.67.89***, in which ***nameofmachine.domain.com*** or ***123.45.67.89*** is the name or IP address of a host that is available as an FTP server.
3. Most FTP servers require a logon id and password, or they will accept anonymous connections. At this point you will be prompted for a logon ID and password.
4. Once you are connected, you can list the files on the remote server by typing **dir**.
5. If you have create privileges on the remote server, you can create a new directory by typing **mkdir**.

6 To download a file, type **get *filename.txt*** where ***filename.txt*** is the name of the file you are downloading.

7 To upload a file, type **put *filename.txt***.

Many other commands are available in FTP. To view them, open an FTP prompt and type **?**. You'll see a screen similar to the one in Figure 5-4.

```
C:\>ftp
ftp> ?
Commands may be abbreviated.  Commands are:

!               delete          literal         prompt          send
?               debug           ls              put             status
append          dir             mdelete         pwd             trace
ascii           disconnect      mdir            quit            type
bell            get             mget            quote           user
binary          glob            mkdir           recv            verbose
bye             hash            mls             remotehelp
cd              help            mput            rename
close           lcd             open            rmdir
ftp>
```

Figure 5-4 When you type **?** at an FTP prompt, you'll see a listing of all available FTP commands.

TFTP

The Trivial File Transfer Protocol (TFTP) is not the same as FTP, even though the two sound similar. TFTP is far less capable than FTP—it doesn't provide user authentication, a directory structure, or commands besides get and put for downloading and uploading files, respectively. TFTP runs over User Datagram Protocol (UDP), which provides an unreliable, connectionless data-delivery service. The application is small, easy to implement, and has minimal overhead. In fact, it is possible to include TFTP in firmware.

One of the most common uses of TFTP is to download firmware images using a network device, such as a printer or a router. TFTP is sometimes used for bootstrapping diskless computers. Figure 5-5 displays the TFTP command structure.

```
C:\>tftp

Transfers files to and from a remote computer running the TFTP service.

TFTP [-i] host [GET | PUT] source [destination]

  -i              Specifies binary image transfer mode (also called
                  octet). In binary image mode the file is moved
                  literally, byte by byte. Use this mode when
                  transferring binary files.
  host            Specifies the local or remote host.
  GET             Transfers the file destination on the remote host to
                  the file source on the local host.
  PUT             Transfers the file source on the local host to
                  the file destination on the remote host.
  source          Specifies the file to transfer.
  destination     Specifies where to transfer the file.

C:\>
```

Figure 5-5 TFTP has far fewer command options than FTP.

HTTP

When people use the Internet, they are usually doing one of two things: sending electronic messages or browsing the Web. Hypertext Transfer Protocol (HTTP) is the application behind Web browsing. It enables users to download Hypertext Markup Language (HTML) files in addition to text, graphics, sound, and multimedia files. Web browsers use a Web browser application, such as Microsoft Internet Explorer, shown in Figure 5-6, which downloads files and immediately displays them in the browser window.

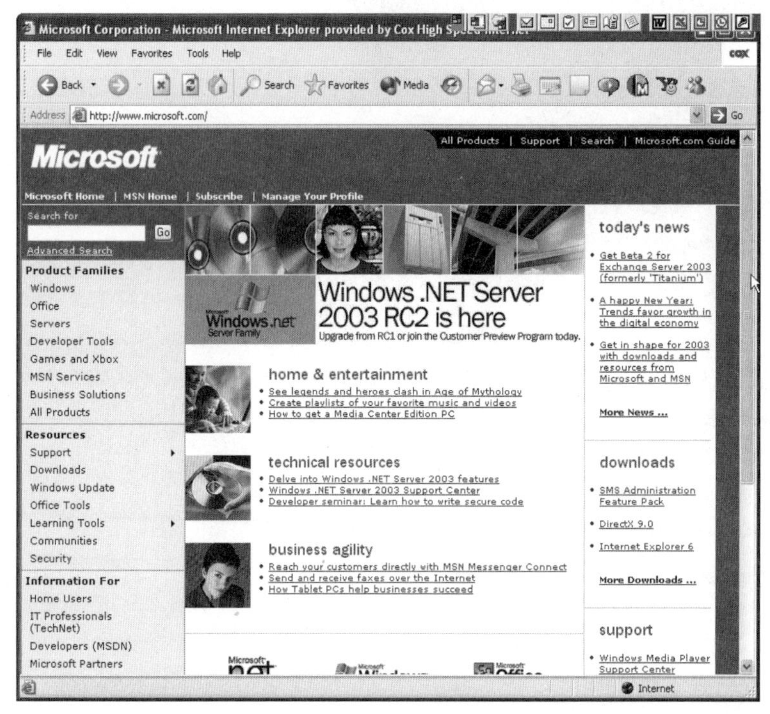

Figure 5-6 Web browser windows display HTML files after retrieving them via HTTP across a TCP/IP network.

Essentially, a Web server runs an HTTP service (on UNIX it is called a daemon) or program that waits for HTTP requests. Upon receiving a request, the Web server transfers the requested file. A client machine uses the Web browser, along with the Uniform Resource Locator (URL) of the file the user attempts to open. The browser creates an HTTP request and transmits it to the server's IP address, which is mapped to the URL. The server's HTTP service processes the request and returns the file. HTTP uses the TCP port 80, unless the administrator assigns a different port. If a different port is used, the user must include the port number with the URL. For example, *http://www.myservice.com:8994* or *http://199.5.88.43:8994*, where the numbers 8994 represent the new port being used by HTTP.

Because of the graphical display of a Web browser, HTML files and their clickable references to other files, along with the ability to add scripted components to the files, the HTTP protocol appears to be the most interactive application of the TCP/IP protocol suite. One difference between the way that HTTP works in comparison with FTP is that no bidirectional session is open on the server. The appearance of interactivity is an illusion. In fact, HTTP consists of the single request by the client, followed by the response from the server. Once the file is opened in the client's browser, it runs solely on the client's machine and connectivity to the server is closed.

You can use Telnet to a Web server to download the source for a Web page. For example, you can Telnet to *www.2test.com* or any other Web site, for that matter, on port 80. Upon a successful connection, you can type **get** / and press Enter twice. You can now see the HTML source code received by your browser and interpreted into a graphical display. This demonstrates that you can use different applications to access information provided by other services, exposing vulnerabilities that you should consider when applying security to a server.

HTTPS

As you browse the Internet, you'll probably find online stores, such as Amazon.com, from which you can purchase items. When you're ready to place your order, you'll be directed to an order form Web page. If you check the address field in the browser, you should see that it begins with https://. In addition, the bottom of the browser will show a graphic of a padlock to indicate that the page is secure, as shown in Figure 5-7. When you click on the command to place your order, the browser will encrypt the data you transmit to the server. The returned acknowledgment is also encrypted.

Hypertext Transfer Protocol over Secure Socket Layer (HTTPS or HTTP over SSL) is a version of HTTP developed by Netscape to incorporate encryption into Web browsing. HTTPS communicates over port 443 instead of port 80. SSL, the encryption protocol, provides for a 40-bit encryption key using an RC4 encryption algorithm by default.

HTTPS is capable of authenticating users. The most widespread use of HTTPS involves online purchasing. In addition, HTTPS is used for the exchange of private information. For example, a Web site might have a "Members only" section that provides encrypted proprietary information to people who are authenticated.

Figure 5-7 Web browsers use https:// to precede URLs and display files downloaded via HTTPS with a padlock in the bottom bar.

POP3/IMAP4

Post Office Protocol 3 (POP3) and Internet Message Access Protocol 4 (IMAP4) are two application-layer protocols used for electronic messaging across the Internet. POP3 is a protocol that involves both a server and a client. A POP3 server receives an e-mail message and holds it for the user. A POP3 client application periodically checks the mailbox on the server to download mail. POP3 does not allow a client to send mail, only to receive it. POP3 transfers e-mail messages over TCP port 110.

IMAP4 is an alternate e-mail protocol. IMAP4 works in the same way as POP3, in that an e-mail message is held on a server and then downloaded to an e-mail client application. Users can read their e-mail message locally in their e-mail client application, but they can't send an e-mail message using IMAP4. When users access e-mail messages via IMAP4, they have the option to view just the message header, including its title and the sender's name, before downloading the body of the message. Users can create, change, or delete folders on the server, as well as search for messages and delete them from the server. To perform these functions, users must have continued access to the IMAP server while they are working with e-mail messages.

With IMAP4, an e-mail message is copied from the server to the e-mail client. When a user deletes a message in the e-mail client, the message remains on the server until it is deleted on the server. POP3 works differently in that an e-mail message is downloaded and not maintained on the server, unless configured otherwise. Therefore, the difference between POP3 and IMAP4 is that IMAP4 acts like a remote file server, while POP3 acts in a store-and-forward manner in its default configuration. (You can configure POP3 clients to leave copies of messages on the server, if you prefer.)

Both Microsoft and Netscape Web browsers have incorporated POP3. In addition, the Eudora and Microsoft Outlook Express e-mail client applications support both POP3 and IMAP4.

SMTP

Simple Mail Transfer Protocol (SMTP) is the application-layer protocol used for transmitting e-mail messages. SMTP is capable of receiving e-mail messages, but it's limited in its capabilities. The most common implementations of SMTP are in conjunction with either POP3 or IMAP4. For example, users download an e-mail message from a POP3 server, and then transmit messages via an SMTP server. Figure 5-8 displays an e-mail configuration screen from Outlook Express that specifies both SMTP and POP3 servers.

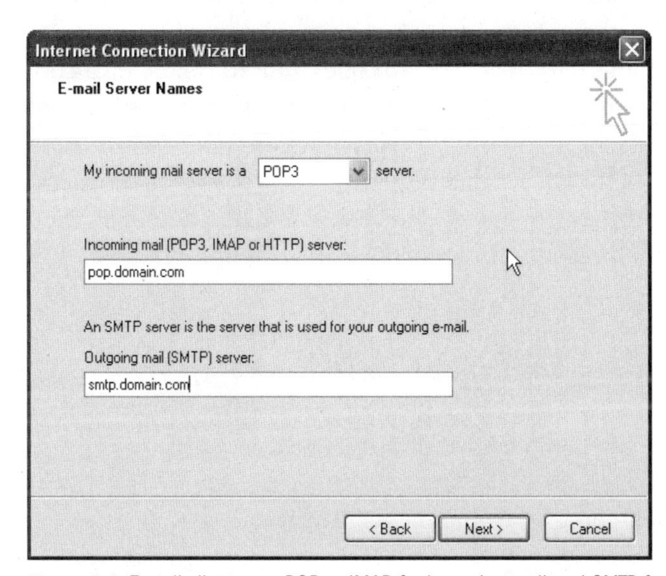

Figure 5-8 E-mail clients use POP or IMAP for incoming mail and SMTP for outgoing mail.

SMTP transmits across TCP port 25. It is supported in Eudora and Outlook Express, in addition to many other e-mail client applications. In Europe, X.400 is often used in place of SMTP.

NTP

Time management is a detail issue in managing network servers, but it can be a cumbersome problem. For example, users might be confused if they receive an e-mail message with a time stamp that occurs in the future. Some distributed applications are driven by time and require servers and network computers to be reasonably synchronized to function correctly. One of the most critical applications that depends on correct time is the air traffic control system. Just imagine what would happen if the clocks were not synchronized in the air traffic control tower, and two different air traffic controllers accidentally gave permission to airplanes to land and take off at the same time. Whoops!

The problem is compounded by the fact that computers' internal clocks are not precise. They function either somewhat faster or slower than actual time measurements. As the days, weeks, and months pass, these computers become increasingly out of sync with actual time.

Network Time Protocol (NTP), developed by David Mills at the University of Delaware to help manage time synchronization, is now an Internet standard. NTP synchronizes computers' internal clocks with a central time source. The central time source can be a server on the Internet, such as *time.nist.gov* or *time.windows.com*. NTP uses Universal Time Coordinate (UTC) to synchronize time within a millisecond. UTC time can be delivered using radio and satellite systems, in addition to time servers.

Ideally, a network administrator will establish one or more central time servers on the network, which will obtain the correct UTC time from a public time source. Then, all time clients, meaning both servers and computers on the network, can obtain the current time from one of the time servers on the network.

Windows XP includes the NTP time client natively and can automatically synchronize to time servers across the Internet. To configure the Windows XP time client, follow these steps:

1. Double-click the time shown in the lower right corner of the screen. (Alternatively, you can open the Date And Time properties in Control Panel.)
2. Click the Internet Time tab.
3. Type in the name of the time server, and click Apply.

SNMP

Network management requires the gathering of information and tracking it over time. Some conditions can indicate a problem on the network. When those conditions are met, the network administrator needs to be alerted.

The Simple Network Management Protocol (SNMP) was developed to manage and monitor network devices. SNMP is typically incorporated into a network-management software package as the underlying protocol. This package can alert a network administrator via pager or e-mail message when an alert condition is met. SNMP applications must be configured in order to work properly.

SNMP uses a Management Information Base (MIB) to define the objects that it will manage. When an administrator uses an SNMP application to define alerts for a type of equipment on the network, the SNMP references its MIB to determine the various events that can occur with that type of object. Then, the administrator configures alerts and assigns the IP address of the equipment to monitor. Many of the SNMP based applications are graphical with these actions taking place behind the scenes. They make it easy for an administrator to simply browse the network, point and click on the objects, and quickly assign the configurable alerts. For new network equipment, a manufacturer will create an MIB specifically for the object so that it can be incorporated in existing SNMP applications.

Transport

The transport layer offers two protocols: TCP and UDP. Application-layer services select either TCP or UDP to deliver data, but never both simultaneously.

TCP

TCP is a connection-oriented protocol that guarantees that the data being sent will be delivered (as long as the other node is running, as well as all underlying protocols). When transmitting data, the TCP protocol keeps track of packets—through the use of port numbers that are assigned to the application—making certain that they'll be delivered to the correct application on arrival at their destination. TCP also ensures that each packet is reassembled in the correct order through the process of sequencing. In sequencing, the packets are given a sequence number so that they can be put back in succession. Sequence numbers have a secondary use as well: they identify duplicate packets so that any duplicates can be discarded.

Data sent via TCP is guaranteed through bidirectional communication using acknowledgments (ACKs). For example, if you are browsing the Web via HTTP, you type in a URL and download an HTML file. When you download that file from the Web server, the server's TCP protocol divides the file into multiple packets, applies sequence numbers to them, and identifies port 80 for the HTTP application to deliver the data to. Each packet is forwarded individually to the next layers and eventually transmitted across the internetwork.

If multiple routes across the network are in place, some packets might travel different routes than the others, and as a result might arrive at the destination in a chaotic shambles. The TCP protocol at the client machine must then reassemble the packets in order. While doing so, the protocol sends an ACK packet to the server and tells it which sequence numbers arrived. Let's say that all packets arrived except for number eight, which was delayed. The protocol sends the ACK but does not acknowledge number eight, so the server transmits number eight again. Because number eight was delayed, the client will receive two copies of packet number eight. TCP will discard the second copy and will send an ACK that the packet has arrived. It reassembles the data and sends it to the HTTP protocol on the client.

The connection between the client and the server is maintained until the message is fully received. In the HTTP example, that would be until the entire HTML file was received on the client machine. If a session-oriented protocol such as Telnet is used, the connection would be maintained until either the client or the server disconnected.

TCP offers a reliable, end-to-end datastream connection over an unreliable internetwork; it doesn't depend on the network architecture, its topology, or even its speed to function correctly.

When two computers establish a connection, the applications used at each end are called ports. Each computer can have up to $2^{16} - 1$ open ports (equivalent to 65,535), which enables it to maintain many open connections with one or more network computers simultaneously.

UDP

If an application doesn't use TCP, it will use UDP, which offers a limited data-delivery service, to communicate. UDP doesn't provide any acknowledgment that data will arrive at the intended destination. If an application uses UDP, it usually has the ability to ensure data delivery without depending on a lower-layer protocol to provide that service. UDP doesn't create the logical connection between two nodes, either. For this reason, UDP is called a connectionless, unreliable data-delivery mechanism.

UDP's lack of overhead ensures a speedy delivery of data. For some network applications, UDP is preferable because the application needs to transmit only small units of data.

Like TCP, UDP uses ports to identify upper-layer applications. These port numbers are identical to those used by TCP. Many applications can use either TCP or UDP, depending on conditions.

Network Layer Protocols

The network layer provides for addressing of data so that it can be routed across an internetwork. The network layer protocols for the Open Systems Interconnection (OSI) reference model are equivalent to the internetwork-layer protocols in the TCP/IP model. Several protocols function at this layer; nearly all applications, however, use only Internet Protocol (IP).

ARP

The Address Resolution Protocol (ARP) maps an IP address to the physical address of an interface. In IPv4, the IP address is 32 bits in length. It is a logical address and is used only when a device looks at protocols above the physical and data-link layers. This is an important concept. When data travels along an Ethernet segment, it is looked at by all the computers on that segment. The computers look only at the data's physical address to determine whether they are meant to look at it. If so, they open up the packet further and look at the data's internetwork layer header.

This means that you can send a data packet from Joe's computer in one building to Lisa's computer in another building using the IP address. In this scenario, one router sits between Joe and Lisa. The data from Joe puts the router's physical address in the Ethernet header, with Lisa's IP address in the IP header. When the data is transmitted, the router will look at the data because its own physical address was in the data packet. The router will forward the data to Lisa by stripping out its own physical address from the Ethernet header and inserting a new header that contains Lisa's physical address.

But what happens when the router doesn't know Lisa's physical address? That's where ARP comes in. The router will maintain a table called an ARP cache, which will contain a mapping of IP addresses to physical addresses of all known hosts. When the router receives the data for Lisa's computer, it will look at the IP address and know which segment Lisa's computer is supposed to be located on. The router will then attempt to learn Lisa's physical address by sending out a broadcast via ARP. The ARP broadcast will ask each computer whether it has Lisa's IP address. Lisa's computer will recognize its own IP address and reply to the broadcast. Then ARP will update the ARP cache for future reference. Plus, the router will then be able to strip off the header and put in a new header with Lisa's physical address in it and forward it to Lisa's computer.

ICMP

If you have ever needed to find out whether a computer was up and running on a TCP/IP network, you probably used Packet Internet Groper (ping) to see if it could communicate with you. Ping is a simple application that sends an Echo command to a remote host via its IP address and waits for a response. Ping will indicate whether a computer is online and functioning, as well as how much time it takes to communicate with the remote host. Ping is entirely dependent on the Internet Control Message Protocol (ICMP).

ICMP is the protocol responsible for message control and error reporting. ICMP uses IP datagrams to communicate. While ping is the most common application that uses ICMP, other applications use ICMP to report errors or control messages between two IP hosts.

IP

IP is a connectionless protocol that provides a data-delivery service to all upper-layer protocols. Connectionless means that no continuing connection exists between the sending and receiving machines. (This doesn't prevent other protocols such as TCP from creating a connection, however.)

Every application in the TCP/IP suite communicates over IP. It provides addressing of both networks and hosts so that data can be routed among multiple segments.

If you've sent a message via SMTP to your SMTP server, the data will be segmented into smaller packets. Each of these packets contains the transmitting computer's IP address (your computer's address) as well as the destination computer's IP address (the SMTP server's address).

Data delivered across an internetwork can be transmitted via different routes, even if the packets are sent from the same computer and to the same destination. Some packets might be delayed and, as a result, received out of order. Keep in mind that IP just delivers the packets. It depends on TCP to put them in the correct order.

Well-Known TCP and UDP Ports

The TCP/IP protocol suite has hundreds of upper-layer services that can communicate across an internetwork, many of which are commonly used. For example, SMTP services many of the e-mail messages sent across the Internet. To make it easier for people to manage ports, the services that are used universally have been assigned port numbers. These are called *well-known ports*. When an application uses a port for TCP, it uses the same number port for UDP. For example, since SMTP uses port 25 over TCP, it also uses port 25 over UDP. Table 5-5 lists some of the well-known ports that you should become familiar with.

Table 5-5 Well-Known Ports

Service	Abbreviation	Port
Domain Name System	DNS	53
Simple Mail Transfer Protocol	SMTP	25
File Transfer Protocol	FTP	20 and 21
Telnet	TELNET	23
Network Time Protocol	NTP	123
Network News Transfer Protocol	NNTP	119
Hypertext Transfer Protocol	HTTP	80
Hypertext Transfer Protocol over Secure Sockets Layer	HTTPS	443
Trivial File Transfer Protocol	TFTP	69 (UDP)
Bootstrap Protocol	BOOTP	67 (UDP) and 68 (UDP)
Post Office Protocol 3	POP3	110

Naming and Addressing

IP addresses and IP names were originally assigned by a network administrator to each and every computer on the internetwork. Not only did the network administrator have to assign the IP address and the IP name, but the admin also had to let all the other computers know what every other host's IP address and IP name was for communication to take place. Three problems evolved from this system:

- Assignments of IP addresses were sometimes duplicated, causing errors on the internetwork.
- Management of IP name to IP address mappings became extremely cumbersome, with many computers having mismatched and out-of-date HOSTS files.
- Management of NetBIOS name to IP address mappings also became cumbersome because the IP address mappings depended on LMHOSTS files, which are similar to HOSTS files, and which also became mismatched and out of date.

DHCP, DNS, and WINS were the upper-layer services developed to fix these dilemmas.

DHCP

Dynamic Host Configuration Protocol (DHCP) provides IP addresses to network computers without an administrator manually assigning them. DHCP lets network administrators establish a central DHCP server and configure it to deliver IP addresses to DHCP clients.

Many Internet Service Providers (ISPs) deliver IP addresses through DHCP. In addition, the DHCP protocol can deliver extended information, such as the subnet mask and a default gateway address, to each DHCP client. If an administrator makes a configuration change to IP addressing, it can all be done through the DHCP console.

DHCP provides for *leases* of IP addresses. A lease is simply a limited amount of time during which the IP address will be valid for a particular computer. The lease time can vary, depending on the administrator's needs. For a busy network in which users are changing frequently and where there are limited numbers of IP addresses, a shorter lease duration can provide sufficient IP addressing to all network hosts.

When configuring DHCP, you should always consider the function of a computer before assigning a dynamic IP address that might change on a periodic basis. For example, a server should have a static IP address, especially when that server is providing Web services across the Internet. Routers should also have statically assigned IP addresses, as should any device that provides DHCP or DNS services to clients.

DNS

The Domain Name System (DNS) provides for a hierarchical naming system of computers on the Internet. Domain names are used to locate the DNS servers in a hierarchical fashion. A DNS server will forward a request to servers either directly above or below themselves in the hierarchy when they do not contain a listing for a particular domain name to IP address.

Take Figure 5-9, for example. The root Internet DNS server is represented by a dot (.). Below this are the top-level domains, including .EDU for educational facilities, .MIL for military facilities, .ORG for nonprofit organizations, and .COM for commercial enterprises. DNS servers exist for each of these domains, and they contain a mapping to every DNS server sitting within their boundaries. .COM includes the DNS server mappings for Microsoft.com as well as Amazon.com and many other .coms. But the .COM DNS server does not contain a mapping to DNS servers for AKC.org.

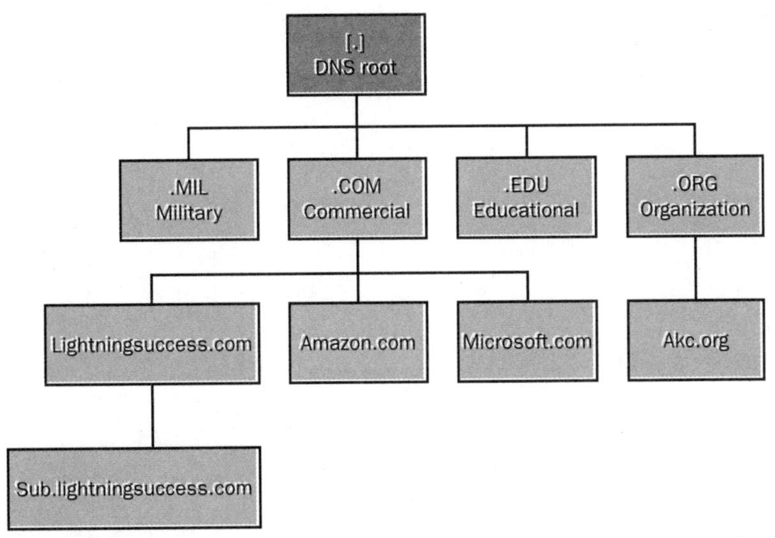

Figure 5-9 DNS servers pass requests to higher-level servers when they do not contain an IP name to IP address listing. At the upper levels, a request is passed down to servers in subdomains until a server has the dot (.).

When you communicate across the Internet, you use the domain name to communicate in the form of a URL. If you decide to browse to *www.microsoft.com*, you will not use the IP address. But the IP address is required to transmit data, so your computer asks its own DNS server for the address. If the DNS server doesn't have that address, your request will be passed up the line to the most common upper DNS server. If you were using a computer at mycompany.com, your request will be passed up to the .com DNS server, which would then pass it to the Microsoft.com DNS server to obtain the IP address. If you were using a computer at mylibrary.org, your request would go up the .ORG tree until it reached the root (.) server and then would travel back down the .COM tree. Once the Microsoft.com server is reached, the response will contain the IP address that your data will use. All of these transactions take place behind the scenes.

WINS

Windows Internet Naming Service (WINS) is used on networks that have NetBIOS names. A pure TCP/IP network would not need WINS; however, legacy Windows networks and OS/2 networks use NetBIOS naming and need some method to map between NetBIOS names and IP addresses.

In WINS, one or more Windows servers are placed on the network to maintain mappings of all IP addresses and NetBIOS names. WINS clients use a WINS server address to look up NetBIOS names.

Key Points

- IPv4 provides for 32-bit long IP addresses and is currently the common version of IP used on the Internet.
- IP addresses are written in dotted decimal format with four numbers between 0 and 255 separated by decimal points. An example is 123.45.67.89.
- There are three classes of IP addresses. Class A addresses begin with an octet range of 1 to 126. Class B addresses begin with an octet range of 128 to 191. Class C addresses begin with an octet range of 192 to 223.
- Each class of IP address has a default mask. The class A default mask is 255.0.0.0. The class B default mask is 255.255.0.0. The class C default mask is 255.255.255.0.
- An IPv6 address is 128 bits in length. It is written in colon hexadecimal format in which each 16-bit portion of the address is separated by a colon. An example is: 2A0F:1234:ABCD:DDD2:0112:0003:78E9:FFF5.
- Telnet is a protocol providing a remote session into another computer.
- FTP provides file transfer services with authentication.
- TFTP provides file transfer services without authentication. TFTP is used only when there is no need for security, such as in firmware downloads.
- HTTP provides Web-browsing services.
- HTTPS is a secure form of HTTP.
- POP3 and IMAP4 are e-mail protocols that allow a person to read e-mail messages.
- SMTP is an e-mail protocol that is used to transmit e-mail messages.
- ICMP is used to send control messages. Ping requires ICMP.

Chapter Review Questions

1 You've been asked to establish a subnetting scheme on a network with five network segments, each of which has 20 or fewer hosts. You are given a class C address. Which subnet mask should you use?

a) 255.255.255.0

b) 255.255.255.128

c) 255.255.255.192

d) 255.255.255.224

Answer d, 255.255.255.224, is correct. By shifting the network mask 3 bits, you will create six subnets with 30 host addresses each in classful addressing and 8 subnets with 30 hosts in CIDR. Answer a, 255.255.255.0, is incorrect because this is the default mask. Answer b, 255.255.255.128, is incorrect because this is an invalid subnet mask in classful addressing where adding one bit is not allowed. Under CIDR, it provides only 2 subnets. Answer c, 255.255.255.192, is incorrect. When you use this address, you create two subnets of 62 hosts in classful addressing and only four subnets in CIDR.

2 You have been told that your address is 178.3.115.88/24. What is the subnet mask?

a) 255.255.255.0

b) 255.255.0.0

c) 255.255.128.0

d) 255.255.255.128

Answer a, 255.255.255.0, is correct. The address is a class B address with 24 bits assigned to the subnet mask. Answers b, c, and d are all incorrect because none of these masks uses 24 bits.

3 Which of the following is a valid IPv6 address?

a) 123.45.67.89

b) 0FA:23:8888:GG:HH99:I111:L99A

c) ABC1:2345:30DD:8877:6691:AFFF:EE77:1234

d) 01-AB-32-88-77-CA

Answer c, ABC1:2345:30DD:8877:6691:AFFF:EE77:1234, is the correct format for an IPv6 address. It is a 128-bit address written in hexadecimal, with each 16-bit section separated by a colon. Answer a, 123.45.67.89, is an IPv4 address. Answer b is incorrect because it includes characters that are not a part of hexadecimal. Answer d is incorrect because this address does not have the appropriate bit length or format.

4 You've been called in to troubleshoot a network. The network administrator added a new router to the network that connects to the Internet. The old router used to connect to the Internet but now only connects to a small office. The administrator says that the new router will not let anyone connect to the Internet. What is the problem?

a) The new router is using IPv6.

b) All of the network hosts are still using the old router's IP address as the default gateway.

c) The old router is on a private network.

d) The new router is on a private network.

Answer b is correct. The most likely problem is that all the network hosts are using the old router's IP address as their default gateway. When the old router receives these packets, it doesn't know what to do with them and discards them. Answers a, c, and d are all incorrect because IPv6 and the use of private network addresses would not affect the network this way.

5 You have a network with 583 computers. You have registered a single class C address. Which of the following can let you assign individual IP addresses to every host on the network with plenty available for growth of the organization?

a) IPv4 addresses

b) A default gateway

c) SNMP

d) NAT with a private address

Answer d is correct. Using network address translation (NAT), along with a private IP address on the internal network, each host can use its private IP address to communicate and map to the registered class C address pool when communicating across the Internet. Answer a, IPv4, is incorrect because classful addressing uses IPv4. Answer b, a default gateway, is incorrect because it is used for mass routing of data. Answer c, SNMP, is incorrect because it is a network-management protocol.

6 Which of the following applications provides for time synchronization?

a) Telnet

b) TFTP

c) NTP

d) DNS

Answer c, NTP, or Network Time Protocol, is used to synchronize time to a central time source. Answer a, Telnet, is incorrect because it is a remote session protocol. Answer b, TFTP, is incorrect because it provides file transfer services. Answer d, DNS, is incorrect because DNS is a hierarchical service that maps IP names to IP addresses.

7 You are configuring a workstation's e-mail application to send e-mail messages to the network servers. Which of the following protocols must be enabled on a server for the client to be able to download an e-mail message and read it?

a) POP3

b) SMTP

c) SNMP

d) NTP

Answer a, POP3, must be enabled on the server from which the client downloads e-mail messages and reads them. Answer b, SMTP, is incorrect. SMTP is used to transmit e-mail messages. Answer c, SNMP, is incorrect because SNMP is a network-management protocol. Answer d, NTP, is incorrect because NTP is used for synchronizing time.

8 How can you be certain that an Internet transaction is using Hypertext Transfer Protocol over Secure Sockets Layer?

a) You can't be certain because it uses UDP and is unreliable.

b) You should be prompted for a user ID and password.

c) You should see https:// in the address and a padlock in the lower part of the browser.

d) You should be required to use port 80 in the transaction.

Answer c is correct. When using HTTPS, you should see an https:// in the address field and a padlock icon in the lower bar of the browser. Answer a is incorrect; UDP has no effect on HTTP over SSL. Answer b is incorrect; while HTTPS supports authentication, it is not required for it to function. Answer d is incorrect; port 443 is used for HTTPS, not port 80.

9 Which of the following is used by computers to receive an IP address?

a) DHCP

b) DNS – IP name to IP address mapping

c) WINS

d) ARP

Answer a, DHCP, is used by DHCP clients to obtain an IP address. Answer b, DNS, is incorrect because DNS provides IP name to IP address mapping. Answer c, WINS, is incorrect because it provides NetBIOS name to IP address mapping. Answer d, ARP, is incorrect because it provides IP address to physical address mapping.

10 Which of the following can be used to determine whether TCP/IP is functioning correctly?

a) Telnet

b) FTP

c) HTTP

d) All of the above

e) None of the above

Answer d is correct. If you can run Telnet, FTP, or HTTP applications successfully, you can be certain that all underlying protocols are functioning correctly.

Chapter 6

Wide Area Networking

Wide area networks (WANs) provide data communications that traverse a broad geographic area. It's typical for a WAN to utilize the transmission infrastructure provided by a third party such as a telephone company. Renting the ability to transmit data across another company's network makes sense, because it would be prohibitively expensive to lay cable across miles of territory, much less receive the rights to do so, for most companies.

Using the Internet to provide WAN services is a growing area for small offices and home networks. People use local Internet Service Providers (ISP) that offer services such as Integrated Services Digital Network (ISDN), Digital Subscriber Line (DSL), cable modems, or even the plain old telephone service (POTS). Once connected, they implement Virtual Private Networking (VPN) to transmit private data to another location anywhere in the world via the Internet.

A WAN usually involves using a third-party network, and WAN technologies function at the lowest three layers of the OSI reference model. WANs consist of physical media and connectors at the physical layer, physical addressing and media access at the data-link layer, and routing at the network layer.

Point-to-point is an important concept in wide area networking. A point-to-point connection in a WAN is not a typical direct connection. Rather, it's a pre-established path from one site to another that passes through a carrier network, as shown in Figure 6-1. A point-to-point connection, including wiring and hardware, is usually rented from a carrier and thus called a *leased line*.

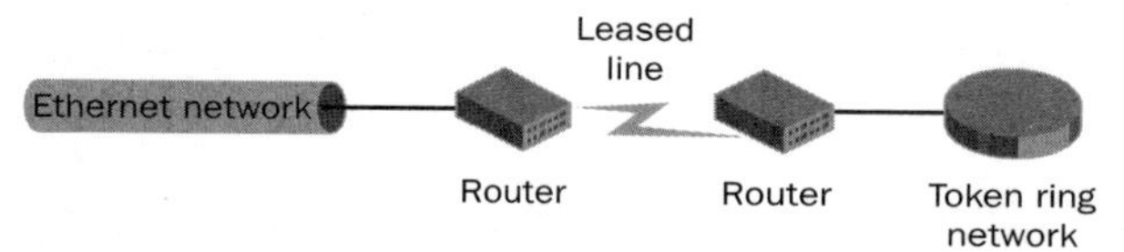

Figure 6-1 A leased line creates a point-to-point connection between networks.

Switching Methods

Switching refers to the routing process used to move data throughout the WAN. A protocol only uses one type of switching. Switching methods influence how quickly routing takes place. The three switching methods we will review are the following:

- Circuit switching
- Packet switching
- Cell switching

Circuit Switching

Circuit-switched networks are the oldest type of communications network. The Public Switched Telephone Network (PSTN) is a circuit-switched network, for example. Circuit-switched networks use temporary paths created through the network along which to transmit data. There are two types of virtual circuits: permanent virtual circuits (PVC) and switched virtual circuits (SVC).

A PVC is a dedicated path between two points used when data communications must take place 24 hours a day, 7 days a week. A PVC is similar to a telephone call that never ends.

SVCs are established on demand and ended when the communication is complete. SVCs are often called a dialup service because the SVC is established just like a telephone call and terminated in the same way. SVCs are used when the need to transfer data is sporadic.

When selecting between a PVC and an SVC, remember that the cost of PVC is higher because the telecommunications carrier must provide constant circuit availability for a PVC, while that for an SVC is provided on an as-needed basis. On the other hand, some additional overhead is involved with an SVC because every time data must be transferred, the circuit must be established, followed by the transmission of the data and finally the termination of the circuit.

Packet Switching

Packet switching normally functions where individual networks connect into a telecommunication carrier's network. The carrier creates *virtual circuits*, a system that specifies a path between specific sites, within its network. The path is not dedicated, however, and data can flow across different routers but still arrive at the same destination, which explains why it is called virtual. Instead of enabling a constant data stream to transfer across the network, the virtual circuit is opened long enough to send a single chunk of data. Packet switching enables a telecommunications carrier company to share its network among several customers. Because the carrier can share the network, the cost to customers is much lower than a dedicated point-to-point link. The shared section is often called a *cloud*.

As indicated by its name, the data transmitted through a packet-switched network is sent as a unit called a *packet*. The source computer sends the data to the destination, but it first encounters a router, which is sometimes called a Packet Switching Exchange. When the router receives the packet, it looks at the destination address. Then it looks at its routing directory to determine which address to forward the packet to next. The packet is then forwarded along the virtual circuit until it eventually reaches the destination computer.

Each packet is transmitted in a store-and-forward process. When a router receives a packet, it stores the packet temporarily. After reviewing the packet's address information, the router forwards the packet. The time that it takes to store the packet causes a slight latency in the transmission.

Packets are kept to a certain size to ensure a high data-transmission speed. It's a simple fact that when there is less data to be stored, the transmission speed is higher. For this reason, the packets that are created at the transport layer of the OSI reference model are usually divided into a smaller size at the sending node.

A packet-switched network offers several advantages. Because packet-switched protocols provide for error checking and flow control, packet switching is very efficient. The virtual circuit minimizes the connection time between any two systems, reducing the load on the network. Because virtual circuits are closed after each packet is forwarded, a router is available to accept information from a router for a different virtual circuit. Figure 6-2 displays how a packet-switched network functions.

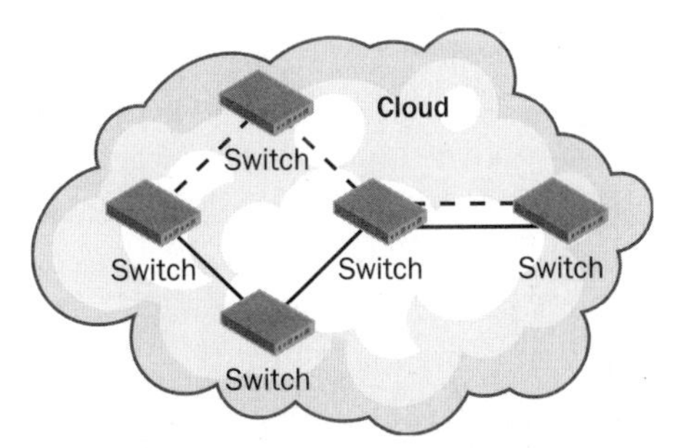

Figure 6-2 Packet switching enables routers to accept packets on multiple different virtual circuits.

Voice over IP (VoIP) is one of the packet-switching technologies that enables a person to place phone calls across any Transmission Control Protocol/Internet Protocol (TCP/IP) network. If you look at how a normal telephone call is made, you can see the value of using a packet-switched network for a VoIP call. In normal telephone calls across the plain old telephone system, a circuit is created during the call. With each router in the circuit dedicated to a single phone call, the circuit must wait until the call has been completed before being used for another call. By contrast, a packet-switched router used for VoIP calls can provide several times the number of phone calls in the same time period.

Cell Switching

Cell switching is a form of packet switching. The main difference between a packet-switched network and a cell-switched network is the size of the cell. Cells are extremely small and do not vary in size. Their size makes them fast and provides for a network with a low latency.

An example of a cell-switched network is Asynchronous Transfer Mode (ATM), which we will discuss in more detail later in this chapter. The cell in an ATM network is 53 bytes in length, including the data portion. In a packet-switched network, the packets vary in length and can be several times as long as a cell. Because a cell does not vary in size, each router in the cell-switched network knows how much data to expect with each cell and is built to take advantage of it. The tiny cell is small enough to be stored in random access memory, whereas a packet-switching router must store a packet to disk. Because the router need only switch the cell in and out of its fastest memory, there is little latency in a cell-switched network.

Using the Digital Telephone with ISDN

One of the WAN technologies telephone companies embraced early on was Integrated Services Digital Network (ISDN). It was developed as the digital offering by the telephone companies to run over the existing telephone copper wiring. ISDN allows subscribers to transmit data, voice, and multimedia.

The two types of ISDN services are as follows:

- **Basic Rate Interface (BRI):** Used for home and small office connectivity. BRI services include two B channels and a single D channel. A B channel offers 64 Kbps and carries user data. A BRI D channel operates at 16 Kbps and carries control and signaling information. Through these two channels, a home connection can reach 128 Kbps of data throughput.

- **Primary Rate Interface (PRI):** Used for WANs and runs across leased lines. The PRI service is composed of 23 B channels at 64 Kbps each for user data, along with a single D channel, also operating at 64 Kbps to handle control information. Overall, the PRI service provides a throughput rate of 1.544 Mbps.

A computer with an ISDN line is able to connect to any other computer that also uses ISDN simply by dialing its ISDN number. (This is similar to modems and telephone numbers except that the number is in a different format.)

The ISDN specification includes several types of equipment, as listed below:

- **Terminal adapter (TA):** Also called an ISDN modem, this is either an internal or external adapter to connect equipment to an ISDN line.
- **Terminal equipment type 1 (TE1):** Terminals with built-in ISDN adapters.
- **Terminal equipment type 2 (TE2):** Terminals that require a terminal adapter to connect to an ISDN line.
- **Network termination type 1 (NT1):** Connects the ISDN line between the customer's location and the telephone company's local loop.
- **Network termination type 2 (NT2):** Used for digital private branch exchanges (PBXs), providing addressing and routing services.

Using Fiber Optics with FDDI

Fiber optics offer great advantages over copper wiring. Light signals are capable of traveling long distances, and fiber optics aren't subject to either electromagnetic interference or radio frequency interference. As a result, fiber optics offer an enormous number of benefits to both WAN and LAN technologies. Fiber Distributed Data Interface (FDDI) is the American National Standards Institute (ANSI) specification for a 100-Mbps token-passing dual ring network over fiber optic cable and is often used as a backbone for a campus network because it can attain distances up to two kilometers with multimode fiber or farther with single-mode fiber.

Both multimode and single mode fiber optics transmit light signals. Multimode fiber uses a light emitting diode (LED) to transmit the signals, while single mode fiber depends on lasers.

FDDI's dual ring structure enables traffic to flow on each ring, but in opposite directions—one runs clockwise and the other counterclockwise. One of the rings is considered primary, the other secondary. The primary ring is used for data transmission. Because fiber optic cables are brittle, the secondary ring acts as a backup in case of a break in the primary ring.

Most network nodes connect only to the primary ring for data transmission. These are called single attachment stations (SAS). Single attachment concentrators (SAC) also connect only to the primary ring. Dual attachment stations (DAS) connect to both rings, as do dual attachment concentrators (DAC).

SAS and SAC do not affect data communications when they are turned off, but the DAS and DAC nodes can. Each DAS and DAC has two ports to connect to each ring. When there is a break in the primary ring, it moves into a *wrapped* state. When the ring is wrapped, the DASs on either side of the breakage pass data through each of their ports to create a single ring, as shown in Figure 6-3. A DAS has an optical bypass switch that prevents ring segmentation during a wrapped state by using optical mirrors to pass signals along if the DAS fails or is powered off. The dual ring technology can tolerate only a single break. If two breaks occur, the single ring becomes two separate rings that cannot communicate with each other.

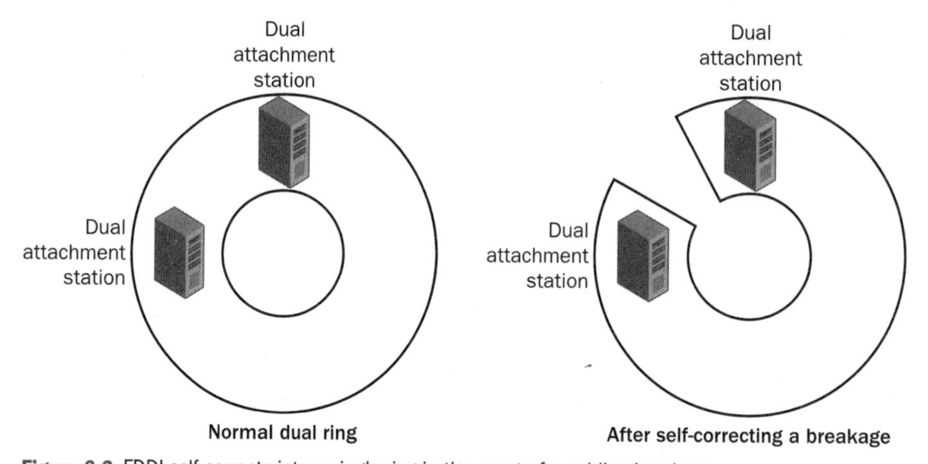

Figure 6-3 FDDI self-corrects into a single ring in the event of a cabling breakage.

FDDI is a physical and data-link layer protocol. Upper-layer protocols such as TCP/IP and IPX/SPX can run across a FDDI ring. The FDDI frame is similar to a token ring frame format. There are two frames: the token frame and the data frame. The FDDI data frame can become as large as 4500 bytes in length. The frame fields, shown in Figure 6-4, are as follows:

- **Preamble:** Identifies the incoming data as a new frame.
- **Start delimiter:** Specifies the beginning of a frame.

- **Frame control:** Includes control information such as the size of the address fields.
- **Destination address:** A 6-byte–long physical address of the destination device.
- **Source address:** A 6-byte–long physical address of the source device.
- **Data:** Variable-length field containing the transmitted data.
- **Frame Check Sequence (FCS):** The value of the cyclic redundancy check (CRC), which is an algorithm used to determine whether the frame has an error. In the case of an error, the frame is discarded.
- **End delimiter:** Indicates that the frame is completed.
- **Frame status:** A field that, when returned to the source node, shows whether the frame was received or whether an error occurred so that the frame can be retransmitted.

Preamble	Start delimiter	Frame control	Destination address	Source address	Data	FCS	End delimiter	Frame status

Figure 6-4 The FDDI data frame includes a variable-length data field.

Gaining WAN Speed with ATM

WANs are typically slow. They are transmitted across long distances over networks that traditionally were unreliable. As a result, the oldest WAN technologies were loaded with failsafe measures, such as error checking, and were clocked to match the slowest expected link. As public networks gained in reliability, WAN technologies became mismatched. Demand for greater data throughput was one of the driving forces behind Asynchronous Transfer Mode (ATM).

Demand for bandwidth is driven by multimedia, distributed databases, and voice across digital lines. These require speeds in the megabits-per-second range for reasonable performance. I remember when a 56-Kbps link was not only expensive, but also considered more than adequate for a WAN link to serve offices with as many as 100 users. Today a cable modem can provide speeds at several megabits per second (depending on the contract you have with the cable provider) for a fraction of the cost.

Bandwidth demands have crossed over from the local area network (LAN) to the WAN. Traditionally, the WAN and LAN served different purposes. LAN technologies enabled people to share resources such as server storage, printers, and databases. WAN technologies were limited to intermittent data transmissions that could withstand the delays inherent within slow WAN technologies. Internet innovations that have compressed the bandwidth requirements for multimedia components in concert with the increased bandwidth available from high-speed technologies like ATM have blurred the lines between WAN and LAN technologies. Voice, video, shared database applications, and other types of data are now transmitted across both LANs and WANs.

Answering the Need for Speed with ATM

The International Telecommunications Union-Telecommunication Standardization Sector (ITU-T) and ANSI developed ATM for high-speed data transfer within public networks. ATM is also capable of performing as a backbone technology for a private network, or even as a LAN technology across all segments of the network.

ATM is a network made up of a series of ATM switches (also called ATM routers) and ATM nodes. The ATM nodes can be routers that connect the ATM network to another type of network, such as an Ethernet LAN, as shown in Figure 6-5. Any ATM node can communicate with any other ATM node by transmitting data across the ATM network.

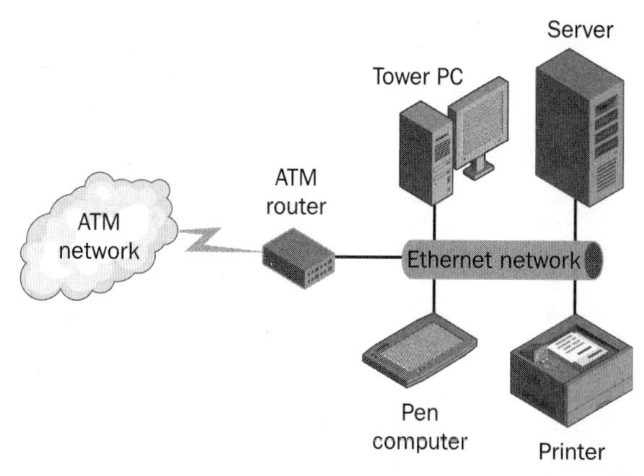

Figure 6-5 ATM routers can connect an ATM network to other types of LANs.

Two types of connections can operate in an ATM network. The User to Network Interface (UNI) is a connection between a node and the ATM network; the Network to Network Interface (NNI) is the link between two ATM switches. LAN emulation (LANE) standards are used to connect ATM networks to LANs running Ethernet or token ring protocols.

ATM is a connection-oriented protocol in which the two ATM end points must establish a connection before transferring data. A signaling request packet is sent through the ATM network from the source node to the destination node, including the Quality of Service (QoS) requirements for the connection. QoS enables the node to request priority for the connection, which is required for voice or multimedia transmissions. If no ATM switch along the path can supply the QoS required, the switch will reject the signaling request and reply with a message to that effect. If all switches and the destination can provide the QoS, an accept message is returned to the source node, including the virtual path and virtual channel identifiers. This creates the connection.

Each ATM cell consists of 53 total bytes, including the data. The header is 5 bytes in length, with 48 bytes of data. The ATM header includes the virtual path identifier (VPI) and virtual channel identifier (VCI) to identify how to forward the cell along the path to its destination.

The ATM Model

Three layers in the ATM networking model map roughly to the physical and data-link layers of the OSI reference model:

- ATM physical layer
- ATM layer
- ATM adaptation layer

The ATM physical layer is concerned with transmission of bit-stream data on the physical media. This layer is divided into logical sublayers. The physical medium sublayer commands the sending and receiving of the bit stream and uses timing information to synchronize the transmitted data. The physical medium sublayer is dependent on the type of cabling used. ATM can transmit across SONET and T3/E3, which we will discuss later, as well as fiber optics, shielded twisted pair (STP), and unshielded twisted pair (UTP). The transmission convergence sublayer controls cell delineation, which is simply maintaining the boundaries between cells. The transmission convergence sublayer also generates and validates the data by checking the header error control information. Where necessary, the transmission convergence sublayer will insert or suppress unassigned cells to match the rate of cells to the capacity of the system; it will also adapt the cells into frames that match the medium's requirements.

The ATM layer creates the virtual connection between the sending and receiving node and then switches the ATM cells through the virtual path. When two virtual connections temporarily share the same path, the ATM layer will multiplex the cells to be able to transmit them as a single bit stream, and then demultiplex the cells when the virtual connections split off into two different directions. This is shown in Figure 6-6.

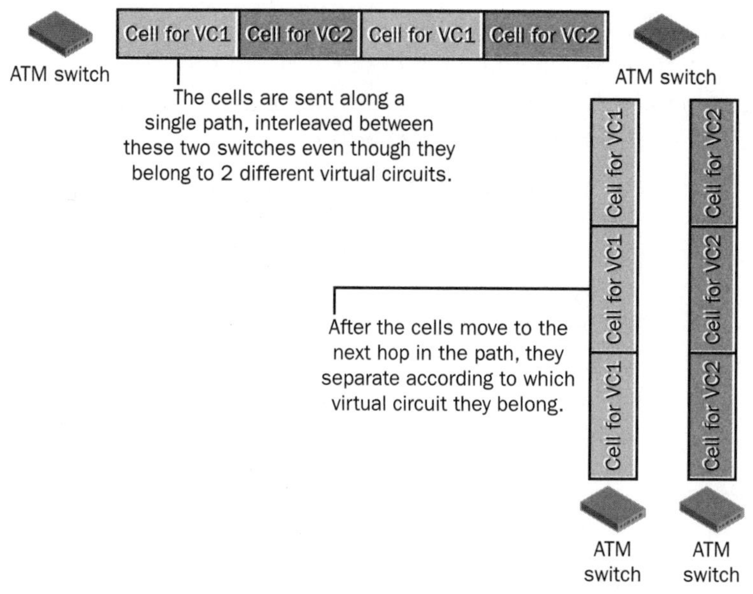

Figure 6-6 ATM layer will multiplex and demultiplex two or more virtual connections.

The ATM adaptation layer receives the packets from upper-layer processes and translates them into ATM cells. These upper-layer packets can come from AppleTalk, TCP/IP, or IPX/SPX, for example.

Frame Relay

Frame relay is a widely applied WAN protocol that uses packet switching. It was originally designed to run across ISDN interfaces. The frame relay protocol provides an efficient data transmission, even though the packets vary in length. The variable length of the packets makes data transmission very flexible.

Even though frame relay was designed to transmit across ISDN, it has since been updated to transmit data across a variety of different protocols. Most of the data transmission takes place within the carrier's network, also called a cloud, which provides for congestion signaling, physical media, and switching.

Frame relay supplies data-link and physical layer services, as shown in Figure 6-7. Both permanent virtual circuits (PVC) and switched virtual circuits (SVC) can be used. It is most common to find frame relay with PVCs.

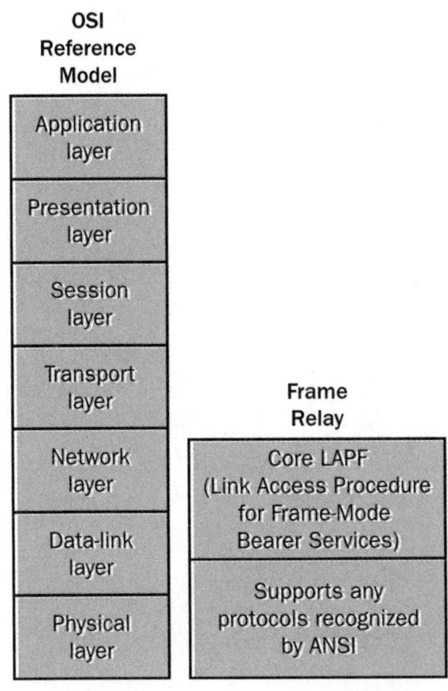

Figure 6-7 Frame relay is a physical and data-link layer protocol specification.

Although all customer networks connect to the same carrier network, the maximum rate of data throughput is determined by the customer's point-to-point link into the frame relay cloud. Each link can be a different data rate, as depicted in Figure 6-8.

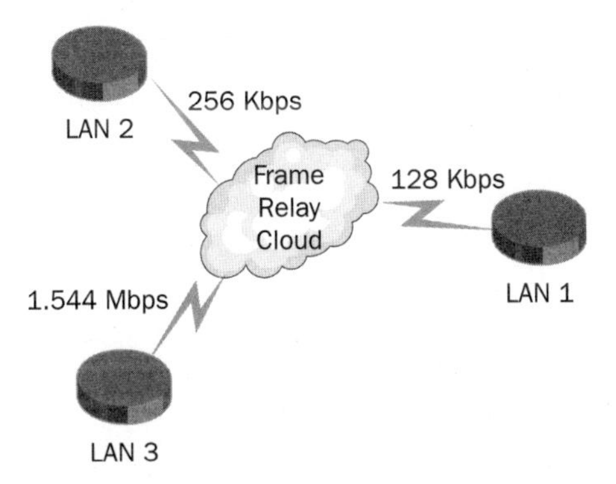

Figure 6-8 Different networks can connect to a frame relay cloud at different rates.

Data Link Connection Identifiers (DLCI) are key to frame relay connections. The DLCI identifies the point-to-point link that begins the virtual circuit to the cloud. It is a logical connection, because a single interface can transmit to different DLCIs. It's easiest to think of a DLCI as a phone number. When transmitting data across frame relay, the interface into the cloud will specify the destination by its DLCI in the same way that a telephone call identifies the recipient by that recipient's telephone number.

Controlling Congestion

Congestion is one problem that can occur in a cloud network shared by multiple customers. Because the switches can participate in any number of virtual circuits, they can become overloaded by traffic. The frame relay protocol includes 3 bits in the address of each frame relay frame header, shown in Figure 6-9, to manage congestion. The first bit is called the Forward Explicit Congestion Notification (FECN). When this bit is set to a one (1) value, it means that the frame encountered congestion. A router that receives a frame with the FECN bit set to 1 will send a reply frame with a different bit named the Backward Explicit Congestion Notification (BECN). Both the FECN and BECN bits are used to notify upper-layer protocols of congestion so that they can initiate flow-control mechanisms to reduce data transmissions.

Flags	DLCI	Extended address	C/R: not defined	FECN	BECN	DE	Data	FCS

Figure 6-9 The frame relay header includes bits to manage congestion.

The third congestion control bit, the Discard Eligibility (DE) bit, is assigned to unimportant frames. Frames with the DE bit set to 1 are discarded when there is congestion on the network. This is an additional method to help manage high-traffic situations within the frame relay cloud.

Given that congestion does exist in the frame relay network, the carrier provides a guaranteed minimum data transmission speed: the committed information rate (CIR). A benefit of frame relay is that bursts of speed up to double the CIR level are generally allowed by the carrier network, but only for short periods of time. Bursting is available only during noncongested time periods when extra bandwidth is available.

Not only is frame relay flexible because of its ability to burst extra traffic, but its data transmission rates are also dependent solely on configuration. A frame relay link can be installed and configured at a lower speed (and at a lower cost) using the same equipment that is used for a much faster link. If a link needs to be upgraded because of an increased need for bandwidth, or downgraded because bandwidth is no longer required, the carrier needs only to make a configuration change.

One reason that geographically dispersed companies select frame relay is that each location requires only a single link into the carrier's cloud. Using point-to-point links increases the number of connections geometrically with each new office added. For example, a WAN with five offices in frame relay requires only five links. However, if point-to-point links were in use, this WAN would need 10 links for all of the offices to be able to communicate with each other. This is illustrated in Figure 6-10.

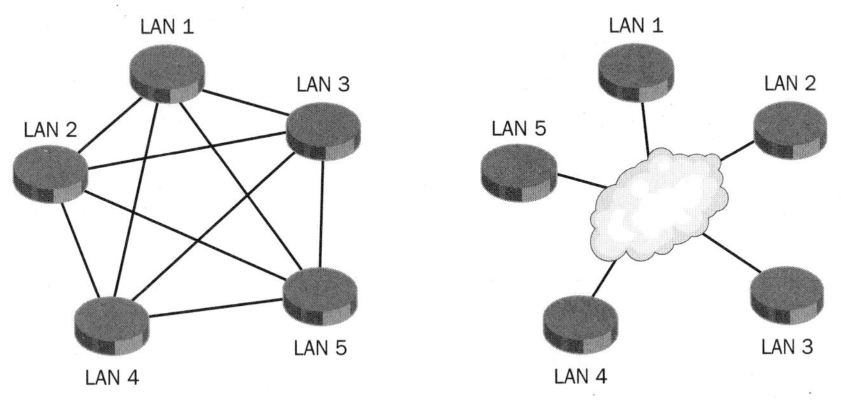

Figure 6-10 A point-to-point WAN requires more links than does a frame relay cloud.

Configuring a Cisco Router with Frame Relay Cisco routers are commonly used to connect a LAN to a WAN. When configuring a Cisco router, you'll need to Telnet into the router from a network host, or through a computer that is directly connected via a serial cable to the router. Once you've connected and logged into the router, you must move into interface configuration mode to begin. The commands are similar to the following, with the words preceding the pound sign (#) serving as the router's prompt. After entering the commands, you complete the process by pressing the CTRL and Z keys on the keyboard.

```
Router# config terminal
```

This places the router in configuration mode.

```
Router(config)# interface serial 0
```

This begins interface configuration mode.

```
Router(config-if)# encapsulation frame-relay
```

This applies the frame relay protocol to the serial interface.

```
Router(config-if)# frame-relay interface-dlci 42
```

This command identifies the DLCI for the link. Other commands are available on Cisco routers to configure frame relay or view its statistics. These commands are available on Cisco's Web site at *http://www.cisco.com*.

SONET/SDH

The Synchronous Optical Network (SONET), or Synchronous Digital Hierarchy (SDH) as it's known in Europe, offers the ability to construct large-scale, high-speed IP networks over fiber optics. The SONET topology can be either a dual ring architecture or a star. The dual ring is preferable because it can be reconfigured in case of a break in the fiber optic cable to ensure the network's survivability. SONET is often used for Internet and large internetwork backbone services. Using time division multiplexing (TDM), SONET is capable of providing high-bandwidth capacity for data transmission as well as voice traffic and even cable television.

SONET is a global standard focusing on synchronous communications that are multiplexed. In synchronous networking, all the clocks are synchronized to the same time. The time division multiplexing enables signals from slower networks to be intermingled directly with SONET signals as they are moved onto the SONET network. This is also partially because of the advanced network management and maintenance features inherent in SONET.

Four layers comprise the SONET protocol stack. The bottom layer, called the photonic layer, provides the conversion from electrical signals to optical signals. The section layer handles the transport of frames across the media. The line layer provides synchronization and multiplexing for the path layer, which maps services between the equipment at each end of a path. An upper-layer service could be a T1 or T3 signal, which we'll discuss further on, as well as video. Note that these services can be much slower than the optical rates available on SONET.

Optical Carrier Signaling (OCx)

The signals used to transmit across SONET are framed as Synchronous Transport Signals. When SONET transmits across fiber optics, it calls these signals Optical Carrier signals.

SONET's basic transmission rate, Synchronous Transport Signal level 1 (STS-1), also considered Optical Carrier 1 (OC1), is 51.84 Mbps. The SONET multiplexing scheme can transmit at rates that are multiples of 51.84 Mbps. Table 6-1 displays the rates. It's common for SONET to be implemented up to the level of STS-192.

Table 6-1 SONET Data Rates

SONET Signal Level	OC Level	Data Rate
STS-1	OC-1	51.84 Mbps
STS-3	OC-3	155.52 Mbps
STS-12	OC-12	622.08 Mbps

Table 6-1 SONET Data Rates

SONET Signal Level	OC Level	Data Rate
STS-24	OC-24	1.24 Gbps
STS-48	OC-48	2.488 Gbps
STS-192	OC-192	10 Gbps
STS-256	OC-256	13.271 Gbps
STS-768	OC-768	40 Gbps

T-Carrier System

The T-carrier system is a series of data transmission formats developed by Bell Telephone for use in the telephone network system in North America and Japan. The base unit of a T-carrier is DS0, which is 64 Kbps. The T-carrier system uses in-band signaling, a method that actually robs bits from being used for data and uses them instead for overhead. This reduces the transmission rates used for T-carrier signals. The E-carrier system used in Europe doesn't perform bit-robbing and as a result has a higher throughput rate.

T1/E1

T1 and E1 lines are each multiples of DS0 signals. The T1 line provides 1.544 Mbps, while the E1 line provides 2.048 Mbps. The difference in data rates results from the T-carrier system's method of bit-robbing.

Customers can purchase fractional-T1 lines, which are actually multiples of the DS0 signal. With Frac-T1, the customer rents a number of the 24 channels within a T1 line. The remaining channels go unused. For example, a Frac-T1 line can be 128 Kbps, 256 Kbps, 512 Kbps, and so on.

T3/E3

T3 lines are digital carriers, equivalent to 28 T1 lines, that can transmit at the rate of 44.736 Mbps. E3 lines provide 16 E1 lines, with a transmission rate of 34.368 Mbps.

Key Points

- Circuit switching is a method that creates dedicated circuits across which networks will communicate.
- Packet switching is more efficient because it uses virtual circuits that are dedicated only for a packet's transmission. As soon as the packet is transmitted, the switch is released and can be used to transmit other packets.

- Cell switching is a form of packet switching that uses small, fixed-length cells. The small size enables cell switching to take place in memory, reducing latency and increasing speed.
- ISDN is a digital telephony method that offers two types of interfaces. The Basic Rate Interface (BRI) provides up to 128 Kbps. The Primary Rate Interface (PRI) offers up to 1.544 Mbps and is transmitted across a T1 line.
- FDDI is a networking method that transmits across fiber optics using a dual ring architecture.
- FDDI's dual ring is self-correcting in the event of a breakage in the cabling.
- ATM is a high-speed WAN protocol that uses cell switching to transmit data. ATM can transmit across fiber optics as well as copper media.
- Frame relay is a WAN protocol that offers up to 2.048 Mbps transmission speeds.
- Networks connecting to a frame relay cloud can be established at different rates of transmission and still communicate.
- The DLCI is used in much the same way as a telephone number, identifying a destination network within a frame relay cloud.
- SONET is a high-speed WAN interface typically deployed over fiber optic media configured in a dual ring. Using SONET, WAN protocols can achieve theoretically up to 40 gigabits per second.
- The T-carrier system was developed for telephone carrier networks. T1 lines offer 1.544 Mbps. T3 lines offer 44.736 Mbps.

Chapter Review Questions

1 Which of the following is a WAN protocol?

a) TCP

b) RIP

c) ICMP

d) ATM

Answer d is correct. ATM is a WAN protocol. Answer a, TCP, is incorrect because it works on LANs and WANs. Answer b, RIP, is incorrect because it is a routing protocol. Answer c, ICMP, is incorrect because it is a network-layer protocol providing control messages.

2 You've been called in to help design a network's WAN connection. The network has two offices that are currently not connected to each other. The offices will be sharing e-mail and data applications across the WAN. Data transmissions will be intermittent and will occur on a small scale. Which WAN specification is the best choice?

a) ISDN

b) ATM

c) Frame relay

d) X.25

Answer a, ISDN, is the best choice because the networks will be exchanging intermittent data in small amounts. Answer b, ATM, is incorrect because there is no need for high-speed transmission. Answer c, frame relay, is an option, but given the intermittent data exchange, ISDN is still a better choice. Answer d, X.25, is incorrect because it is a legacy WAN protocol used mainly for text data.

3 When a virtual circuit is created by one network and temporarily connects to the other network, what type of virtual circuit is it?

a) Dialup

b) Cell circuit

c) SVC

d) PVC

Answer c, a switched virtual circuit (SVC), is created when one network connects temporarily to another network. Answer a, dialup, is incorrect because there is no such thing as a dialup virtual circuit. Answer b, cell circuit, is incorrect because there's no such thing. Answer d, PVC, is incorrect because a permanent virtual circuit is not temporary.

4 Which protocol uses cell switching?

a) FDDI

b) ATM

c) Frame relay

d) ISDN

Answer b, ATM, uses cell switching. Answers a, c, and d are incorrect because none of these protocols use cell switching.

5 Which of the following specifications offers 1.544 Mbps?

a) E1

b) E3

c) T1

d) T3

Answer c, a T1 line, offers 1.544 Mbps transmission speed. Answer a, E1, is incorrect because it offers 2.048 Mbps. Answer b, E3, is incorrect because it offers 34.368 Mbps. Answer d, T3, is incorrect because it offers 44.736 Mbps.

6 You are designing a WAN solution for a network. The network has several offices located in a large metropolitan area, which has a SONET ring. These offices have used frame relay and are disappointed by the slow transmission rates. What WAN protocol would you suggest?

a) Fast Ethernet

b) ATM

c) ISDN

d) Token ring

Answer b, ATM, can offer high-speed data transmission across the SONET ring available in the metropolitan area. Answer a, Fast Ethernet, is incorrect because it is not a WAN protocol. Answer c, ISDN, is incorrect because it will not provide an increase in speed. Answer d, token ring, is incorrect because it is not a WAN protocol.

7 You are configuring a frame relay connection. The carrier has given you two DLCIs for your two offices, 14 and 19. When you have completed the configuration, the connection doesn't work. Why?

a) You assigned the DLCIs to the wrong connections.

b) Frame relay requires a backup dialup connection to function.

c) You forgot to enter the number to dial.

d) The fiber optic cable was unplugged.

Answer a is correct. You must assign the DLCIs to the correct connections or the frame relay links will not function. Answers b, c, and d are all incorrect because these situations are not applicable.

8 Which of the following use fiber optics?

- **a)** ISDN
- **b)** Packet switching
- **c)** T1
- **d)** FDDI

Answer d is correct. FDDI uses fiber optic media. Answer a, ISDN, is incorrect because it uses regular telephone lines. Answer b, packet switching, is incorrect because it isn't media-dependent. Answer c, T1, is incorrect because T1 lines use copper wiring.

9 How fast is an OC-48 link?

- **a)** 2.488 Gbps
- **b)** 48 Mbps
- **c)** 48 Kbps
- **d)** 48 Gbps

Answer a is correct. An OC-48 link is 2.488 Gbps. Answers b, c, and d are all incorrect.

10 If you require a connection that transmits at or above 51 Mbps, which of the following will be adequate?

- **a)** T1
- **b)** T3
- **c)** OC-1
- **d)** ISDN PRI

Answer c is correct; an OC-1 link offers 51.48 Mbps. Answer a is incorrect; a T1 line offers 1.544 Mbps. Answer b is incorrect; a T3 line offers 44.736 Mbps. Answer d is incorrect; ISDN PRI offers 1.544 Mbps.

Chapter 7

Networking with Remote Clients and Servers

Telecommuting was once only a theory. Today people telecommute on a regular basis. Some can conduct their entire jobs away from the office. Others are able to work remotely when traveling for business. Many people depend on telecommuting to stay current with various data tasks at home after hours. Telecommuting is largely dependent on users being able to access a business network from remote locations.

Remote access grew from early use of remote terminals connecting to mainframe computers across telephone lines. Terminals would connect to individual applications and then disconnect. Eventually, users demanded the same access to network-connected personal computers and server-based applications.

In this chapter, we'll review the two main methods of remote computing: remote node and remote control. We'll look at the protocols involved in each of these methods, including thin client computing. Finally we'll review the process of configuring remote-computing connections.

Remote Node

Traditionally, remote nodes were computers that connected to a network via a dialup line. Today remote nodes connect via ISDN, DSL, cable modem, and Virtual Private Networking (VPN) across the Internet, in addition to dialing up using the plain old telephone system (POTS). When connecting via remote node, a computer makes a connection through a public network to a remote access server (RAS). The remote access server then acts as a router, exchanging traffic between the remote computer and the network. This enables the remote computer to act as though it is a network node, able to transfer files, access database information and applications, and print to network printers.

Remote node computing is notoriously slow, primarily because dialup and other remote access connections are far slower—averaging between 56 Kbps to 128 Kbps—than the typical LAN connection, which is usually 10 Mbps or 100 Mbps. When the remote node accesses applications from a network location, the application must first download to the remote node before it is processed. Updates made to data must be uploaded across that slow link as well. It can be frustratingly slow to execute applications remotely. In fact, running an application across a dialup connection might seem slower than trying to suck peanut butter through a straw.

Aside from the limitations placed on speed by the remote link, the remote node is not much different from a local node on the network. A server handles requests the same way regardless of whether the node is local or remote. As you can see in Figure 7-1, the path to the network from the remote node is simply longer (and slower) than the path from a local node.

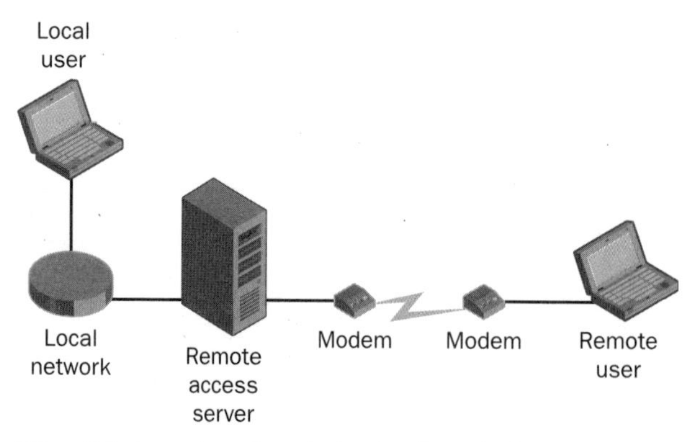

Figure 7-1 A remote node and a local node are both connected to a network, just via different paths.

Remote node computing is simply a point-to-point link. The remote node connects directly to a remote access server via an intervening network. Figure

7-2 shows two ways a remote connection can be made—directly via dialup or across the Internet.

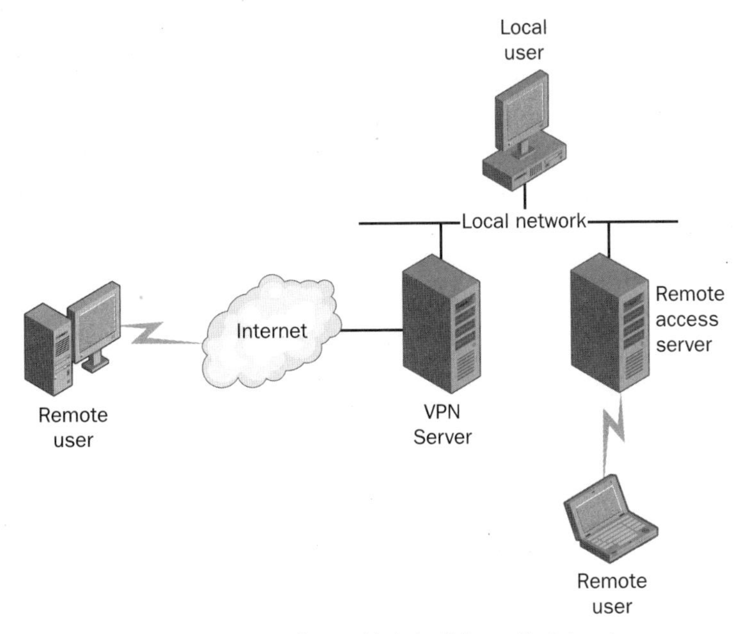

Figure 7-2 Remote access can be provided via dialup or the Internet.

Test Smart Remote nodes differ from local nodes in only one way: Data travels through a modem to access the network in a remote node when it uses a dialup connection. Data travels through a network interface card (NIC) to access the network in a local node. Therefore, a remote node simply treats its modem as though it were a NIC.

Advantages of Remote Node Computing

Connecting via a dialup method can be so slow that users generally limit their sessions to small transfers of data. Where possible, users run applications on the local machine. For example, a user can spend several minutes downloading an application from a remote server that is a couple of megabytes in size, and only then can the user access the data. By contrast, the user can run an application nearly instantly from the local hard drive and spend far less time downloading and accessing data on the network.

Given the problem with slow connectivity, why would anyone bother to use remote node computing? Well, one of the major reasons is ease of configuration. Because you need only a remote PC with a modem, a remote access server with a modem, and a common protocol, the setup is fairly simple.

The nice thing about remote node computing is that it can be run across any dialup line, cable, DSL and so on. Telephones are available in most locations, and modems that can access the cellular telephone network are growing in popularity. The fact that users can connect to the network from nearly any location provides a geographic extension of a business's network. In addition, the ability to telecommute extends working hours and offers other benefits as well.

Another advantage to using remote node computing is that a wide variety of operating systems natively support remote access. Not only do network servers of all types support common remote access protocols, but so do workstation operating systems. A user who has a different operating system on his or her home machine can still access the network and its data. This can bring up a problem with application compatibility, but for common applications, such as e-mail programs or office suites, the data can still be accessed and used. Therefore, even in a situation in which the user has a different home machine, basic services can still be available to the client.

When remote node clients dial up directly into RAS, there is little likelihood that the connection will be interrupted. However, when the remote node client connects across the Internet, there is a definite need for security. This is why remote access computing across the Internet requires special security protocols, which we'll discuss in a later section.

Still another benefit to remote node computing is that there are no graphics restrictions. When a remote node runs an application, accessing data from a network server, the remote node will use whatever graphics display it is set to. If a remote node needs to view a graphic in great detail, there is no barrier except for the speed at which the graphic downloads. This is an important feature to consider in comparing remote node to remote control sessions (also discussed later in this chapter).

The final benefit of using remote node computing is that the server's hardware requirements for supporting a large number of remote clients are minimal. There is no need to maintain a one-to-one relationship between local machines and remote nodes. If you implement remote node computing over the Internet, there is no need to maintain a pool of modems, either. A single server can support many remote nodes, so administration of the remote access system is centralized, reducing the effort that it takes to diagnose and troubleshoot.

Disadvantages of Remote Node Computing

The major disadvantage of remote node computing is the issue of slow speed, which we've already discussed. The more data a remote node must access, the slower its response. For some applications, even if the application's executable is located on the remote node, the slow speed is intolerable. Some legacy applications interact so heavily with their data that the data must be downloaded in its entirety into the client's memory. If that data is in a large database, the client might be so slow that it could appear to be unresponsive. Whenever updates are made to the data, they must also be transmitted back to the server, which further slows down the remote node. Unfortunately, there is no good way to handle legacy applications like these if you're tied to remote node computing. The best way to cope with it is to try higher-speed access methods such as ISDN, DSL, or cable modems, which, even though they're faster than dialup, are still much slower than local connections to the network. Faster access methods are more expensive as well.

Another drawback to remote node computing is driven by the need to use certain applications. Users might not own personal computers with the hardware requirements specified by the organization's applications. In these cases, the user must upgrade at a personal cost, or be supplied with another remote computer at a cost to the organization. It can be simply prohibitive to supply two computers to every network user just because they must be able to work remotely as well as locally. The best solution to this dilemma is to provide users with laptop or tablet computers, even though their cost is somewhat higher than that of a desktop.

When users must install applications on remote nodes, another problem for network administrators arises—licensing. Most software requires that every computer running that application must be licensed to do so. For applications that cost hundreds of dollars per license, remote node computing can be extremely expensive. In addition, tracking the licenses to ensure compliance can be nearly impossible.

Finally, support issues can be horrendous. The organization has no control over the configuration of a remote node owned by the user. The user might install applications that are in conflict with the organization's applications, or that have some other incompatibility. Add that to the fact that users expect the help desk to support them and their incompatible machines, and the cost of remote node computing can shoot through the roof.

Creating a Point-to-Point Connection with PPP

When you create a dialup connection to RAS, you must use a protocol to communicate. The protocol most often used to create the point-to-point connection across the telephone network is the aptly named Point-to-Point Protocol (PPP).

PPP is based on an earlier protocol, the Serial Line Internet Protocol (SLIP). In fact, to understand PPP better, we must explore SLIP. SLIP is a legacy UNIX protocol used for basic remote node connectivity. It works at the physical layer to provide a serial connection between two computers, between a network and a remote computer, or even between two networks. Because SLIP works across the RS-232 interface, which is universally incorporated into personal computers, it has become popular. But SLIP also has a few drawbacks. UNIX is an operating system that originally used only Transmission Control Protocol/Internet Protocol (TCP/IP) natively. Because of this, SLIP worked only on TCP/IP networks, so it was unusable on networks with other protocols. Because it's a physical-layer protocol, SLIP doesn't provide higher-level services such as security or multiprotocol support.

As you can probably guess, the Internet Engineering Task Force (IETF) developed PPP in 1991 with an eye toward fixing the problems inherent in SLIP. First of all, PPP functions at both the physical and data-link layers. Because it operates at the data-link layer, PPP can provide encryption, security, error control, and multiprotocol support. PPP supports authentication through both clear text and encrypted protocols. PPP works over more than just a serial line; it also can transmit data across ISDN and WAN links.

PPP uses three types of frames; one is a data frame that transports data. The header for this frame is shown in Figure 7-3. PPP also uses a Link Control Protocol (LCP) frame that establishes the link and then terminates it, and a Network Control Protocol (NCP) frame that interfaces with the upper-layer protocols, such as TCP/IP or Internetwork Packet Exchange/Sequenced Packet Exchange (IPX/SPX).

Address	Control	Protocol	Data	FCS

Figure 7-3 The PPP data frame header transmits data across dialup connections.

PPP offers several advanced capabilities. When it is used to connect with a remote network, it encapsulates the upper-layer protocols. This process enables a remote node to appear to be connected locally. PPP's link-control ability indicates when a connection is poor, providing for automatic termination and redialing. PPP supports both Password Authentication Protocol (PAP) and Challenge Handshake Authentication Protocol (CHAP), which both prompt users to log on to establish a connection using encryption or clear text passwords.

PPP offers an excellent solution for connecting two remote networks as well. As shown in Figure 7-4, a dialup connection can be configured between two servers, which then route information between the two networks. While dialup connections were once common as the only connection used between remote networks, today they are more often used as backup links in case a higher-speed WAN link fails.

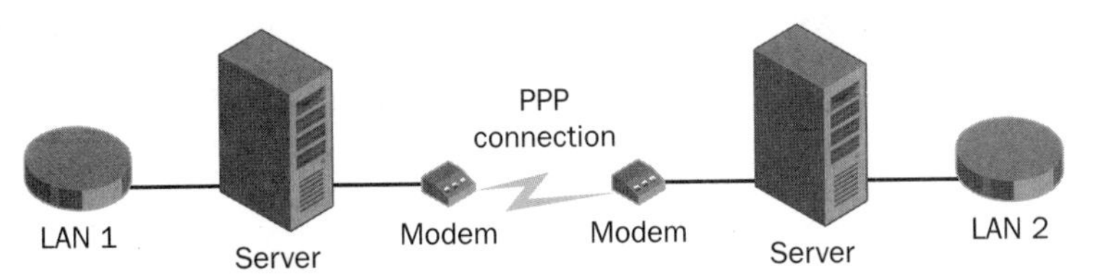

Figure 7-4 A PPP connection can be created between servers or routers on two different networks to route data between the two.

Using DSL for Remote Node

Digital Subscriber Line (DSL) is growing as a dialup technology for remote node. DSL uses the same lines as regular telephone calls, but doesn't interrupt those phone calls. There are several types of DSL to be aware of:

- **ADSL:** Asymmetric Digital Subscriber Line is the form of DSL that most home users are familiar with. The term "asymmetric" refers to the mismatched rates of upload speed to download speed. Most of the bandwidth in ADSL is dedicated to download speed, which makes it ideal for Web browsing and remote node connectivity. These two activities are mainly used to retrieve data, graphics, text, e-mail messages, and so on and use little upstream bandwidth for user requests and responses. ADSL provides up to 6.1 Mbps downstream and 640 Kbps upstream rates. ADSL uses a splitter or filter to enable telephones to function even while using that same telephone wiring to run the DSL connection.
- **G.Lite:** G.Lite is also known as DSL Lite and splitterless ADSL. It's a much slower form of ADSL that doesn't require a filter because the telephone company splits the signal at the central office. The downstream rate is approximately 1.544 Mbps, but it can go as high as 6 Mbps. The upstream rate ranges between 128 Kbps and 384 Kbps.
- **HDSL:** High bit-rate Digital Subscriber Line is an early form of DSL that provides symmetrical rates for uploading and downloading data. HDSL is used for organizations, because they typically have equal amounts of upstream and downstream traffic. HDSL provides a 1.544-Mbps rate.
- **VDSL:** Very high data rate Digital Subscriber Line is the fastest version of DSL. Short distances are the key to achieving VDSL. At approximately 1,000 feet, VDSL can deliver more than 51 Mbps. VDSL is dependent on Fiber to the Neighborhood (FTTN), a program in which fiber optic cabling is installed short distances from subscription points. Copper cabling can be installed for the final distance. The longer the length of the copper cabling, the slower the speed. For example, a distance of 4,000 feet delivers VDSL speeds of approximately 13 Mbps.

Remote Access Service (RAS)

When you dial into a network as a remote node, you log into a remote access server. This is often the same server that provides remote node services across the Internet, via tunneling protocols. A variety of remote access servers is available. In general, organizations use a dedicated server to provide remote node services because it can maintain security better and offer higher performance.

RAS server placement is key to performance. When a RAS server is placed across a slow WAN link from the data and applications that users want access to, the performance is slower than it would otherwise be. When a RAS server provides services to remote nodes across the Internet, it should be placed near the Internet connection so that it acts as a doorway to the network. Optimal placement of RAS servers is demonstrated in Figure 7-5.

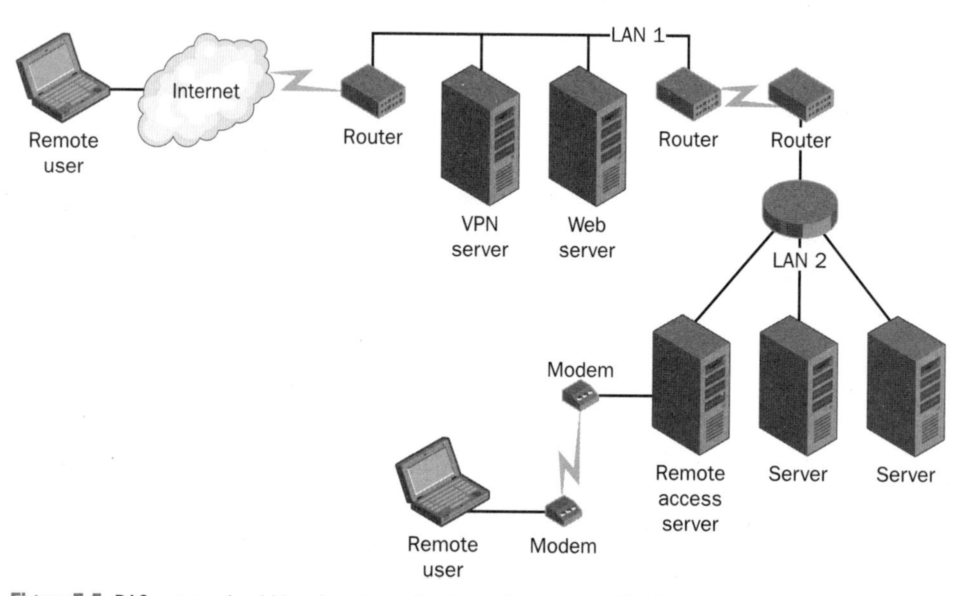

Figure 7-5 RAS servers should be placed near the Internet connection if VPN services are provided. Dialup RAS servers should be placed closest to the data being accessed.

Tunneling to a Virtual Private Network

Virtual Private Networking (VPN) describes remote nodes that access a network via the Internet in a secure fashion. That security is provided by tunneling protocols, along with encryption. Many encryption schemes can encode data with strengths up to 128 bits, an encryption strength that virtually prevents decryption altogether.

VPN is available to clients who connect to the Internet through nearly any type of link. Whether the client connects via ISDN, DSL, cable modem, or dialup line, a VPN session can usually be created. The times when this is not possible are entirely because of the use of a proprietary VPN solution that doesn't recognize different types of Internet connections.

VPN creates a virtual point-to-point connection to the RAS. Tunneling is driven by the need to protect that virtual point-to-point link from being interrupted or eavesdropped upon. Tunneling works by encapsulating data within IP packets in an encrypted format. Figure 7-6 shows a typical VPN link.

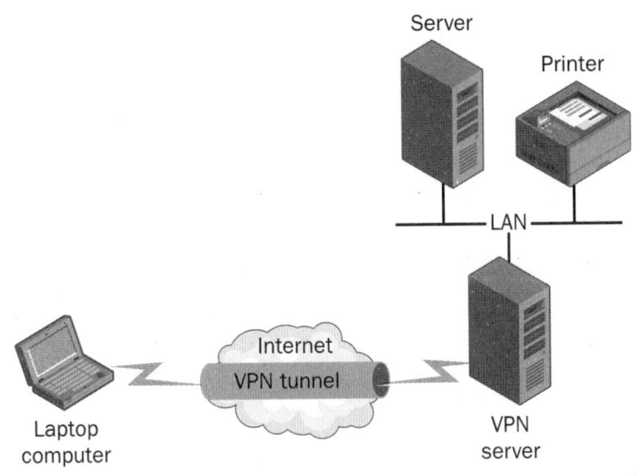

Figure 7-6 VPN connections tunnel data through the Internet.

Point-to-Point Tunneling Protocol

Point-to-Point Tunneling Protocol (PPTP) takes its name from PPP because it uses PPP frames in its tunneling process. PPTP encapsulates PPP frames within IP datagrams, which are then transmitted across the Internet. Microsoft developed PPTP, which was then used to form the basis for Layer 2 Tunneling Protocol (L2TP), developed by the IETF, which we'll discuss in the next section.

By deploying PPTP, you'll provide two services on the network—a RAS providing PPP and PPTP service. Of course, the only way you can use these services is through a remote access client that is equipped with PPTP. First the client creates a PPP session with the RAS and then initiates a link to the PPTP service. When the client connects, PPTP first encrypts and compresses PPP frames. These PPP frames are then encapsulated by PPTP into IP datagrams. The IP datagrams travel across the Internet, routed to the PPTP server. This server discards the IP header and then decrypts the PPP packet and forwards it to its correct destination.

Layer 2 Tunneling Protocol

The Layer 2 Tunneling Protocol (L2TP) was developed to establish a viable alternative to PPTP as a standard. Like PPTP, L2TP is an extension of PPP that supports multiple protocols. Two servers provide an L2TP tunnel: the first is an L2TP access concentrator (LAC), which is simply a RAS. The second is an L2TP network server (LNS), which provides the L2TP service.

One way L2TP clients can share a single server is through the assignment of a virtual access interface. When a client connects via L2TP to the LNS, the client is assigned a unique virtual interface.

L2TP clients follow a similar process to PPTP clients when connecting to the VPN. First the remote client connects to the LAC on the Internet via PPP. Next the LAC authenticates the client. When the client requests connectivity to the LNS, the LAC and LNS authenticate each other and the tunnel is created. With the L2TP session beginning, the L2TP client presents its authentication information via an authentication protocol. If the authentication information is correct, the client can then participate on the remote network as a remote node. L2TP is most often used in concert with Internet Protocol Security (IPSec), which offers both authentication and encryption services. Even when using IPSec, L2TP uses a separate authentication protocol, such as CHAP, discussed later in this chapter. L2TP supports header compression and tunnel authentication, which PPTP does not support.

Remote Control

Now that you have an understanding of remote node computing, you'll be able to compare and contrast it with remote control. Early on in network computing, the value of being able to work from a remote location proved to be very high. Working hours were extended and productivity increased. Remote control was an early remote networking technology that enabled users to run applications on the network with fair performance. The user would create a remote control session with a computer that was connected directly to the LAN. On the remote computer, a window would appear with the remote computer's desktop within it. All application processing and data remained on the LAN; the only data that traveled to the LAN from the remote computer were keyboard and mouse clicks. The graphical user interface contained the data traveling back to the remote computer.

Remote control computing overcame some of the issues with remote node computing. Because the application ran only on the LAN-connected computer, the remote computer didn't need to be compatible with the network applications, nor did it require any additional hardware. A remote control computing example is shown in Figure 7-7. In this setup, application performance was fairly good because so little data had to traverse the connection between the remote computer and the local node. Finally, there were no issues with licensing because the application ran on the local node regardless of where the user was located.

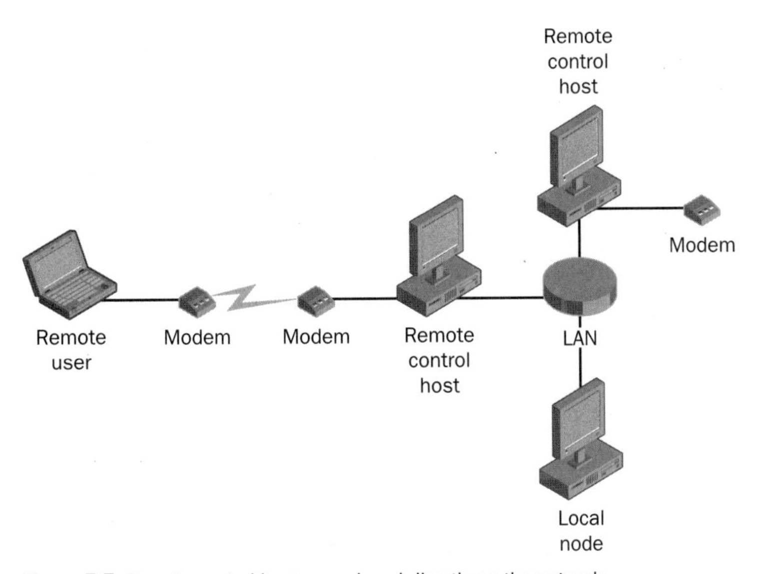

Figure 7-7 Remote control hosts are placed directly on the network.

It wasn't long before administrators realized the benefit of remote control computing. With Windows servers, an administrator could execute a remote control session from any location on the network, or even from home, and be able to manage the server remotely. Enterprise packages such as Systems Management Server (SMS) were useful in assisting users. When a user had a problem, the administrator could access the user's desktop remotely through a remote control session and provide assistance from any location in the network.

Problems arose when remote control computing became popular. Users needed to compute remotely, and additional computers were added to the network in response to demand. I remember walking into an office once to work on a server, and as I was being led back to the server room, I passed a room full of shelves upon shelves of computers, all whirring away, which I was told were the remote control computers. With a cost at the time of about U.S. $2,000 each, those shelves represented well over U.S. $250,000 in equipment.

The ability to run multiple remote control sessions on a single computer overcomes this problem, allowing a single server to take over for several remote control hosts. X Windows is an example of a multisession remote control technology on UNIX. Citrix MetaFrame is a version of the technology available for

32-bit Windows applications. Citrix Corporation came out with multiuser remote control technology for Windows machines. Eventually, its technology was integrated with Microsoft Windows NT 4.0 Terminal Services Edition, and later it was incorporated natively into Microsoft Windows 2000. These types of remote control sessions are now known as *thin clients*.

Thin client technologies combine the best of both remote node and remote control computing. A single server can provide remote control sessions to multiple remote computers, as shown in Figure 7-8. These remote computers don't have to be compatible with the network applications. Sessions can be executed via dialup as well as across the Internet. Thin client performance has been optimized to function over dialup links, so with a faster connection the performance is exceptional. Keep in mind when you consider Citrix MetaFrame that you must add the cost of Citrix on top of the cost of Windows 2000.

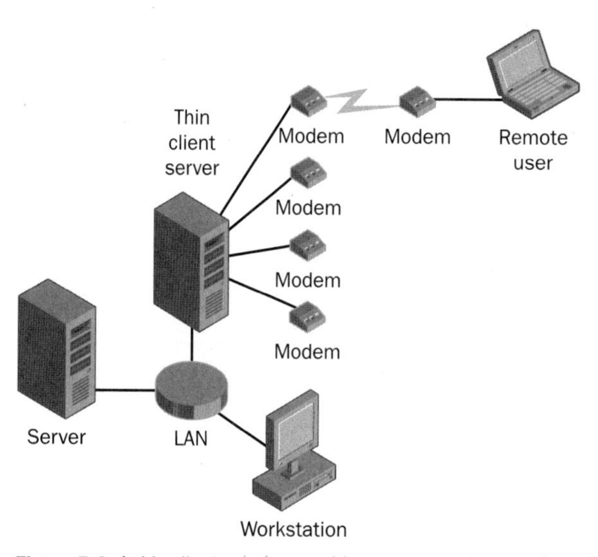

Figure 7-8 A thin client solution provides many remote control sessions from a single server.

Graphical display can be a drawback to using remote control computing. Because graphics are a large part of the data that is exchanged between the local node and the remote computer, remote control programs will limit the graphics in the number of colors and screen size in an attempt to improve performance. Administrators can usually configure a session to limit the graphics for the same reason. For users who must access detailed graphical data remotely, this can be unacceptable.

Citrix MetaFrame provides thin client technologies via its Independent Computing Architecture (ICA) protocol. Microsoft provides thin client sessions across Remote Desktop Protocol (RDP). We will review each of these protocols in the following sections.

Independent Computing Architecture from Citrix

Citrix developed its ICA protocol to facilitate remote control sessions. The protocol runs within the upper layers of the OSI reference model, including the application, presentation, and session layers. It establishes the session, maintains it, and terminates it. During the session, ICA carries keystrokes, mouse clicks, and graphical data in the form of compressed draw commands. ICA is highly optimized in that it will update only the graphical data that has changed on the screen. The protocol also allows file transfers between the local and remote computers.

An additional benefit that ICA offers is the ability to conduct a *shadow* session. (This is also available with the latest RDP client.) When shadowing a session, one person is able to see the same session that another person is using on an entirely different computer. Depending on the security controls, the person who is shadowing the session can actually interact with the session. Shadowing is especially useful in assisting a person with a problem, and help desks often use it. But it's a security risk, and enabling it is often forbidden on servers in highly secure environments.

ICA is supported by the latest Web browsers. A thin client session can be invoked directly from a Web browser. Because browsers support ICA natively, there's no need for users to install an additional ICA client. This is highly beneficial for providing sessions to the public across the Internet, such as in demonstrating a software product without providing the software executable.

Citrix has developed ICA clients for many different operating systems. These include DOS, Windows 3.1*x*, Windows 9*x*, Windows NT, Windows 2000, Windows XP, Macintosh, Windows CE, EPOC32, Linux, SCO UNIX, IBM AIX, HP/UX, and more.

ICA was developed to function in two ways: First an ICA client can dial directly into a Citrix MetaFrame server and run a pure ICA session across the direct connection. An ICA client can run across any network connection, even if it's a remote node connection. The protocols that it will run across include TCP/IP, IPX/SPX, and NetBEUI. When ICA runs across IPX, which is a connectionless protocol, it is flexible enough to add connection-oriented services to its stack. When ICA runs across TCP or SPX, which are both connection-oriented, ICA doesn't incorporate connection-oriented services.

ICA requires very little bandwidth and can provide solid performance over a 20-Kbps connection. This means that the average computer using a 56-Kbps modem connection will experience exceptional performance with an ICA session.

Microsoft Remote Desktop Protocol

Remote Desktop Protocol (RDP) offers much the same type of service as ICA. It supplies the transport for keystrokes, mouse clicks, and display data for a server providing sessions to a thin client application. RDP version 4.0 was the original protocol used by Windows NT 4.0 Terminal Services Edition. Since then, it has been replaced by RDP version 5.0 on Windows 2000 Server.

RDP clients are available for Windows for Workgroups 3.11, Windows CE, Windows 9*x*, Windows NT, Windows 2000, Windows XP, and a special client named the Terminal Services Advanced Client (TSAC) for Internet Web browsers. As you can see, RDP clients are basically provided for Microsoft operating systems. If you require remote control services and are using a non-Microsoft client, you must use Citrix MetaFrame.

RDP is limited in the protocols it will run across; there's no direct dial method. RDP clients will operate only across a TCP/IP network. If you need to run remote sessions across a network that is solely IPX/SPX or NetBEUI, you would be required to use Citrix MetaFrame.

RDP is also the protocol used in Windows XP's two native remote applications. One is the Remote Assistance application, which provides to a help desk an interactive session with a user's desktop. The other is the Remote Desktop Connection application, which provides a single remote control session to a Windows XP computer.

RDP provides graphical data through an RDP display driver. A unique and separate RDP display driver is assigned to each user session, along with the Win32 kernel. The Graphic Display Interface (GDI) on the Windows 2000 Server sends commands to the RDP display driver and passes them on to the Terminal Services device driver. This data is then encoded and transmitted via RDP to the remote computer, which updates its display according to the new information.

RDP captures every keystroke and mouse movement generated on the client, encodes it, and transmits it to the server. At the server, the RDP packet is decoded and processed as though it took place locally on the server within that session. As keystrokes and mouse movements change the screen, the graphical updates are transmitted back to the client to reflect the user's changes.

RDP offers three encryption levels for the communication between client and server. The RC4 encryption algorithm ensures security. Designed by Rivest for RSA Security, RC4 is a variable–key-size stream cipher. The variable-length key can be up to 2048 bits, and it guarantees extremely tight security. The three levels of security offered by RDP are as follows:

- **Low security:** Though this level of security is low, data is still encrypted whenever it is sent from the client to the server. It is not encrypted in the reverse direction. RDP 5.0 uses a 56-bit encryption key.
- **Medium security:** This level of security is a 56-bit encryption key that encrypts data traveling in both directions between the client and server.
- **High security:** This level of security uses a 128-bit key to encrypt data traveling between the client and server in either direction.

Installing the Terminal Services Client If you're using Windows XP, you can use the Remote Desktop Connection application to connect to a Windows 2000 Terminal Server. However, if you're using another Microsoft client, you'll need to install a client, via disk or network installation.

To create a diskette installation, you'll use the Terminal Services Client Creator application located on the Windows 2000 Terminal Server Start menu under Administrator Tools. Select either the 16-bit or 32-bit installation, and insert diskettes as prompted.

A network installation method can work even when remote clients connect to the network. On the server, you'll need to share the C:\Winnt\system32\clients\tsclient\net directory. Remote clients can connect to the network, access the Win32 (or Win16, if you're using a 16-bit Windows operating system) directory, double-click SETUP.EXE, and run the installation over their connection. After the Terminal Services client installation, the remote client can begin running an RDP-based session.

If a user wants to install the Windows CE Terminal Services client, the installation must be done on a workstation with Windows CE services already installed on it. Then the user can run the installation from the Windows 2000 CD under the D:\valuadd\msft\mgmt\mtsc_hpc directory, assuming that the CD-ROM is drive D. Upon the next synchronization, the CE device will have the Terminal Services client.

Configuring a Connection

Remote node is much more prevalent than remote control applications. In fact, even when you install thin clients, you'll probably also need to configure a remote node connection. It's fairly easy to establish remote node computing, and it's usually highly appreciated.

A typical VPN deployment begins with a remote client that needs to access the local network that is already connected to the Internet. You might discover that the user already has an Internet connection—possibly a high-speed one. If this is the case, offering a remote node link via a dialup link won't allow the user to take advantage of the resources already available. In addition, when you deploy VPN connections, you end up saving money because there's no need to set up a rack of modems and purchase additional dial-in telephone numbers.

About 90 percent of home users already use some version of Windows. This makes it fairly easy to configure a dial-up networking connection to an Internet Service Provider (ISP) using PPP. When the client has connected to the ISP, the client can transmit data across the Internet, but that doesn't enable the client to get inside the network. To do this, the client must make a second call across the Internet over the existing PPP connection. This is also done using dial-up networking. The second call uses either PPTP or L2TP, which encapsulates PPP frames within the IP datagrams sent across the Internet. The call connects to a RAS providing PPTP or L2TP services and creates a tunnel. After the user authenticates to the network, the user can access resources by mapping drives and printers, transferring files, and exchanging e-mail messages through applications installed on the user's computer.

Test Smart Even though the Network+ certification exam is not specific to an operating system, you'll be tested for general configuration knowledge.

Server-Side Configuration

Many types of proprietary servers are available to provide RASs. Plus, most server operating systems offer some form of RAS natively. We'll use Windows 2000 Server as an example.

Windows 2000 servers are equipped with Routing and Remote Access Service (RRAS), which, as its name implies, provides both routing services between two networks and RASs. Logically, this is a sensible choice, because RASs are simply the routing of data between a network interface and a modem interface or a network interface and a tunnel interface, while routing services are the routing of data between two different network interfaces. There is very little difference in the processes except in the type of interface.

Windows 2000 RRAS is able to authenticate directly to the Active Directory service, providing users the advantage of a single secure logon. You can place multiple RRAS servers anywhere throughout in the network, which can then provide any combination of routing, VPN, and dialup remote node services. It's a very flexible system.

You can distribute dialup RRAS servers throughout the network to reduce dialup telephone costs, given that users can dial a local phone number to reach these servers. However, when you install a VPN RRAS server, you should place the server close to the Internet connection, as shown in Figure 7-9.

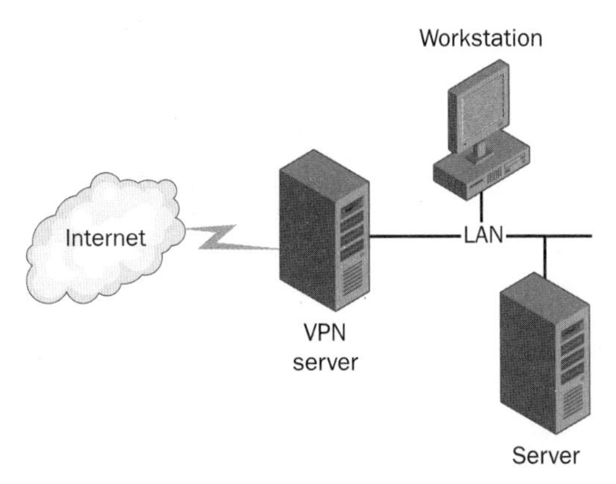

Figure 7-9 VPN servers should have a network interface that leads to the Internet and another that leads to the private network.

You can configure either PPP connections in RRAS. The default dialup connection in Windows 2000 is PPP. The following instructions will help you set up a RAS in Windows 2000 Server:

1. To accept incoming dialup connections, you should have one or more modems installed in the server before you start.
2. Click the Start menu and point to Programs and then Administrative Tools. Click Routing And Remote Access.
3. You'll need to add the server first. Click the Action menu, and select Add Server. Select This Computer, and click OK.
4. Now you can select the RAS in the left pane of the RRAS window by clicking on it. With the server selected, click the Action menu and select Configure And Enable Routing and Remote Access.
5. A wizard will start. Click Next.
6. To accept dialup connections, click Remote Access Server in the window shown in Figure 7-10. Then click Next.

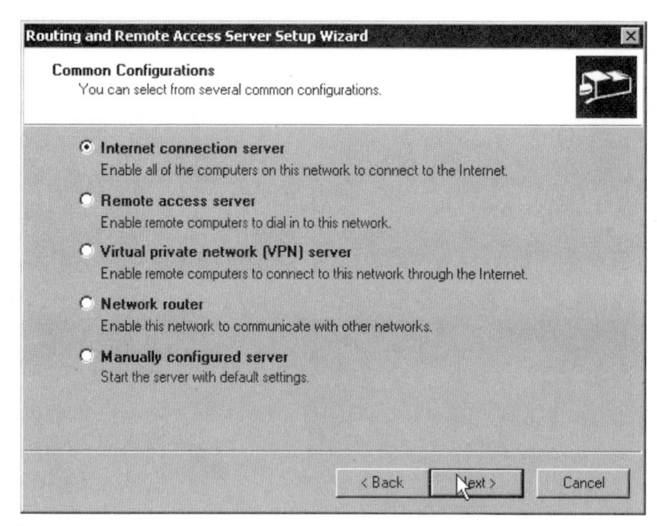

Figure 7-10 You can select the type of service to configure in this RRAS wizard.

7 In the next screen, shown in Figure 7-11, select Set Up an Advanced Remote Access Server so that it will be able to utilize the Active Directory. Click Next.

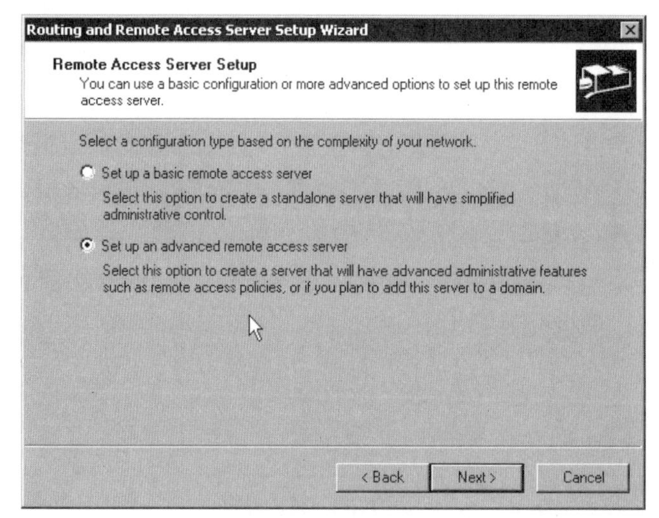

Figure 7-11 If you're using a server that belongs to an Active Directory domain, select Set Up an Advanced Remote Access Server.

8 This screen will prompt you to make certain that the correct protocols are installed. If you don't have the protocol installed on the server that you will provide to dialup users, you'll need to add it. If the server is already functioning on the network, you probably have the necessary protocol stack installed. Select Yes, all of the required protocols are on this list, and click Next.

9 When users dial in to the network using TCP/IP, they will require an IP address. This screen, shown in Figure 7-12, will allow you to automatically assign IP addresses, or select from a preassigned pool of IP addresses. Select Automatically (for this exercise), and click Next.

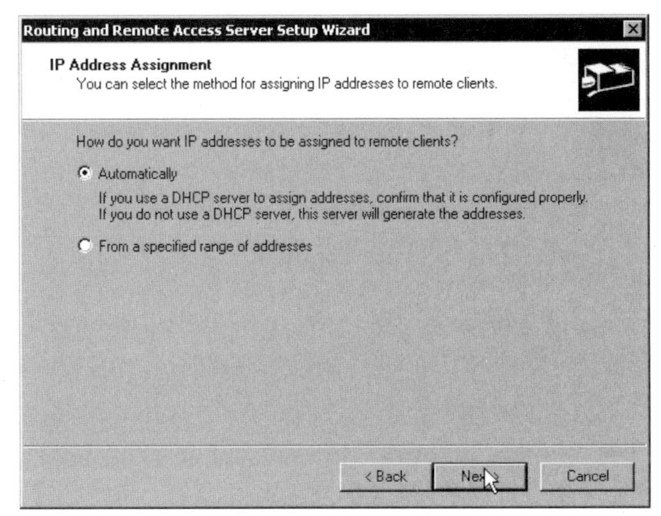

Figure 7-12 When you provide TCP/IP to dialup users, you'll need to assign IP addresses.

10 You'll be prompted about whether to use Remote Authentication Dial-In User Service (RADIUS), as depicted in Figure 7-13, which is a method of providing a single point of authentication to multiple remote access servers. Select No, and then click Next.

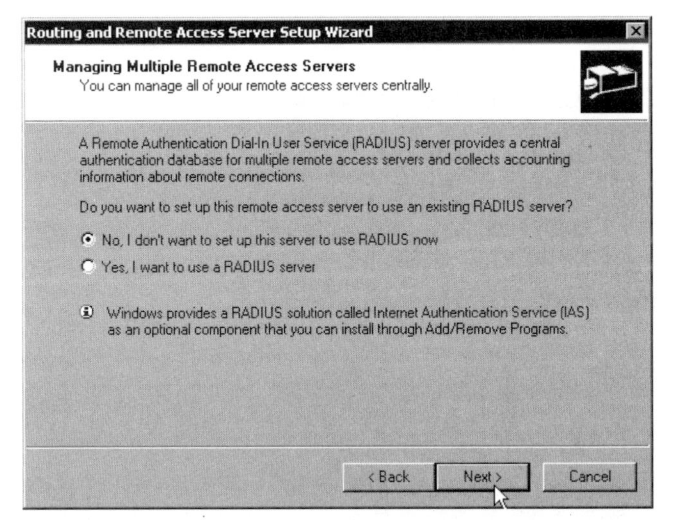

Figure 7-13 RADIUS can be used to manage multiple remote access servers.

11 Click Finish to end the wizard. The RAS is now available for dialup users. The RRAS screen will appear as displayed in Figure 7-14.

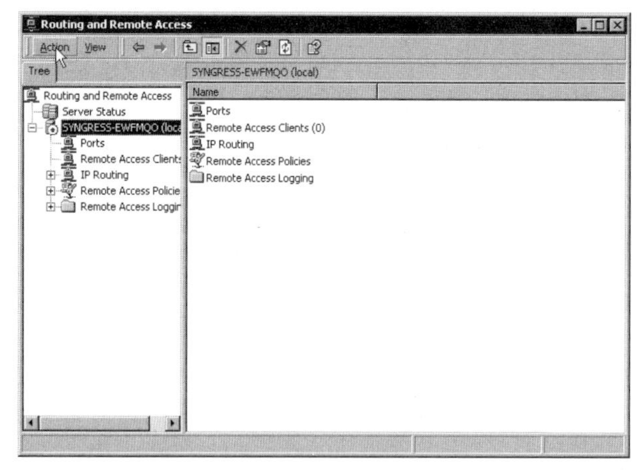

Figure 7-14 RRAS services have multiple components on Windows 2000.

The process to configure VPN services is similar to configuring dialup connections in Windows 2000 RRAS. At the first screen of the wizard, however, you should select Virtual Private Network (VPN) server. At the screen that prompts you to ensure that the protocols are correct, make certain that TCP/IP is available, given that both PPTP and L2TP must run across TCP/IP. The wizard will have an additional screen that will prompt you to select the network interface connection that connects to the Internet, which is shown in Figure 7-15. Your server should have at least two NICs, one that connects to the private network and the other that connects to the Internet.

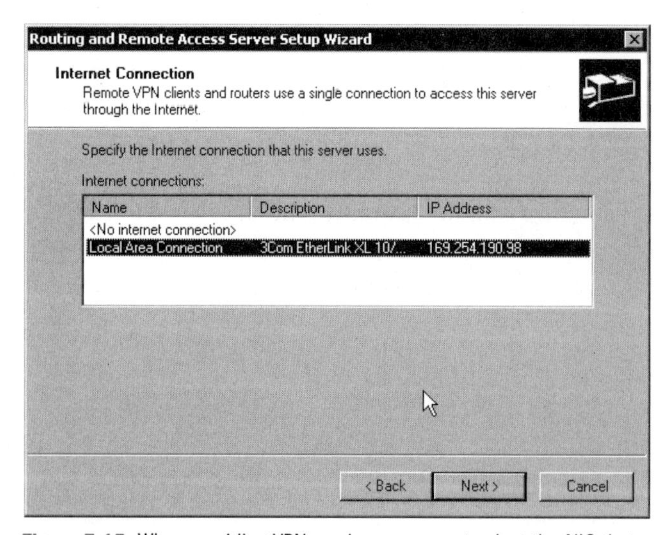

Figure 7-15 When providing VPN services, you must select the NIC that connects to the Internet.

VPN services on Windows 2000 Server provide both L2TP and PPTP connections by default. Figure 7-16 shows the types of connections that you might see. You can delete the connections for the protocol that you do not intend to provide.

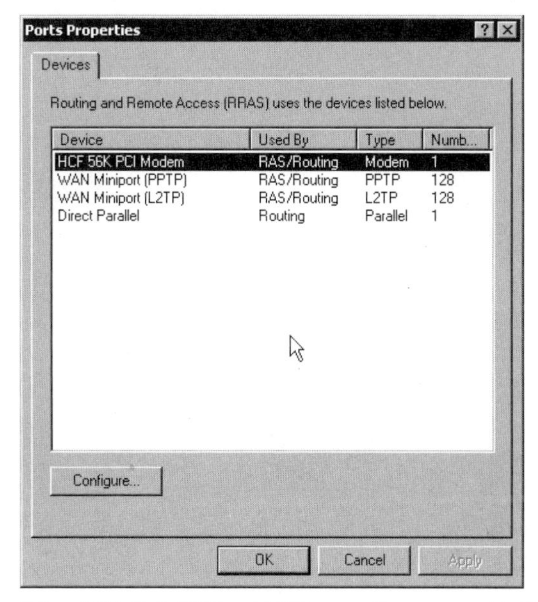

Figure 7-16 Both PPTP and L2TP connections are provided by default in Windows 2000 Server.

Client Configuration

Once you've configured a RAS, the second step is to configure a client to dial up or to connect via VPN to the server. The default client within Windows 2000 and later computers can connect to any RAS that provides PPP connections, and to any VPN server that provides L2TP or PPTP services. Older versions may require additional client software to connect via VPN. (VPN was not supported in Windows 3.*x*.) If you're using a RAS with proprietary protocols, use the proprietary client application to dial up or connect to the server.

It's common for users to connect via DSL, so we'll walk through the configuration of a Windows XP client using DSL to connect to the Internet and VPN to connect to a private network.

To start installing an internal DSL adapter (external DSL adapters connect to a NIC in the computer), you'll need to install a DSL adapter physically into the computer. After the computer is powered on, you must install the drivers so that the new adapter appears to be a network interface. You'll be able to see the adapter in the Network and Dial-Up Connections screen when you are finished. Make certain that you install and configure TCP/IP for the adapter. Also make sure that the filters are placed on the telephone jacks so that the telephone lines function without data interruptions, unless your telephone company provides a dedicated pair of wires or offers filtering at the central office.

To begin creating a VPN connection in Windows XP, follow these steps:

1. Open Control Panel and select Network And Internet Connections category.
2. Under Pick A Task, select Set Up Or Change Your Internet Connection, as shown in Figure 7-17.

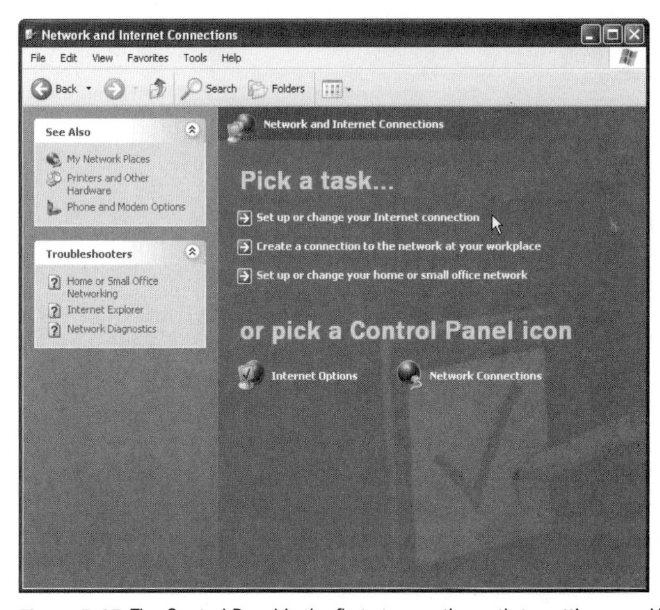

Figure 7-17 The Control Panel is the first stop on the path to setting up a VPN connection.

3. In the dialog box that appears, click the Add button. A wizard starts, as depicted in Figure 7-18.
4. Select Connect To A Private Network Through The Internet, and then click Next.
5. In the following screen, select the connection that represents the DSL adapter (or if you're using a dialup, select the dialup connection). If you're connected directly to the Internet and do not need to dial up a connection, select Do Not Dial The Initial Connection.

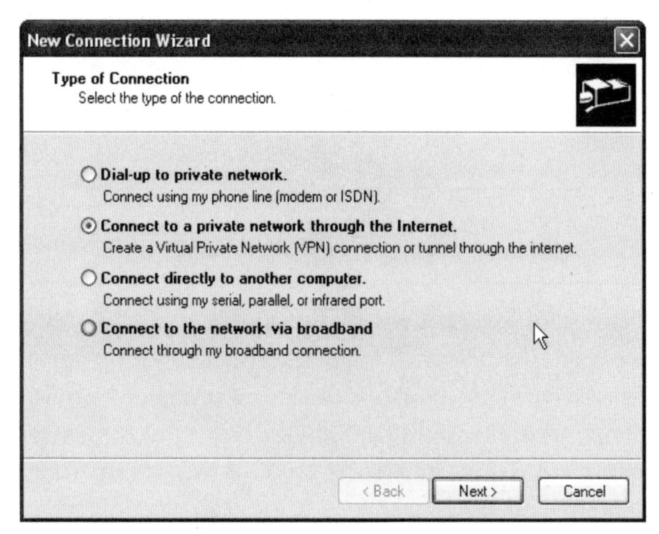

Figure 7-18 For VPN, you connect to a network through an Internet link.

6. In the next wizard screen, type the fully qualified domain name (FQDN) or IP address of the VPN server. For example, if you're connecting to a server named EMF34-RRAS and it's located in the server.xxxyyyzzz.com domain, the name you put in this box is EMF34-RRAS.server.xxxyyyzzz.com. If the RRAS server's IP address is 200.22.33.44, you'd put the IP address in the box.
7. Finally, provide a name for the connection, such as VPN to Work, and click the Finish button.

A dialog box appears that allows you to customize the way that the VPN connection functions. You can click the Properties button to change security, protocols, and advanced options. The default VPN connection will attempt to connect via L2TP in Windows XP, and if that fails, then via PPTP; so you can typically accept the default properties to connect to a VPN server.

Authentication

Whenever a network is made accessible to the public, whether through dialup connections or an Internet connection, there is an associated risk. Sensitive information can be accessed if an outsider manages to breach one of these connections. As you can imagine, a RAS can be tantalizing to hackers. To prevent an attack that could have devastating consequences to the organization, you should ensure that all users are authenticated before allowing them access to the network.

Many standard security protocols have been developed for remote access security. With tunneling protocols, the data is encrypted as it travels between the client and the server. This helps to ensure that no one can eavesdrop on the conversation, but it wouldn't work if anyone could open a tunnel into the network undetected. An authentication protocol ensures that only those people with permission to access the network will be able to.

Test Smart Authentication protocols will be tested on the exam. You should know which protocols provide authentication and which are encrypted.

PAP uses a list of user names and passwords on the server, and compares what the remote user sends to that list. PAP is not encrypted and is very vulnerable; an eavesdropper could easily find out a user's name and password by simply looking at passing network traffic at the same time that a user is logging in.

CHAP is a fully encrypted authentication protocol. It uses a key from the RAS to encrypt the username and password as they are transmitted across the wire. The RAS then decrypts the information and compares it to the list of approved usernames and passwords. CHAP's encryption key is dynamic—the remote server transmits a different key for each connection.

Key Points

- Remote node solutions are made up of remote clients and at least one RAS providing access to the network via a dialup or a VPN interface.
- Remote control solutions are composed of remote control hosts that are attached directly to the organization's network and remote clients. These hosts await connections from remote control clients, and upon connection provide a session, making the remote control client appear to be connected locally.
- In remote control solutions, applications run on the remote control host with only graphical data, keystrokes, and mouse movements transmitted across the wire.
- The Point-to-Point Protocol (PPP) is used to create a dialup connection to a remote host.
- Serial Line Interface Protocol (SLIP) is a legacy UNIX-based protocol that will connect only across TCP/IP.
- PPTP and L2TP are two types of tunneling protocols used to create VPN connections.
- Thin client solutions are servers that offer multiple remote control sessions to clients.

- Independent Computing Architecture (ICA) and Remote Desktop Protocol (RDP) are the two protocols used in thin client sessions.
- When placing a RAS on the network, you should position it near the information that is being accessed.
- RASs that provide VPN connections should be placed close to the Internet connection. They need at least two network interfaces, one that leads toward the Internet connection and the other toward the private network.
- Both PAP and CHAP provide authentication services. PAP information is vulnerable because it is transmitted in clear text. CHAP information is encrypted.

Chapter Review Questions

1 You've been asked to design a remote access solution for a company. The network has 300 users, all working out of a single office building. The six servers all run Windows NT and Windows 2000, but the administrator says only four of them are being used. Users have asked to work from home and need access to e-mail messages and office files. Which solution will both be cost-effective and meet the users' needs?

a) Remote control via pcAnywhere

b) Thin client via Citrix MetaFrame

c) Remote node via Windows 2000 RRAS

d) Remote node via a proprietary RAS solution

Answer c is correct. Remote node via Windows 2000 RRAS will be the most cost-effective, given that the network already has Windows 2000 servers available for use as RRAS and that the users' needs are appropriate for remote node services. Answer a, remote control via pcAnywhere, will not be cost-effective, because it could require the organization to purchase additional PCs. Answer b, thin client via Citrix MetaFrame, is not cost-effective; in addition, it offers more services than are needed by the users. Answer d, remote node via a proprietary RAS solution, is not correct because it won't be cost-effective.

2 Which solution provides multiple remote control sessions from a single server?

a) Novell NetWare

b) Windows NT 4.0

c) Windows 98

d) Windows 2000 Terminal Services

Answer d is correct. Windows 2000 Terminal Services offers a thin client solution, which is essentially multiple remote control sessions from a single server. Answer a, Novell NetWare; answer b, Windows NT 3.51; and answer c, Windows 98, are all incorrect because none offers a thin client solution.

3 What equipment may be required to install an ADSL connection? Select all that apply.

a) DSL adapter

b) Network adapter

c) Filter

d) Modem

Answers a and c are correct. You'll need to physically install the DSL adapter and the filters for the telephone jacks when you set up an ADSL connection. Answer b, network adapter, is correct because it can be connected to an external ADSL adapter. Answer d, modem, is incorrect because modems are used for regular telephone dialups.

4 Which of the following protocols can be used for a VPN connection? Select all that apply.

a) PPP

b) PPTP

c) L2TP

d) IP

Answers a, b, c, and d are all correct. You can create a VPN connection that will use PPP to connect to the Internet. It will then tunnel through the Internet with PPTP or with L2TP and transmit IP datagrams.

5 If a user dials up directly to a network computer and sees a screen that looks like the desktop of that network computer, what type of connection has the user made?

a) Remote node dialup

b) Remote node VPN

c) Remote control

d) Dialup to an ISP

Answer c is correct. The user has made a remote control connection. Answers a, b, and d are all incorrect because remote node connections don't display sessions showing a network computer's desktop.

6 You are asked to design a remote access solution for a network. The network has offices located all over the country. Users have asked to use e-mail applications and office files remotely. The administrator wants to have a solution that can be managed from the central office. The network is connected to the Internet from the central office, and all users given remote access privileges have a connection to the Internet available to them. Which is the best solution?

a) VPN solution

b) Remote node dialup to the central office

c) Remote node dialup to several offices

d) Thin client solution

Answer a, a VPN solution, is correct. Given that the network is connected to the Internet and users have Internet access, the VPN solution can be centrally installed and managed without requiring long distance calls. VPN is the optimal solution. Answer b, remote node dialup to a central office, is incorrect because it will require long distance calls from users not located at that office. Answer c, remote node dialup to several offices, does not meet the administrator's request for a solution that is located at the central office. Answer d, a thin client solution, offers more services than the users need.

7 Which of the following remote access protocols does not support multiple protocols and encryption?

a) PPP

b) SLIP

c) IP

d) IPX

Answer b is correct. SLIP is a remote access protocol that supports only IP and does not provide encryption. Answer a, PPP, is incorrect because it supports multiple protocols and encryption. Answer c, IP, is incorrect because it is not a remote access protocol. Answer d, IPX, is incorrect because it is not a remote access protocol.

8 You've been called by a user to help troubleshoot his VPN connection. He is using Windows 98, and he installed the default VPN client through dialup networking. His Internet connection is provided by a cable modem. The network server provides VPN using L2TP and IPSec. What is likely the problem?

a) The user's PC has a corrupted hard drive.

b) The user's modem is not connecting to the phone line.

c) The ISP does not support L2TP.

d) The default client in Windows 98 uses PPTP, not L2TP.

Answer d is correct. The default client in Windows 98 uses PPTP, not L2TP. You can either reconfigure the client or add PPTP services to the RAS. Answer a is incorrect because a corrupt hard drive would also manifest itself in other areas. Answer b is incorrect because a cable modem doesn't connect to the phone. Answer c is incorrect because L2TP travels within IP datagrams, so the ISP only needs IP support.

9 You've been asked to design a remote access solution for a network. The network already has a remote control solution in place, with over 50 dedicated remote control hosts that wait for dialup connections. These machines are becoming outdated and need to be either upgraded or replaced. The users must use remote control because the company doesn't want to distribute a proprietary application outside the network. What solution will best serve the company's needs?

a) Thin client solution

b) Remote node via dialup

c) VPN solution

d) Remote control solution upgrading existing computers

Answer a is correct. The thin client solution can provide the company's needs and will be cost-effective given that a single server can replace the entire set of remote control hosts. Future upgrades would apply only to that server. Answers b, remote node via dialup, and c, VPN solution, are both incorrect because the application cannot be installed on computers outside the network. Answer d, a remote control solution upgrading the existing computers, is incorrect because it is not as cost-effective or easy to manage as a thin client solution.

10 Which of the following is an authentication protocol?

a) L2TP

b) PPP

c) PAP

d) ICA

Answer c is correct. PAP provides authentication services. Answer a, L2TP, is a tunneling protocol for VPN. Answer b, PPP, is a remote access protocol. Answer d, ICA, is a thin client protocol.

Chapter 8

Security Protocols

Security protocols protect a computer from attacks. To understand how security protocols work, you must first understand what types of attacks they protect against. Networks and data are vulnerable to both active attacks, in which information is altered or destroyed, and passive attacks, in which information is monitored. Attacks that you might encounter include the following:

- **Altering data.** This active attack takes place when data is interrupted in transit and modified before it reaches its destination, or when stored data is altered.
- **Eavesdropping.** This passive attack takes advantage of network traffic that is transmitted across the wire in clear text. The attacker simply uses a device that monitors traffic and "listens in" to discover information. You'll hear this term referred to as sniffing the wire, and sometimes as snooping.
- **IP address spoofing.** One way to authenticate data is to check the IP address in data packets. If the IP address is valid, that data is allowed to pass into the private network. IP address spoofing is the process of changing the IP address so that data packets will be accepted. IP address spoofing can be used to modify or delete data, or to perpetuate an additional type of attack.

- **Password pilfering.** A hacker will obtain user IDs and passwords, or even encryption keys, to gain access to network data, which can then be altered, deleted, or even used to create another attack. This type of attack is usually done by asking unsuspecting users, reading sticky notes containing passwords that are posted next to computers, or sniffing the wire for password information. Sometimes a hacker will attempt to get hired at a company merely to obtain an ID and password with access rights to the network.
- **Denial of service.** This active attack is intended to cause full or partial network outages so that people will not be able to use network resources and productivity will be affected. The attacker floods so many packets through the network or through specific resources that other users can't access those resources. The denial-of-service attack can also serve as a diversion while the hacker alters information or damages systems.
- **Virus.** A virus is an attack on a system. It is a piece of software code that is buried inside a trusted application (or even an e-mail message) that invokes some action to wreak havoc on the computer or other network resources.

The methods for securing a network and its systems are often developed as soon as certain types of attacks occur. For example, before anyone wrote code for the "Internet worm" virus, there never was an anti-virus software. (The first virus, called the Internet worm, was a self-replicating program released on the Internet in 1988 that effectively shut down hundreds of thousands of computers in a matter of hours.) As attacks become more sophisticated, so does the defense against them. Table 8-1 lists security methods and the types of attacks they protect against.

Table 8-1 Security Methods and Attacks

Security Method	Type of Attack	Notes
Authentication	Password guessing attacks	Verifies the user's identity
Access control	Password pilfering	Protects sensitive data from access by the average user
Encryption	Data alteration	Prevents the content of the packets from being tampered with
Certificates	Eavesdropping	Transmits identity information securely
Firewalls	Denial of service (as well as others)	When configured correctly, can prevent many denial-of-service attacks
Signatures	Data alteration	Protects stored data from tampering
Public key infrastructure	Spoofing	Ensures that data received is from correct sender

Table 8-1 Security Methods and Attacks

Security Method	Type of Attack	Notes
Code authentication	Virus and other code attacks	Protects the computer from altered executables
Physical security	Password pilfering	Protects unauthorized persons from having access to authorized users and their IDs and passwords
Password policies	Password pilfering	Ensures that passwords are difficult to guess or otherwise decipher

Security protocols are developed to implement a defense against the many attacks that can be waged on a network. With the exception of physical security, anti-virus software, and password policies, security protocols are the guard dogs that tirelessly protect the doors into the network. In addition to security protocols, two types of hardware-based solutions focus solely on network security: firewalls and proxy servers. In this chapter, we'll review Internet Protocol Security (IPSec), Layer 2 Tunneling Protocol (L2TP), Point-to-Point Tunneling Protocol (PPTP), Secure Sockets Layer (SSL), and Kerberos, in addition to the security properties of firewalls and proxy servers.

IPSec

IP version 4.0 doesn't include a native security method, so IPSec was created. Not only is IPSec considered an addition to IPv4, it also has been incorporated into the IPv6 protocol.

Based on cryptography, Internet Protocol Security, or IPSec, ensures the privacy of network traffic as well as its authentication. IPSec is used for peer-to-peer and client-server communications across a private or public network; secure LAN-to-LAN communications across a WAN; and remote access transmissions via either dialup or virtual private network (VPN). IPSec functions at the Network layer. Upper-layer protocols utilize IPSec to provide secure transport of application data. The fact that IPSec is a network-layer protocol makes its services transparent to applications. A user doesn't need to invoke any special security software to use IPSec. Instead, the administrator configures IPSec, and those services are provided to IP datagrams according to the configuration.

IPSec provides several types of services that prevent different kinds of attacks from interfering with the network. These include authentication and encryption of the data stream. While other security strategies are built to create a perimeter of privacy around a network, IPSec ensures that data cannot be tampered with while it is traversing any part of the network—whether it's within the private network boundaries or across the public Internet.

The inner workings of IPSec, as defined by the Internet Engineering Task Force (IETF), are based on its authentication header (AH) and encapsulated security payload (ESP). As you can probably guess, AH provides authentication for data integrity and covers the entire packet, while ESP provides encryption for confidentiality. IPSec uses User Datagram Protocol (UDP) port 500. When you use IPSec across a firewall, you must ensure that UDP port 500 is open, along with IP type 50, which is used by ESP, and IP type 51, which is used by AH. Because ESP provides encryption, it negatively affects the size of the packet. To reduce the size of the packet, IPSec includes IP payload compression (IPcomp), which compresses the packet before it is encrypted by ESP.

Internet Key Exchange (IKE) is the part of IPSec that provides the key. A security key is agreed upon using the Diffie-Hellman Technique is and shared by the sender and recipient, ensuring that data is not changed during transmission. The Diffie-Hellman Technique is a public key cryptography algorithm, which is a fancy name for an equation, that enables two computers to communicate and agree upon a shared key. The process begins with each of the computers exchanging their public key. Then each one combines its private information along with the other's public information to generate a shared secret value.

An IPSec session is initiated when the IP protocol receives data from upper layers. IPSec works with the destination computer to agree upon the shared key and then encrypts the data at the sending host. This process ensures that the data packets are unreadable while en route to their destination, where they are then decrypted using the shared key.

IPSec supports two types of encryption modes—transport and tunnel. The transport mode encrypts only the data part of the packet, not the header. Tunnel mode encrypts the entire header and data.

When data is sent using IPSec, the AH follows the standard IP header. After AH, the ESP header is next. These headers are added before the transport layer headers, whether those are UDP or TCP, which means that the transport-layer header is encrypted.

Test Smart IPSec is used with L2TP to create a tunnel. IPSec offers the tunnel an authentication and encryption method.

VPN Protocols

Virtual private networking is a system of creating a private network connection that travels through a public network. Even though the VPN exists within a public network, the connection is intended to be as safe as a private network. One of the top considerations for using a VPN is to reduce costs. When you compare them, the costs of leasing a line to create a private WAN are far higher than the costs of using local Internet services to create a VPN. A second reason that organizations use VPNs is to overcome the costs of computing to multiple remote locations. When many individuals need access to the network, an administrator has the choice of installing one or more dialup sites with many modems and telephone lines (along with the accompanying support and telephony costs) or using the existing Internet connection with a VPN server.

L2TP

L2TP is a VPN protocol used along with IPSec to ensure confidentiality of the data transmission. L2TP grew out of the combination of two prior protocols—Point-to-Point Tunneling Protocol (PPTP) courtesy of Microsoft and Cisco's Layer 2 Forwarding (L2F) protocol. L2TP borrows many of the qualities of PPTP, especially in that it extends the Point-to-Point Protocol (PPP) used for remote access.

In a typical L2TP scenario, a remote user dials up to the Internet via a local Internet Service Provider (ISP). The Internet link uses PPP, and an L2TP access concentrator (LAC) tunnels through the Internet via PPP to an L2TP network server (LNS) located at the private network. This is shown in Figure 8-1. The tunnel is created by a message between the LAC to the LNS through UDP port 1701, followed by an authentication procedure. Keep in mind that the LAC and LNS are services that are transparent to the user. The remote user's PPP frames are encapsulated within IP packets that include an L2TP header. At the LNS, the L2TP headers are stripped off and the client is then logged into the network just as though the user were working locally.

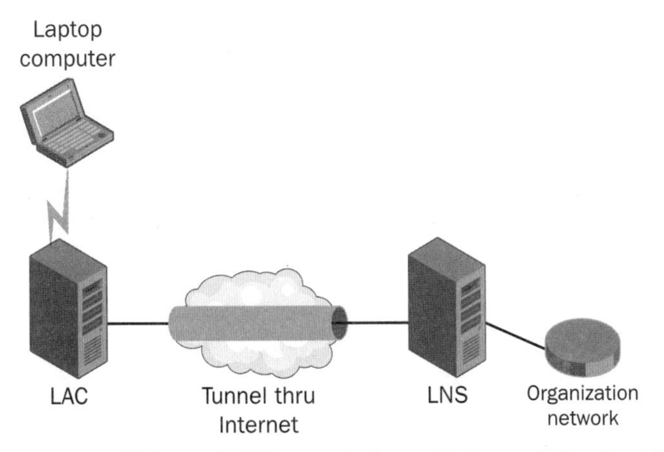

Figure 8-1 L2TP extends PPP across the Internet between a LAC and an LNS.

L2TP on its own lacks the security basics required by a true VPN. This is why it's combined with IPSec to create a secure environment. In the L2TP/IPSec configuration, L2TP creates the tunnel while IPSec encrypts and ensures the privacy of the data.

Why Use L2TP Instead of PPTP?

Whether a network is using L2F or PPTP, it can easily switch over to L2TP. But it's never a good idea to upgrade just because an upgrade is available. You should really understand why making a change might be a good idea. It helps to know what the differences are.

The main reason that an administrator would consider using L2TP over L2F is that the L2TP client is included in Windows 2000 and later operating systems. This means that installation is unnecessary, and configuration is fairly simple for remote users, reducing administrative tasks. In addition, Windows 2000 servers understand L2TP, enabling a network administrator to combine Cisco routers and Windows servers to establish tunnels among multiple networks.

PPTP is natively supported by more Windows OSs than L2TP. (You can download an L2TP client from Microsoft.com for some Windows OSs, however.) But its other features make L2TP a more attractive choice than PPTP. For example, L2TP supports both Cisco TACACS+ and Remote Authentication Dial-In User Service (RADIUS) authentication, while PPTP does not. (TACACS+ by Cisco, which is a form of RADIUS, and RADIUS itself, are both remote access systems that manage authentication and accounting for other remote access servers [RASs].) PPTP is supported by Windows servers only, while L2TP was developed to be a standard that is already natively supported by Cisco routers and Windows 2000 servers. Because L2TP is usually implemented along with IPSec to create a true VPN, it offers a much higher level of security than PPTP. And finally, L2TP offers a wider variety of protocols than PPTP—supporting not only TCP/IP but also IPX/SPX and Systems Network Architecture (SNA), which is used by mainframe computers, and other types of computers.

Secure Sockets Layer (SSL)

SSL is the abbreviation for Secure Sockets Layer, but users will likely be more familiar with its manifestation as the HTTPS:// that precedes the URL of a secure Web site. SSL is a protocol that uses a public key to encrypt the data transmitted across the Internet. It is commonly used to provide privacy for sensitive information such as credit card numbers.

SSL runs transparently to applications, because it sits below upper-layer applications and above the IP protocol, as shown in Figure 8-2.

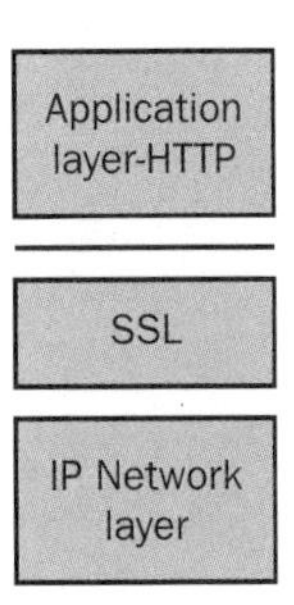

Figure 8-2 SSL is transparent to users because it works below the application layer.

Working on behalf of upper-layer protocols, the SSL server authenticates itself using a certificate and public ID to an SSL-enabled client, which includes both Netscape Navigator and Microsoft Internet Explorer Web browsers, and others. The SSL client ensures that the server's certificate has been issued by a trusted certificate authority (CA), it authenticates itself back to the server using the same process, and an encrypted link is created between the two. A CA is a server that issues certificates to validate hosts within certain domains. This process ensures that both the client's and the server's identities are confirmed. During the ensuing data transmission, SSL enacts a mechanism to ensure that the data is not tampered with before it reaches its destination. The two subprotocols within SSL enable this entire process. The SSL handshake protocol exchanges the series of messages at the initiation of an SSL connection to establish a link. The SSL record protocol specifies the format that will be used to transmit the data.

SSL is able to use several different types of ciphers:

- **Data encryption standard (DES) and Triple DES.** DES is a private key exchange that applies a 56-bit key to each 64-bit block of data. Triple DES is the application of three DES keys in succession.
- **Key Exchange Algorithm (KEA).** KEA enables the client and server to establish mutual keys to use in encryption.
- **Message Digest version 5 (MD5).** This cipher creates a 128-bit message digest to validate data.
- **Rivest-Shamir-Adleman (RSA).** This is the most commonly used key exchange for SSL. It works by multiplying two large prime numbers, and through an algorithm determining both public and private keys. The private key does not need to be transmitted across the Internet but is able to decrypt the data transmitted with the public key.
- **Secure Hash Algorithm (SHA).** SHA produces a message digest of 160 bits using the SHA-1 80-bit key to authenticate the message.

A network administrator can enable or disable ciphers for the network. When an SSL client initiates contact with an SSL server, the SSL handshake protocol is used to agree upon the cipher to use. They identify the strongest ciphers that the two share, and then use those for the session. In general, the stronger the cipher used, the slower the transmission. An admin should balance the need for security with the transmission speed required, as well as the international export laws regarding which ciphers can be used across country boundaries.

Test Smart The longer the encryption key in number of bits, the slower the transmission speed will be because of the extra processing of the encryption.

When a client establishes the trustworthiness of an SSL server, it uses the server's public key, the certificate's (delivered by the CA) serial number and validity period, the server's domain name and the CA's domain name, and finally the CA's digital signature. The steps that a client uses to validate an SSL server are shown in Figure 8-3.

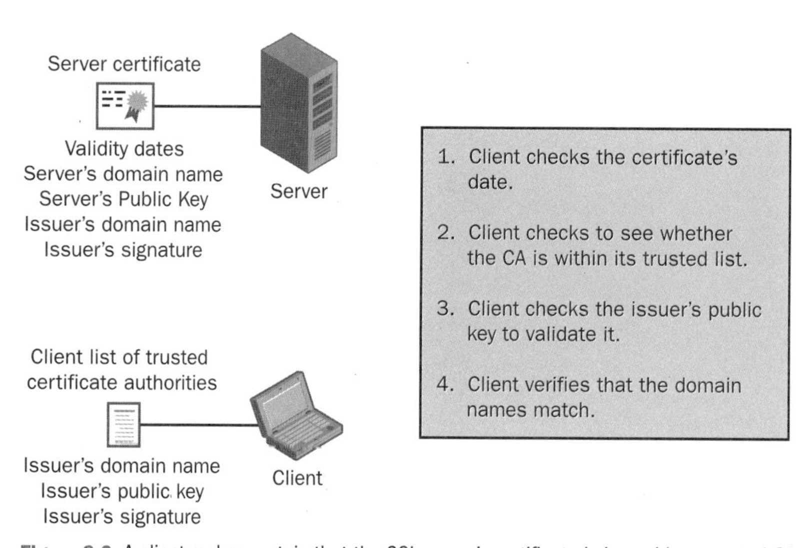

Figure 8-3 A client makes certain that the SSL server's certificate is issued by a trusted CA.

The server goes through a similar process, usually requiring the same type of information from a client during the authentication of the SSL client. This process is depicted in Figure 8-4.

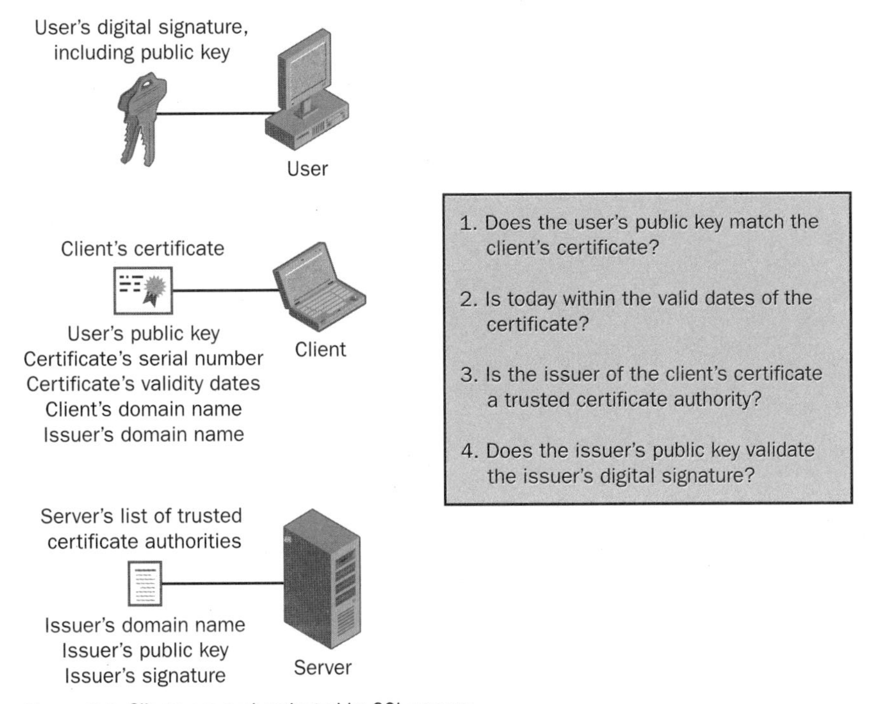
User's digital signature,
including public key
User
Client's certificate
User's public key
Certificate's serial number
Certificate's validity dates
Client's domain name
Issuer's domain name
Client
Server's list of trusted
certificate authorities
Issuer's domain name
Issuer's public key
Issuer's signature
Server
1. Does the user's public key match the client's certificate?
2. Is today within the valid dates of the certificate?
3. Is the issuer of the client's certificate a trusted certificate authority?
4. Does the issuer's public key validate the issuer's digital signature?

Figure 8-4 Clients are authenticated by SSL servers.

Kerberos

Kerberos was the name of the legendary three-headed dog who guarded the gates of Hades. It is an appropriate name for the enormously strong security protocol developed by the Athena project at the Massachusetts Institute of Technology (MIT). As the guard dog of information transmitted across the Internet, Kerberos is an authentication protocol that is used to establish trust relationships between domains and verify the identities of users and network services.

When an entity attempts to access a Kerberos-protected resource and provides correct authentication information, Kerberos issues a *ticket* to it. This method does not require a password for transmission across the network. The ticket is actually a temporary certificate with the information required to identify the entity to the network. The entity uses this Kerberos ticket to request further Kerberos tickets to allow it to access subsequent services on the network. Each process requires a complex mutual authentication, but this is completely transparent to the user.

Specifically, the first request for access to the network is passed to an Authentication Server (AS). The AS creates a ticket-granting ticket, which is an encryption key based on the password, the user name, and a value that represents the requested service.

The ticket-granting ticket must then be sent to a ticket-granting server. This ticket-granting server sends back a new time-stamped Kerberos ticket, which can then be sent to the server requesting a specific service. The final server either accepts the ticket and provides the service or rejects it. The time stamp on the Kerberos ticket enables it to be used only for a limited time.

Because of the use of time-stamping, computers using Kerberos authentication require fairly tightly synchronized time settings. Given the unreliability of computer time clocks, the best way to ensure that computers are synchronized is to use a network time service.

Kerberos Trust Relationships

Kerberos enables trusts to be established between two different UNIX realms, between two Windows 2000 domains, or even between a UNIX realm and a Windows 2000 domain. Trust relationships are established using Kerberos so that authentication credentials can be passed on to network resources in trusted domains or realms. For example, when domain A trusts domain B, the users in domain B are able to access resources in domain A.

Kerberos trust relationships are typically *transitive* and *bidirectional* in nature. Transitive means that if domain A trusts domain B, and domain B trusts domain C, domain A is understood to trust domain C. Bidirectional means that when domain A trusts domain B, domain B automatically trusts domain A. The only time a trust relationship is nontransitive and unidirectional is in the case of connection between a UNIX realm and a Windows 2000 domain, between two UNIX realms, or between domains within two different Windows 2000 Active Directory forests. (A *forest* is a group of related domains considered to be a single Active Directory entity.)

Let's look at how a trust relationship would work in an Active Directory forest of domains. Figure 8-5 shows a forest in which the root domain is named rootdomain.com. There are two subdomains, named one.rootdomain.com and two.rootdomain.com. Rootdomain.com has a Kerberos trust with each of its subdomains. Because it is bidirectional, the subdomains trust the rootdomain.com in return. Because those Kerberos trusts are transitive, it is understood that one.rootdomain.com trusts two.rootdomain.com and vice versa.

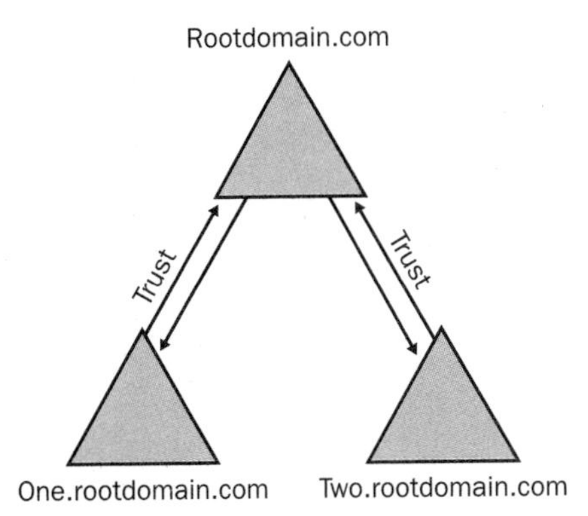

Figure 8-5 An Active Directory forest is related through transitive, bidirectional Kerberos trusts.

Wherever a Kerberos trust exists, the users in one domain will be able to access resources in the other domain as long as the administrator has granted those users access. A network administrator can view Kerberos trust relationships in the Windows 2000 Active Directory Domains and Trusts console, shown in Figure 8-6. To access this console, click the Start menu and point to Programs and then Administrative Tools. Click Active Directory Domains And Trusts On A Windows 2000 Domain Controller.

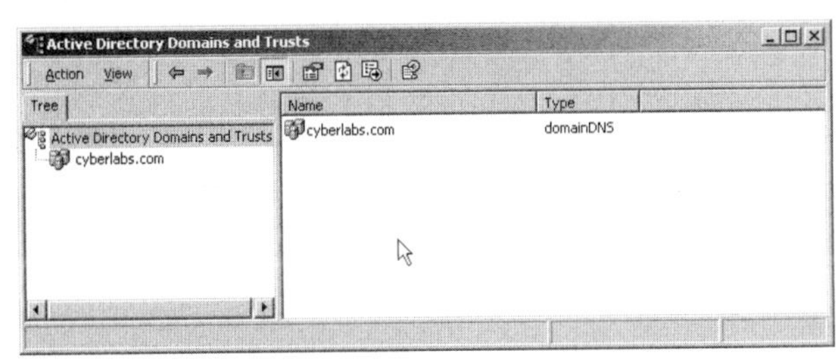

Figure 8-6 The Active Directory Domains and Trusts console shows Kerberos trusts.

Smart Cards

Smart cards are gaining in popularity as a way to ensure secure authentication using a physical key. Smart cards are able to provide an interactive logon, secure e-mail messages, and authenticate access to network services.

Smart cards contain chips to store a user's private key and can also store logon information; public key certificates; and other information, depending on the smart card's usage. When a user needs to access a resource, the user inserts the smart card into a reader attached to the network. After typing in the user's personal identification number (PIN), the user is authenticated and can access network resources. The private key is automatically available for transparent access to encrypted information.

Smart cards require Public Key Infrastructure (PKI), a method of distributing encryption keys and certificates. In addition, each protected resource will require a smart-card reader. Some implementations of smart cards combine the smart card with employee badges so that employees need a single card for building and network access.

Firewalls

When a network administrator initially establishes the perimeter of defense for a network, one of the first pieces of equipment installed is a firewall. This piece of equipment is actually a router with two interfaces—one leading to the public network and the other to the private network. Keep in mind that the "public" network doesn't need to be the Internet—it can be a regular network—while the "private" network can be a portion of the network where the most sensitive data is stored or used. For example, a large software development company that often uses contract labor might decide to place all research and development information within a private network that is bounded by firewalls from the rest of the network.

The simple status of router does not qualify a piece of equipment to serve as a firewall. One of the methods a firewall uses to secure the network is *packet filtering*. This is the process of receiving data packets from one interface and examining them to see which packets meet the rules. For packets that meet firewall rules, they are either permitted or blocked, depending on how the rule is implemented. For example, a person sets up a firewall and implements a rule to block all traffic for Telnet (TCP port 23) coming from the public network. All other traffic can flow through the firewall. A different way of setting up that firewall is to block all traffic and then permit only the traffic that you want to receive. There are other criteria upon which to permit or deny traffic; you can decide that the traffic shouldn't be received from a certain IP address (or sent to a certain IP address) or that a type of traffic should be denied. This can come in handy if a network administrator decides to block all e-mail traffic from a domain where a lot of spam e-mail messages originate.

Whether you block a certain type of traffic and allow the rest, or block all traffic and allow a set of traffic types can make a major difference in how a data packet is handled. For example, if you've decided to block all Telnet traffic on TCP port 23 in an attempt to stop denial-of-service attacks because some of them use Telnet, you won't be successful; a Telnet request can be configured to use a different port. In addition, other methods can use other ports and achieve the very same result—a network outage. On the other hand, if you decide to block all traffic and allow only certain types, you should really understand which data to allow and which not to. Some applications will function only across a specialized port, and using such a policy on the firewall can cause them to fail. However, this is the more secure method of implementing firewall rules.

Test Smart Firewalls provide packet filtering based on the permit and deny commands found in access control lists. Traffic can be filtered based on its source or destination address, as well as its source or destination ports.

Firewalls are useful for protecting the network from unauthorized access to data, as well as for protection against denial-of-service attacks, but they're not foolproof. If you need to use a particular service, that type of traffic must be permitted to pass through. For each packet type permitted to enter the network, there is a risk. However, because the firewall acts as a choke point—the only access in or out of the network—it's easy to use it to log access attempts.

You should know that firewalls can't protect data that doesn't pass through the network. Several years ago, I sent a team to consult with a company that had been victim to unauthorized access to data—including sensitive information that somehow ended up in the local news. A thorough review of the network's security finally revealed that the company's data was accessible to the public because an executive secretary was putting sensitive reports into the garbage can. This information was available to anyone who could sift through a pile of papers. A consistent security policy must be established and adhered to for the entire company, not just the network.

Before configuring a firewall, you should first decide what type of traffic should and what type shouldn't be allowed to pass through it. Figure 8-7 depicts a scenario in which DNS, HTTP, and SMTP traffic is allowed into the network, any other incoming traffic is denied, and all outgoing traffic is allowed. It's always a good idea to end an access control list on a firewall with a deny to all other traffic because this will prevent any attacks through higher ports—which can lead to a software vulnerability or even an intentionally placed back door into an application.

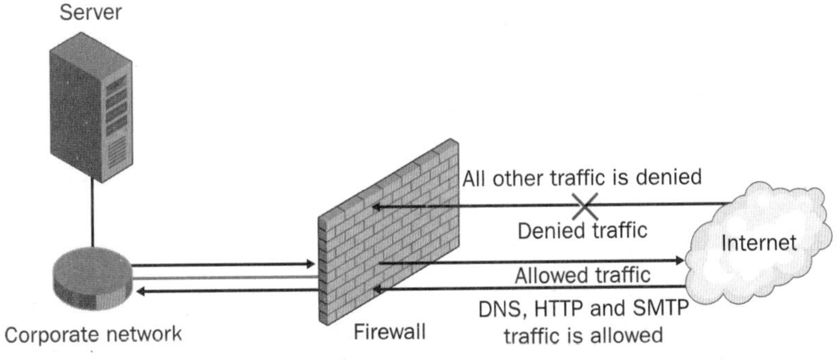

Figure 8-7 A firewall is usually more restrictive to incoming traffic than outgoing traffic.

A firewall uses an access control list for all the commands to execute packet filters. When the firewall implements the access control list, it does so in order of the commands. This means that when a "deny all" command precedes a "permit this" command, the "permit" command is not executed. This process can cause problems in application performance.

Another reason an application might not perform correctly after a firewall is installed is the port number used. Many applications use one or more TCP or UDP ports that are not well known. When implementing a new firewall, you should review every application that must function across the firewall. When the port numbers are known, they must be permitted within the access control list on the firewall.

Demilitarized Zones

A demilitarized zone (DMZ) is an offshoot from a firewall that is not considered part of the Internet, nor is it considered part of the private network. As shown in Figure 8-8, a DMZ is a middle area that offers more freedom of access from the Internet. This configuration places the DMZ between two firewalls.

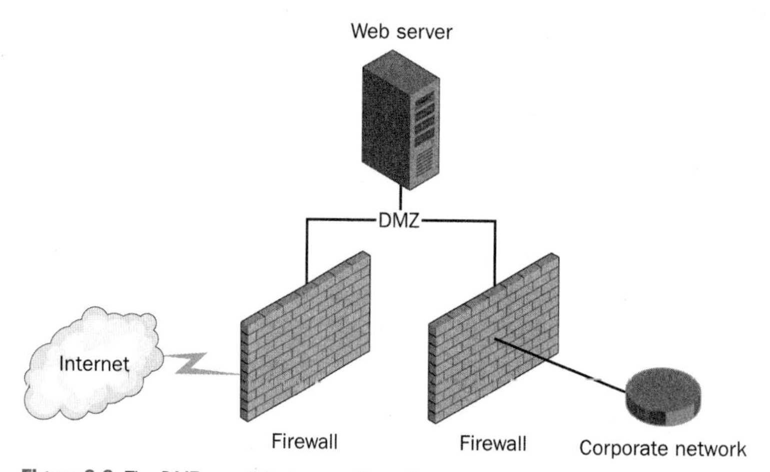

Figure 8-8 The DMZ can sit between a firewall to the Internet and a firewall to the private network.

Alternatively, a DMZ can be an offshoot area such as that displayed in Figure 8-9. In this scenario, the firewall has three interfaces, one that connects to the Internet, a second that connects to the DMZ, and a third that connects to the private network. This configuration is driven solely by access control lists in which the DMZ access is relaxed compared to that of the private network.

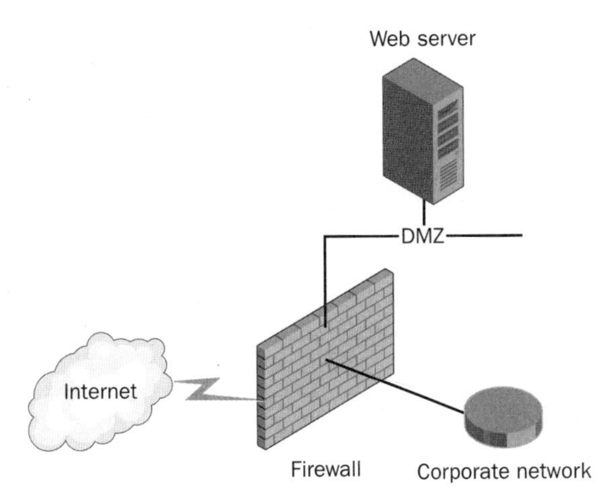

Figure 8-9 The DMZ can be configured as a third interface on a firewall.

One reason to create a DMZ is to provide access to certain servers, such as a Web server or e-mail server, yet still protect the rest of the network from those types of traffic.

Proxy Servers

The traditional firewall acts at the network layer, filtering packets from one interface to the other. This type of firewall will make all permit/deny decisions based on the source or destination addresses, domain names, or ports, which are all in the IP packet header. For a more sophisticated and secure method of blocking and permitting traffic, you need to use a proxy server, which is sometimes called a dual-homed gateway.

A proxy server doesn't permit traffic to pass through it between networks. It does, however, examine each packet up to the application layer and reassemble a new packet for the other network. Because each piece of data is so thoroughly examined, the proxy server is able to log traffic and perform audits. A proxy server configuration is shown in Figure 8-10.

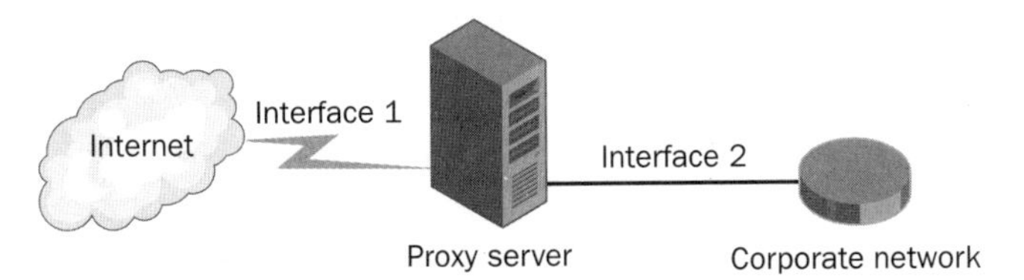

Figure 8-10 A proxy server is called a dual-homed gateway because it has two interfaces to two different "home" networks.

Proxy servers are often preferred over firewalls because they fully prevent traffic from passing through from one network to another. Because the proxy understands the application-layer protocol, it can implement security specific to that protocol.

Key Points

- Security protocols are intended to protect a network from various types of attacks on the network. Most security protocols are capable of protecting against multiple, but not all, types of attacks.
- IPSec is a security system made of subprotocols that are capable of authentication and encryption.
- Tunneling protocols are used to create virtual private networks (VPN).
- Layer 2 Tunneling Protocol (L2TP) was developed using Cisco's L2F and Microsoft PPTP, and has many advantages over both of the older protocols, such as wider usage and additional security through its implementation with IPSec.
- Secure Sockets Layer (SSL) is a protocol that can provide confidentiality for information transmitted across the Internet.
- When SSL is in use, a URL is preceded by https://.
- Kerberos is an authentication protocol that will also create a trust relationship between UNIX realms and with Windows 2000 domains or between Windows 2000 domains.
- Firewalls are routers that perform packet filtering between two interfaces, one connected to a protected network and another connected to a public network.
- Firewalls make filtering decisions given the data available in the IP packet header. This means that the packet can be filtered based on source IP address, destination IP address, and ports.
- Demilitarized zones are networks that are less private and that either sit between a public network and a private network with two firewalls separating them, or off an interface of a router that is also connected to a public network and a private network.
- Proxy servers function at the application layer and can provide protocol-specific security, in addition to performing logging and auditing.

- The extra services provided by a proxy server make it slower than passing data through a packet-filtering firewall.

Chapter Review Questions

1 Which of the following types of attacks will IPSec help guard against? Choose two.

a) Virus attack

b) Denial of service

c) Eavesdropping

d) Password pilfering

Both answers c and d are correct. IPSec, because it provides authentication and encryption, can protect against both eavesdropping and password pilfering. Answer a is incorrect because IPSec will not protect against a virus attack. Answer b is incorrect because IPSec cannot protect against a denial-of-service attack.

2 Which of the following will protect against IP address spoofing?

a) Authentication

b) Encryption

c) Packet filtering

d) Physical security

Answer a is correct. Authentication is a method of security that can protect against IP address spoofing because it ensures that the data received is coming from the correct source. Answer b, encryption, is incorrect because it will not guarantee that the IP address isn't changed. Answer c, packet filtering, is incorrect because IP address spoofing is one of the attacks used to break through a packet-filtering firewall. Answer d, physical security, is incorrect because it can't guarantee that packets come from the address in their headers.

3 Which of the following two protocols are used together to create a virtual private network connection?

a) IPSec

b) Kerberos

c) SSL

d) L2TP

Answer a, IPSec, and d, L2TP, are the two protocols used together to create a virtual private networking link. Answer b, Kerberos, is incorrect because it is used to authenticate users and create trust relationships. Answer c, SSL, is incorrect because it is used to ensure confidentiality of data across the Internet.

4 Which of the following ciphers can be used by SSL?

a) L2TP

b) PPTP

c) MD5

d) DMZ

Answer c, MD5, or Message Digest version 5, is one of the ciphers that SSL can use to ensure confidentiality of data. Answer a, L2TP, is incorrect because it is a tunneling protocol used with IPSec. Answer b, PPTP, is incorrect because it is a tunneling protocol. Answer d, DMZ, is incorrect because a demilitarized zone is a network that is connected to the Internet and protected by a firewall, but is more accessible than the private network.

5 Which of the following services do smart cards enable?

a) Confidentiality

b) Packet filtering

c) Authentication

d) Encryption

Answer c is correct. Smart cards are used to enable a secure authentication of users to the network. Answer a, confidentiality, is incorrect because smart cards do not encrypt or otherwise secure data after the user logs on to the network. Answer b, packet filtering, is incorrect because smart cards do not filter packets. Answer d, encryption, is incorrect because smart cards do not provide encryption.

6 Which of the following is a network that provides greater access to certain hosts than what is available on a private network?

a) DMZ

b) KEA

c) PKI

d) MD5

Answer a is correct. A DMZ is a network that enables greater access to hosts than does the private network. Network admins place servers such as Web servers on the DMZ so that they are able to provide services on the Internet that are not available from the private network for security reasons. Answers b, c, and d are incorrect because these are ciphers rather than networks.

7 Which of the following works at the application layer?

a) IPSec

b) SSL

c) Firewalls

d) Proxy servers

Answer d is correct. Proxy servers function at the application layer, which allows them to provide greater security than firewalls. Answer a, IPSec, is incorrect because IPSec works at the network layer. Answer b, SSL, is incorrect because it works at the network layer. Answer c, firewalls, is incorrect because firewalls function at the network layer.

8 If you access a Web page using a URL that reads *https://www.microsoft.com*, which of the following security methods is in use?

a) Encryption

b) Tunneling

c) Packet filtering

d) Physical security

Answer a is correct. When a URL is preceded by https://, the Secure Sockets Layer (SSL) protocol is in use. SSL offers encryption through public key. Answers b, c, and d are incorrect because none of these methods requires a special URL.

9 You've been called in to troubleshoot a network. A firewall has been installed, but now a critical application won't function across the Internet, though it worked well before the firewall was installed. The administrator states that the TCP port (1899) is allowed on the firewall. What could be the problem?

a) The users need to implement new smart cards.

b) The permit TCP port 1899 command follows a deny all command.

c) A new Kerberos trust should be created.

d) The application is on the private network when it should be on the DMZ.

Answer b is correct. The list of commands in a router's access control list are executed in order; so, if a deny all command is executed, all the permit commands following it are ignored. Answer a is incorrect; there is no reason for smart cards to be used with the application because it worked before the firewall was installed. Answer c is incorrect because a Kerberos trust would not be involved. Answer d is incorrect because the production applications should be in a location that is as protected as possible.

10 Which of the following uses an authentication header and encapsulated security payload?

a) PPTP

b) IPSec

c) Kerberos

d) SSL

Answer b is correct. IPSec consists of an authentication header (AH) and encapsulated security payload (ESP), which provide authentication and encryption. Answer a, PPTP, is incorrect because it is a tunneling protocol. Answer c, Kerberos, is incorrect because it provides authentication and trust relationships. Answer d, SSL, is incorrect because it handles encryption for confidentiality of data.

Part 3

Network Implementation

A large part of administering a network involves the installation of applications, peripheral equipment, workstations, and servers—this is otherwise known as *implementation*. Networking has become a diverse science. Many different types of network operating systems (NOSs) are used on networks, and each is available in multiple versions. As the number of versions increases, more features and capabilities are added, making newer NOSs more complex than older ones.

Client operating systems once required proprietary network client applications to interoperate with a NOS. Today these operating systems typically include all the tools needed to access multiple types of NOSs and even provide services to peer workstations on the network. Home networks offering the ability to share files and printers, and even Internet connections, are gaining in popularity.

In this section of the book, we'll look at UNIX, Macintosh, Windows, and NetWare NOSs; authentication; file and print services; security; and interoperability. We'll review how to establish connectivity using Dynamic Host Configuration Protocol (DHCP), Domain Name System (DNS), and Windows Internet Name Service (WINS). This section also establishes the foundation for small office home office (SOHO) networks and storage systems.

Chapter 9

Looking at Servers and Their NOSs

File and printer sharing are the two main reasons that networks were created in the first place, not to mention simple economics—if 20 people can share one expensive printer using comparatively inexpensive networking equipment, well, that's a no-brainer. Back when a 5-megabyte (MB) hard drive cost U.S. $5,000 or more, storage space was a precious commodity. Sharing storage was a cost-effective solution that provided the basis upon which networking began to grow. Convenience played a part, as well. Sneakernet (carrying files to another person's desk on a floppy disk in order to share that data) was time-consuming and nearly impossible to use over long distances. Networks grew out of two different methods (and have since converged into a combination of them).

- **Client/server.** This networking method is based on a central server that provides services consumed by clients. True client/server applications actually share some of the application processing among machines. In the networking paradigm, however, the server is the main provider and the client is solely a consumer.

- **Peer-to-peer.** In a true peer-to-peer network, all computers are created equal. Each client is able to share files, printers, and other resources with every other client, or *peer.*

Today it's common to find peer-to-peer and client/server systems functioning in tandem on a single network. Every server utilizes a special operating system that enables it to provide services and resources to the other devices (and users) on the network. This comprises the NOS. As a provider on the network, the server shares files, printers, applications, and other resources. Clients are able to act as pure clients when they only access resources from servers, or as peers to other clients by sharing their own resources on the network. As a result, today's client operating systems share many of the same capabilities as server NOSs. For a client to access any resources shared on the network, it must use the same protocol stack as the server. Remember that the protocol stack is like a language that must be spoken by both the client and server for the two to communicate.

The differences between a server's NOS and a client's operating system are beneath the surface. The NOS has a much more secure method of authenticating users and resources—its directory service. In the directory service, user accounts are given access rights to files and printers by a network administrator. Each user is provided with a logon ID and a password with which that user logs in to the network on a client workstation. When the user requests access to a resource, the directory service checks to see what rights are associated with the logon ID and then allows only that level of access to the resource. This is called *user-level security.*

The directory service also provides for groups. Groups simplify security management by placing all user accounts that require the same access to a set of resources into a group, and then granting rights to the group. It's much easier to keep track of the rights provided to a few groups than it is to manage rights assigned to hundreds of individual users.

Every NOS includes utility applications that maintain and monitor the server. You'll usually find tools for administering the directory service, managing files and disk storage, and optimizing server performance. Tools that manage printers and back up the data on the server to tape are also available, usually along with a tool that will log and audit access attempts.

Another way to protect files and printers from unauthorized users is to apply a general password to a resource and then supply that password to anyone who should be allowed to access the resource. This is called *share-level security.* The problem with this method is that it is easily defeated. Because everyone shares the same password, anyone could hand that password to an unauthorized person.

NOSs often include client applications that improve performance and access to the resources they share. File and printer sharing is optimized in a NOS, and security is enhanced. Older NOSs focused on sharing files and printers, while today's NOSs often provide Web services, e-mail applications, database applications, and dynamic addressing.

This chapter represents a crash course in NOSs, in which we'll review server operating systems and their characteristics. We'll look at core concepts in Microsoft Windows, NetWare, Macintosh, and UNIX NOSs. From this chapter forward, you'll occasionally encounter topics that apply to these NOSs, and you might find it helpful to refer back to this chapter at those points.

Microsoft Windows Servers

Windows servers began as Microsoft Windows NT Advanced Server 3.1, a 32-bit network operating system. Back when it was first introduced, everyone was excited about the "new technology," which is what NT stands for. The graphical user interface (GUI), or desktop, looked identical to that of Windows 3.1, which explained why the first version was called 3.1. The NOS was quickly replaced by Windows NT 3.5, which was almost immediately replaced by Windows NT 3.51. While the desktops on these versions all looked identical, improvements were apparent in the increasingly higher performance of file- and printer-sharing capabilities.

Windows NT 3.51 was eventually followed by Windows NT 4.0. This version had a new desktop that looked exactly the same as that of Windows 95, which was released less than a year before Windows NT 4.0. (One thing to keep in mind about any version of Windows NT is that there are two types, one for clients and the other for servers.)

Windows NT Domains

Windows NT uses domains as logical boundaries that house servers, user accounts, and resources. While a domain is similar to a workgroup, it uses a central directory service that authenticates the users when they log on. Servers with a copy of the directory service database (which is called the Security Account Manager, or SAM) are called domain controllers; there is one primary domain controller (PDC), and the rest are backup domain controllers (BDCs). A BDC can't create new users in the domain, because it contains a read-only copy of the SAM. If an administrator makes any changes through an administrative console on any PC or server in the network, the changes appear to take place locally, but in fact they take place only on the PDC. BDCs provide a measure of redundancy, so if the PDC fails, a BDC can provide authentication services so that users can log on and access network resources. In the event of a total PDC failure, a network administrator can convert a BDC into a PDC. The rest of the servers, and the workstations, are members of the domain.

The physical location of the domain's controllers and its members doesn't matter as long as some type of network connection exists between them. Domain planning, however, is almost an art form. Because a measure of bandwidth is consumed between a PDC and its BDCs, and bandwidth is also consumed by domain members when they log on and request access to network resources, the placement of the PDC and BDCs around the network will affect performance. For example, if you decided to place a PDC across a slow WAN link from hundreds of users, performance would be unacceptably slow. If you added a BDC to that location, performance would be drastically improved.

Because the domain structure is a logical organization, more than one domain can exist in a network. Domains can be placed in any geographical location with PDCs and BDCs of multiple domains criss-crossing the entire physical network. This makes it difficult to imagine their logical relationships even if you know where each of the domains' controllers, servers, and members are placed.

Trust Relationships

Because each domain's PDC has a separate SAM, the user of one domain can't access resources owned by another domain. To overcome this flaw, domains can be joined together in a *trust relationship*. Once a trust relationship has been created, a user in a trusted domain can be granted access to a resource in a trusting domain.

To understand trust relationships, imagine them to be political in nature. For example, if the entire network were the planet earth, each domain would be a country and each computer would be a city within the country. Each country has its own security policies, which have no effect on any other country's security policies. So, let's say that a Thai citizen wants to visit New York City. To do this, the U.S. must trust Thailand and then must grant specific access in the form of a visa to the Thai citizen. Without both the trust and the access rights, the Thai citizen can't visit New York. Now, if a U.S. citizen wants to visit a city in Thailand, Thailand must first trust the U.S. and then grant a visa to the citizen to visit the city. Even if Thailand trusts the U.S., the U.S. will not necessarily trust Thailand. Furthermore, even if Thailand trusts the U.S., this doesn't necessarily mean that it will allow an American to visit its cities. Likewise, a trust relationship between two domains is unidirectional: there is one trusting domain and one trusted domain. Two separate trust relationships are required for domains to trust each other. Without the correct trust relationship, a user won't even see the other domain in the logon dialog box. In addition, even when a trust relationship is in place, the network administrator must also grant access rights to the domain's resources to the users in the other domain for the users to be able to use them. This is shown in Figure 9-1.

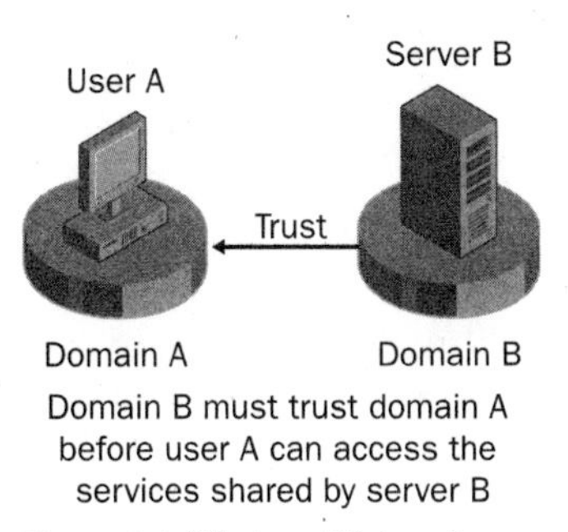
User A
Server B
Trust
Domain A
Domain B
Domain B must trust domain A
before user A can access the
services shared by server B

Figure 9-1 Windows NT domain trust relationships are unidirectional.

You should know one additional fact about trust relationships—they are not *transitive*. Again, consider this in terms of a political relationship. Imagine that Sweden trusts Peru, and Peru trusts the U.S. The state of being nontransitive doesn't automatically mean that Sweden trusts the U.S., or vice versa. So, in a domain trust relationship, if domain A trusts domain B, and domain B trusts domain C, it does not follow that domain A trusts domain C, as depicted in Figure 9-2.

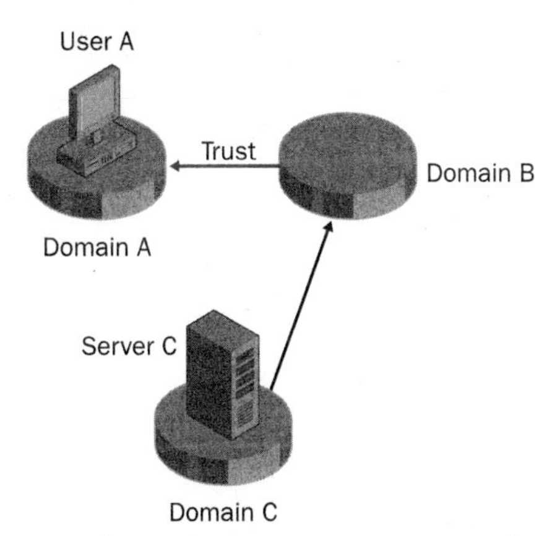
User A
Trust
Domain B
Domain A
Server C
Domain C
User A can't access resources on server C
because no direct trust relationship exists
between domain C and domain A.

Figure 9-2 Windows NT domain trust relationships are nontransitive.

Domain Models

All Windows NT domains have a distinct disadvantage related to the network operating system: they're difficult to rename or redefine. First of all, you can't change a domain controller's name or transform it into a standard server—whether it's a PDC or a BDC—without reinstalling the NOS. In fact, to change the name of a domain, you must reinstall all the domain controllers and then manually change each member of the domain so that they have joined the new domain. Because domain changes are so problematic, you must be certain of your domain plan before you begin implementing it.

Microsoft has provided several domain models as a plan for domains, which are helpful for handling organizational changes such as mergers or migrations. The following models are available:

- **Single domain:** This model consists of a PDC, workstations, and users that are all members of the same domain. There might be one or more BDCs. There are no trust relationships.
- **Master domain:** This model has at least two, and possibly more, domains—a master domain, which contains user accounts, and at least one resource domain, which contains resources such as printers, files, and so on. Every resource domain trusts the master domain, but the master domain doesn't trust any resource domains. This model is depicted in Figure 9-3.

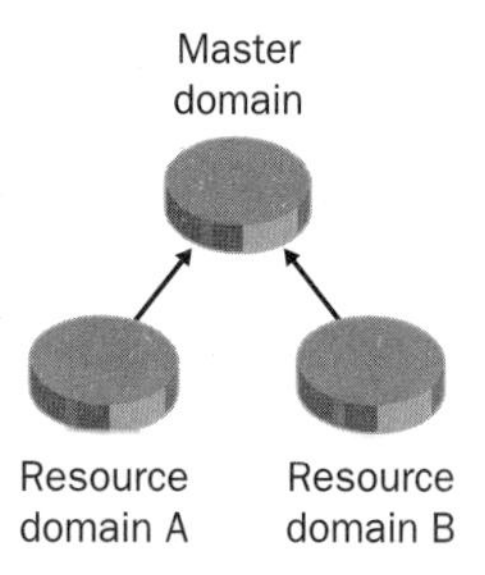

Figure 9-3 The master domain model has only one master domain and one or more resource domains.

- **Multiple master domain:** This model builds on the master domain model in that it has two or more master domains and one or more resource domains. Each master domain contains user accounts and trusts the other master domain(s). Resource domains contain network resources such as files and printers. Each resource domain trusts all of the master domains, as shown in Figure 9-4.

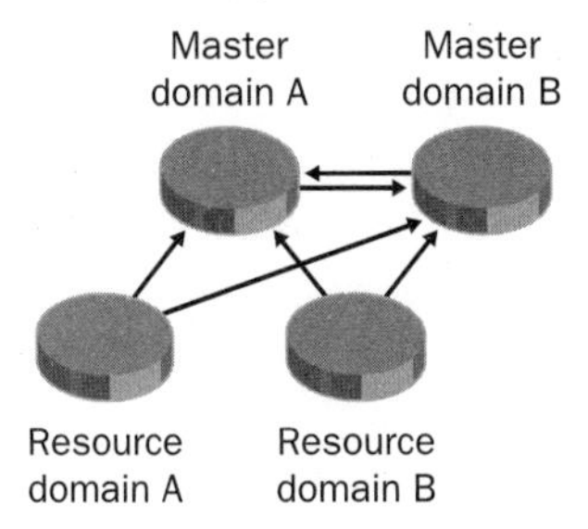

Figure 9-4 The multiple master domain model has two or more master domains and one or more resource domains.

- **Complete trust domain:** This model resembles a spider web, as shown in Figure 9-5. It contains two or more domains in which every domain trusts every other domain. Each domain can contain user accounts or resources, or both.

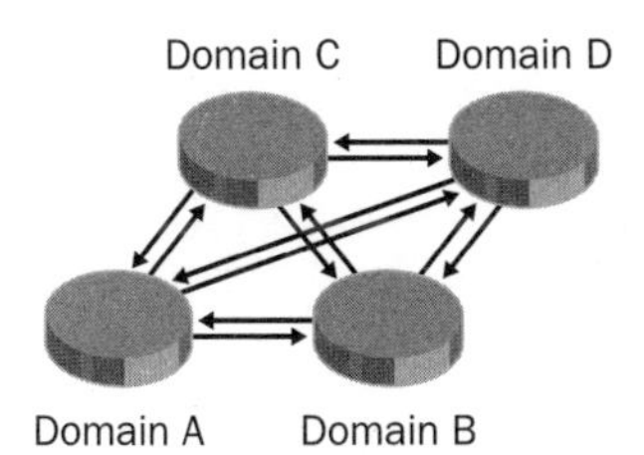

Figure 9-5 The complete trust domain model contains multiple domains that all trust each other.

Organizing Users into Groups

Windows NT domains use special groups to function between domains. There are two types of groups—global and local. A user account created within a domain is actually a global user account, whereas a user account created on a standalone computer is a local user account. Global users function within a domain in the same way global groups do.

The local group is created within a local security database on the server, as are local users. This group is local to the server and is not applied to any other server or workstation on the network. Local user accounts and groups can be granted access only to the resources on the server or Windows NT workstation they reside on.

Global groups and global user accounts are created in the SAM on the PDC. Global groups and users can be made members of any local group on any server or Windows NT workstation in the domain. They can also become members of local groups within any domain that trusts their own domain. When designing security plans for a Windows NT network, you should create global users, place them within global groups, and then make those groups members of local groups. You should apply access rights only to local groups. In this way, you'll find it easy to maintain and manage the network's rights and privileges.

Tip A local group can't move outside its local server. Global groups and global users are like travelers who can visit any local server and use its resources as long as they're granted access to those resources. It's much easier to manage rights and privileges if you grant them only to local groups.

To begin implementing a domain security plan, use the Windows NT User Manager for Domains on a domain controller, and User Manager on a workstation or server. Within this utility, do the following steps:

1. Create a user account within the domain (a global user account).
2. Create a global group.
3. Add the user account as a member to the global group.
4. Create a local group on each server or Windows NT workstation that will be sharing resources. This step and the following step must be performed in the User Manager utility.
5. Make the global group a member of the appropriate local groups.
6. Using Windows Explorer, apply rights to the local groups so that they are able to use files and printers on the local server or Windows NT workstation.

The User Manager for Domains utility also provides the ability to apply password policies. This utility is available on the PDC and each BDC, and you can install the client version on a workstation so that you can administer the network from a client computer. Note, however, that you must always have a network connection available to the PDC to make changes to the domain's SAM.

Another utility you can use is the Server Manager, which provides administrative access to the domain's servers, PDC, BDCs, and member computers. The member computers, however, must be running Windows NT, Windows for Workgroups 3.11, Windows 2000, or Windows XP. Regular Windows 3.1, Windows 95, Windows 98, and Windows Millennium Edition don't function as true domain members and won't show up in this utility. In the Server Manager, you can synchronize the BDCs to the PDC to make certain they all have the most recent copy of the SAM. You can also add or remove domain members, view a list of connected users, monitor shared services, send messages to connected users, and manage services running on domain member computers. Plus, you can promote a BDC to a PDC, which is an extremely important function in the case of a PDC failure.

Windows 2000 and Active Directory

While Windows NT servers are still in use, you'll more often find Windows 2000 servers on a network. Windows 2000 uses a much different method of managing users and computers—the Active Directory, which is a true hierarchical directory service that provides secure authentication for users who want to access resources on the network.

The Active Directory is hierarchical because it uses the Domain Name System (DNS), which is a globally accessible table of domain names and corresponding IP addresses. By using DNS, the Active Directory is able to integrate with the Internet and services that are offered via the Internet. This is a drastic leap for Windows servers from the flat-file database that the SAM offered.

By using DNS, the Active Directory becomes hierarchical at the domain level, and it affects the trust relationships between domains that are within the Active Directory. Yes, the Active Directory still uses the logical organization known as a domain, but it is not named "DOMAIN" with a NetBIOS name. Instead, it is named "domain.com" with a DNS name. The *forest* is a new logical grouping of domains that lies within the Active Directory. Trust relationships are both reciprocal and transitive within an Active Directory forest. When a domain is installed, it can be installed into a new or existing forest. If it's added to an existing forest, it automatically trusts every domain already within that forest, and every domain automatically trusts it. These trust relationships are based on Kerberos, an authentication protocol. However, one thing remains true—the users must be granted access to resources before they can use them, regardless of the fact that there are trust relationships in place.

In the forest shown in Figure 9-6, you can see a root domain (called root.com) and subdomains (called sub.root.com and domain.com). Each different domain name type creates a different domain tree. In this case, root.com and sub.root.com belong to the same domain tree, while domain.com belongs to a different tree. Domains don't need to share a domain name to belong to the same forest.

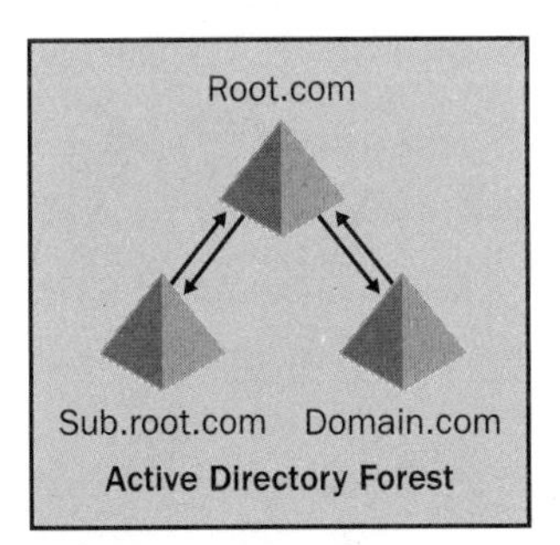

Figure 9-6 The Active Directory forest consists of one or more domains, all of which trust each other but are organized in a hierarchical fashion.

The Active Directory is a distributed directory service, which means that the actual database is distributed throughout the domain controllers on the network. Yes, there are domain controllers, but there is no such thing as a PDC or BDC within the Active Directory. Every domain controller is created equally. This overcomes one of the problems inherent within the old Windows NT domain structure, in which you can't create users if the PDC is down. Because all domain controllers are equal, if one goes down, the others will still be able to provide the same services. In general, when you're planning domains, remember that you should always have at least two domain controllers per domain for redundancy.

Each domain contains objects that represent users, groups, computers, and other network resources, in addition to organizational units that can logically organize the users, computers, groups, and other network resources into a tree-like structure for ease of use. (See Figure 9-7.) The Active Directory is based on the X.500 standard for directory services, which means that it is similar to any other X.500 standards–based directory services such as Novell Directory Services, which we'll discuss later in this chapter.

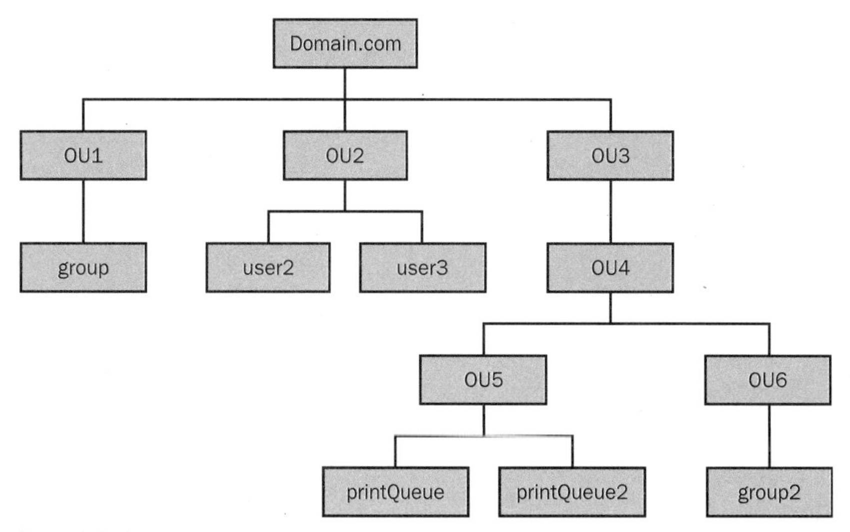

Figure 9-7 Organizational units are used to logically organize the users, groups, and resources within a domain into a tree-like structure.

The Active Directory allows a network administrator to manage the entire network from a single seat. In larger organizations, the Active Directory can be organized so that only specific administrators are capable of performing certain duties. For example, a Help Desk employee might be given the right to change passwords but not to examine files and access applications. This is an important change from Windows NT, in which the administrator had all administrative rights or none, a source of potential problems from a security standpoint on one hand and a productivity standpoint on the other.

The Active Directory database contains two basic types of objects—container objects and leaf objects. A container object can hold other objects whether leaf objects or other container objects, but a leaf object is a single entity and cannot hold a container object. The container objects are used to create the hierarchy, or tree structure, within each domain. The most common container objects are called Organizational Units, or OUs. Objects that represent resources are always leaf objects, which means a user, a printer, and a group are leaf objects.

Authentication

The Active Directory provides a view overseeing one or more domains. Each domain is considered a security and administrative boundary. From a DNS standpoint, domains, called domain trees, often share a contiguous DNS namespace—for example, the tree could have tree.com, branch.tree.com, leaf.branch.tree.com, and twig.branch.tree.com, as shown in Figure 9-8.

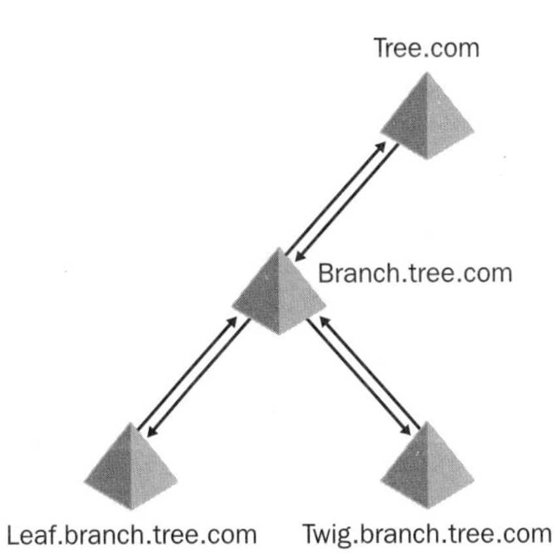

Figure 9-8 A domain tree consists of multiple domains that share a contiguous DNS namespace.

About X.500 and LDAP Standards The X.500 directory service standard has been around since it was first offered in 1988 by the ITU-T (the standardization division of the International Telecommunication Union). X.500 offers a model for linking multiple local directories to form a single, global distributed directory service. In distributing the directory, X.500 allows partitions of the directory to be placed on different servers, which are then linked so that each server has access to the information within all the other servers without necessarily containing that information in their local directory. Servers containing a partition of the directory, along with an index of the directory information, are called Directory System Agents (DSA). When users and administrators look at an X.500 directory using any type of utility, they don't see partitions, they see the entire directory service.

An administrator can alter directory service from any DSA by making changes to the entities representing the users and network resources. Each entity is provided a unique identifier called a Distinguished Name (DN), which includes information about the location of the entity within the X.500 hierarchical tree.

The Lightweight Directory Access Protocol (LDAP) is another Internet standard used for locating and managing resources within directories. Because it is an Internet standard, many different types of directory services support LDAP, which makes it a universal access method for directories. LDAP is able to access DNS, which acts as a mapping system for names to IP addresses and requires that end users know which name to look for. With LDAP, end users can browse through a directory such as DNS and access resources without first knowing the name of what they're searching for.

Messaging systems employ a directory service of their mailboxes, which is usually integrated with LDAP. LDAP is able to browse any X.500-compliant directory, including the Active Directory and Novell Directory Services. LDAP-compliant directories understand naming conventions that use the LDAP Distinguished Name, which identifies a root of the tree, country information containers, organizational unit containers, and the individual resource.

Windows NT File System

The Windows NT File System, NTFS, is used on both Windows NT and Windows 2000 computers. NTFS was developed to surmount the inadequacies within the 16-bit file allocation table (FAT) file system that was used by DOS and earlier Windows operating systems.

NTFS supports long filenames, as opposed to the "8.3" filename system used by FAT in which 8.3 referred to the requirement that each filename be a maximum of eight letters long, followed by a period and then a maximum three-letter extension. NTFS filenames can not only break out 8.3 into long names, but they are also case-sensitive for UNIX programs and case-insensitive for older operating systems that use the FAT file system. NTFS also maintains an 8.3 file name mapped to the long filename so that a client accessing that file can actually open it.

File access in NTFS is provided at the user level. An administrator can assign access rights to both files and directories. This can be performed using Windows Explorer in Windows NT and in Windows 2000.

Both Windows 2000 and Windows XP provide NTFS version 5. NTFS 5 offers file compression to optimize file storage. (This is a property of each of the files.) Another way that NTFS optimizes file storage is by using large partitions. NTFS provides for volumes, which associate a single drive letter with a collection of free space regions that can be spread across several hard drives. A volume can even be a Redundant Array of Inexpensive Disks (RAID), created through the NOS. In Windows 2000 and later, you use Disk Management to manage hard disks. In Windows NT, you use Disk Administrator.

Utilities

In Windows NT and Windows 2000, you use the Network icon in Control Panel to set up the protocols that the server uses to communicate across the internetwork. In Windows NT, NetBEUI is the standard protocol, but TCP/IP is the default protocol in Windows 2000. TCP/IP is required for a Windows 2000 server to participate in the Active Directory.

To configure client services for Remote Access Services (RAS), Dynamic Host Configuration Protocol (DHCP), or even DNS, you use the Network icon in Control Panel. However, to provide RAS to the network, you use the Routing and Remote Access Services (RRAS) utility. There are separate utilities for DHCP and DNS, as well.

When you monitor and manage servers across the network, you use performance monitor (perfmon) and event viewer. The perfmon offers real-time graphs of server performance based on the counters that you select. A counter is the item of service such as the amount of disk space used in bytes, or the amount of traffic being transmitted on a network interface. A perfmon graph is shown in Figure 9-9.

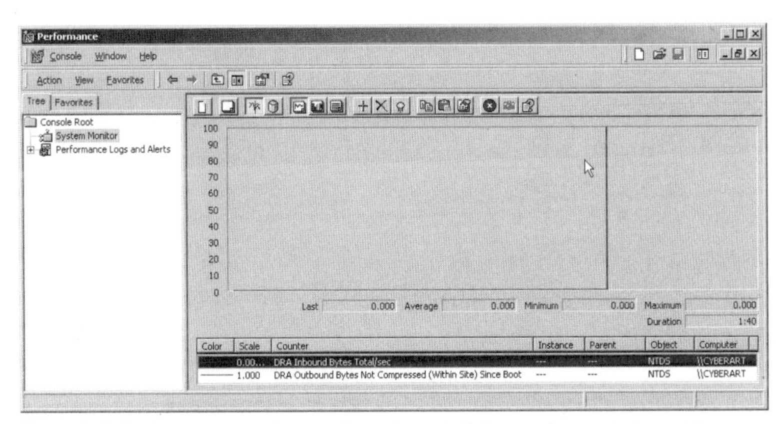

Figure 9-9 Performance monitor graphs various statistics on the server in real time.

To start Perfmon from the command prompt, type **perfmon** and press Enter. You are allowed to select counters related to specific objects, including the following:

- Connections
- Memory

- Physical and logical disks
- Processor
- Server
- Users

Each object has a series of counters that represent that object's measurable characteristics. Perfmon is extensible, which means that vendors can add a new object and associated counters for their own applications or equipment. In machines with multiple components of the same type, you'll see multiple instances of an object. For example, a multiprocessor server will have an instance of the Processor object for each additional processor within the server.

When you have Perfmon opened, you can add counters by highlighting System Monitor in the left pane and clicking on the plus (+) sign on the button bar on the right. You can also right-click within the window and select Add Counters from the popup menu. This will bring up the Add Counters dialog box. Here you can select an object, and then select the counters for that object. If you have questions about a counter, click the Explain button. Once you have added objects and counters, the information will be graphed on the screen with a different color line representing each counter. The color key will be displayed at the bottom of the graph.

Some of the objects and counters you should look more closely at follow:

- Processor
  - % Processor Time. This looks at how much time the processor is utilized executing non-idle threads.
  - Interrupts/sec. Interrupts are used to provide processor time for devices. Interrupts degrade performance because the processor is busy handling hardware rather than applications. If this counter increases drastically, it could indicate a faulty driver or a hardware problem.
- Memory
  - Available bytes. Shows the available virtual memory for applications.
  - Commit Limit. Shows the amount of memory that can be committed without increasing the page file.
  - Pages/sec. Displays the rate at which pages are read from the disk into physical memory. More than 20 pages/sec can indicate a need for more random access memory (RAM).

- Physical and logical disks
  - % Disk Time. Shows the percentage of time that the hard disk is busy. The disk should not be busy more than 90 percent of the time. If it is, you either need a new disk or more RAM because the disk is taking on too many pages/sec.

Event viewer provides audit information that can show alerts for various events that you configure, as well as many default events such as application faults. In event viewer, the icons for various types of messages are different. A red alert icon usually identifies a serious problem that you should examine. Any service, whether it is a fault with a hardware interrupt or a network service failure, will leave its mark as an alert within the event viewer. The event viewer is very helpful for troubleshooting. For example, you can monitor the event viewer to see whether users have attempted to log on to the network and failed to provide the correct password. Event viewer and perfmon allow you to supervise activity on the server and make administrative decisions about it.

The event viewer displays the log files that are stored in %systemroot%\system32\config (%systemroot% is the wildcard used in place of C:\WinNT or C:\Windows, that is, wherever you have installed the Windows operating system). You'll be able to see the following logs:

- **Application log:** Contains the events that are logged by applications.
- **Security log:** Contains the events related to authentication and resource usage. Auditing must be enabled for security events to be recorded.
- **System log:** Contains events related to system components, including services, drivers, and hardware.

Windows 2000 Domain Controllers contain these additional logs:

- **Directory Service log:** Contains the events related to the Active Directory
- **DNS Server log:** Contains DNS events
- **File Replication Service log:** Contains the events regarding replication to other servers

Information is recorded in the logs for any significant event that occurs to the operating system or within an application. Critical events will also elicit a popup message on the console. Each event differs depending on its type, and the Event ID and Description are required to determine what has taken place on the system and how to handle it. The three types of events are as follows, in order of most severe to least:

- **Error.** This message indicates that a significant condition has occurred in the system or application, such as a system shutdown or failed driver. Indicated by a red circular icon with a white X in the middle.
- **Warning.** This event warns of potential problems, such as low disk space or timed-out requests. Indicated by a yellow triangle icon with an exclamation point in the middle.
- **Information.** This message is for your information only. It creates a historical reference of the system's operation, such as successful startup of drivers or services. Indicated by a white balloon icon with a blue I in the middle.

You'll also find that the Security log will provide Success Audit and Failure Audit events. As you can probably guess, the Success Audit shows a successful attempt to log on or access a resource, while a Failure Audit displays the failed logon and resource access attempts.

Each event shows the date and time of the event, the source application or service that reported the event, the event ID, and the name of the computer where the event occurred. When you double-click an event, you can see further information about the event on its property sheet, although this might only be binary information that the vendor of the failed application understands. The event viewer is shown in Figure 9-10.

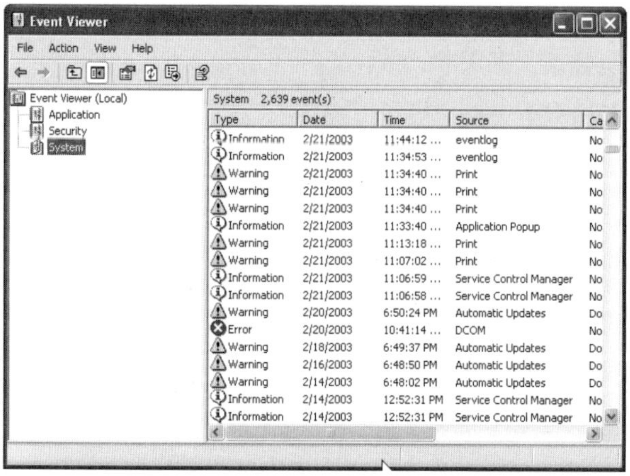

Figure 9-10 There are three types of messages in the event viewer—Error, Warning, and Information.

Novell NetWare

One of the first corporate successes in NOSs was Novell NetWare. Originally developed to provide shared disk storage, NetWare grew into a robust, high-performance network operating system. The oldest versions are rarely used and no longer supported. The latest version is NetWare 6, but you're likely to find versions all the way back to NetWare 3.*x* in use on networks.

Each version includes some of the same core services. These include file services, print services, and applications. Services are provided as NetWare Loadable Modules (NLMs), which are executables within NetWare.

In a Novell NetWare network, the servers are intended to be the only systems that share services, so it provides a true client/server structure. In fact, some services are manageable only from a connected client workstation and are not available from the server's desktop.

NetWare File System

The NetWare file system is transparent to an end user. When users connect to the NetWare server, they see what looks like a local hard drive. The NetWare file system's features are designed to provide files to a variety of different client operating systems.

When you install the NetWare server, or when you configure it with either the Install or Nwconfig NLMs, you'll find that the basic division of a hard disk is a NetWare partition. Within the partition, NetWare provides a logical division called a volume, which can be up to 32 terabytes in size. The volume can be constructed from a single contiguous space on a hard disk or from multiple areas of different hard disks. Once created, the volume is shared out to end users.

Once a volume is created, it is given a namespace that provides the rules according to which the client will interact with the files within the volume. In a volume to which a DOS namespace has been applied, for instance, you'll be able to save files with the 8.3 format. The file system appears to be the same as a local FAT file system. When you use the LONG namespace, which was originally intended for OS/2 clients and grew to include Windows 95 and later clients, files can be saved with long filenames. The MAC namespace is used for Apple Macintosh file support. A namespace is a NetWare-loadable module with a .NAM extension.

The file system offers security for files through trustee assignments applied to files and directories. When you grant a specific access right to a user account or group, you provide it a trustee assignment. When you grant trustee assignments to a directory, the files within the directory, as well as all subdirectories, inherit the same security rights unless there is an explicit denial of access through an inherited rights filter. Inherited rights filters prevent specific rights from being passed down through the file directory tree.

Trustee assignments can be set in the NetWare Administrator program or granted through the Rights command, a DOS command-line utility. Trustee assignments follow in Table 9-1.

Table 9-1 Trustee Assignments

Trustee Right	Meaning	Function
R	Read	Open a file or directory.
W	Write	Add data to an existing file.
C	Create	Create a brand-new file or directory.
E	Erase	Delete a file or directory.
M	Modify	Rename or change the attributes of a file.
F	File scan	View the files or subdirectories of a directory.
A	Access control	Grant or revoke trustee assignments.
S	Supervisor	Automatically have all rights.

NetWare supports both file compression and data migration, both of which must be enabled on a NetWare volume before files can be either compressed or migrated. Data migration is a process whereby the server moves files that haven't been modified or accessed over a period of time to an archival system such as an optical storage system. Compression is an attribute that can be applied to files using the NetWare Administrator program.

Client Support

NetWare was developed to be an enterprise network operating system that can provide services to multiple and distinct client operating systems. These clients include all versions of Windows, OS/2, Macintosh, UNIX, and Linux. The namespace is one of the ways that NetWare provides support for these different client operating systems.

NetWare is capable of providing services using different protocol suites. The oldest versions of NetWare used the IPX/SPX protocol suite natively. In addition, NetWare supports TCP/IP. While older versions can use TCP/IP to communicate, they can't use it as the only protocol suite. NetWare version 5.*x* and later do support TCP/IP as the native protocol, however. TCP/IP is a protocol used by UNIX and Linux clients, and it is often the only protocol suite used by a variety of different client operating systems.

The third protocol that NetWare supports is AppleTalk. AppleTalk is used to communicate with Apple Macintosh clients, and it is seldom used for any other purpose.

Novell Directory Services

Through NetWare version 3.*x*, Novell used a flat-file database to hold user accounts and resources that were specific to each separate server. Users had to have an account on every server that they wanted to access. This database was called the Bindery. NetWare 4.*x* brought about the hierarchical directory service called Novell Directory Services (NDS).

NDS is distributed across multiple servers, which maintain partitions and replicas (copies of partitions) of the database. The administrator creates partitions and replicas to optimize performance of authentication. Partition planning is required in large internetworks with WAN links because of the performance issues that occur across slow WAN links.

NDS is a single hierarchical tree consisting of an organization (O) container that is then structured into organizational units (OUs). Each OU can contain other OUs, user accounts, and network resources, enabling an administrator to manage the entire network from a single seat. The NDS is organized as shown in Figure 9-11.

NDS is the main administrative utility used to manage the NetWare Administrator. This application allows the administrator to view each network resource and its position within the tree, and then modify the properties of that resource as needed—whether that consists of changing a user's password or adding a new file attribute or trustee assignment.

Container units, which are the O and OU containers, act like folders in that they hold other containers and leaf objects. Leaf objects, the resources in the network, can't contain any other objects.

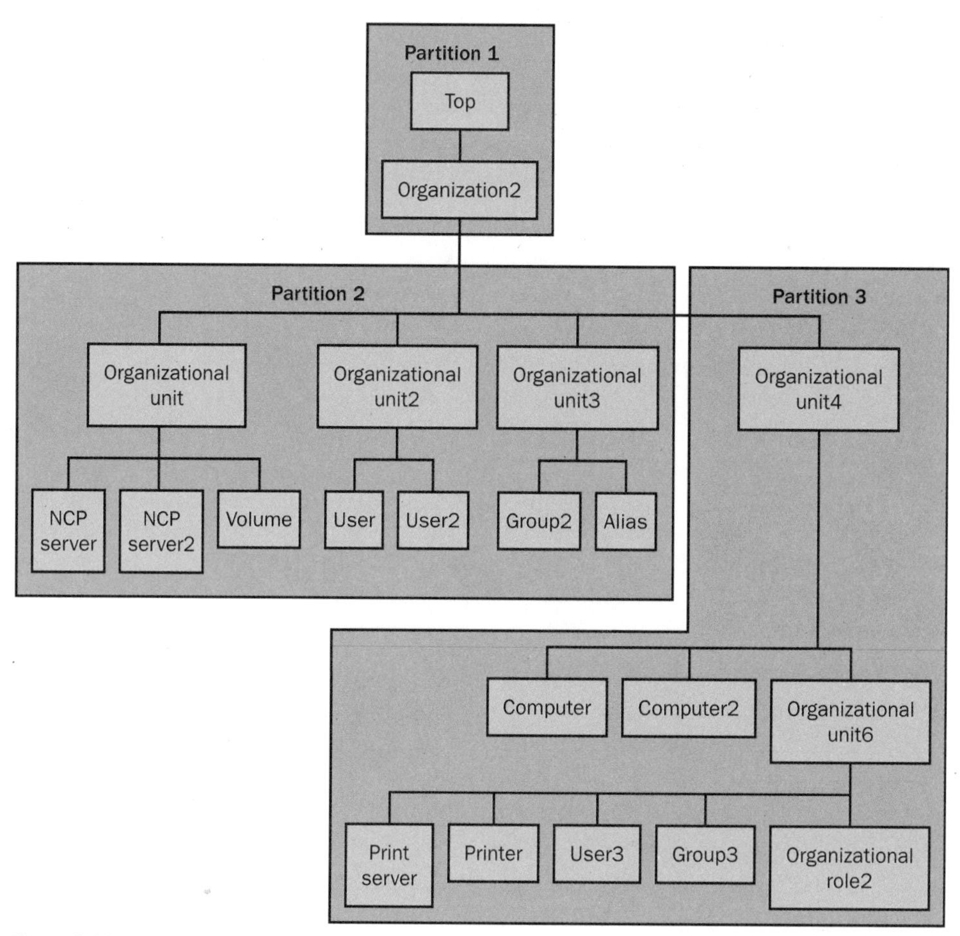

Figure 9-11 NDS is organized into a hierarchical tree that is distributed through partitions and replicas.

NetWare 4.*x* and earlier servers all provided character-based server console applications. NetWare 5.*x* and later servers provide a Java-based graphical console in addition to character-based console applications. To change between screens on the server, you can press the Alt key and the Esc key. Many utilities are available on a NetWare server. They are shown in Table 9-2.

Table 9-2 NetWare Server Utilities

NetWare Utility	Function
Console One	Available on NetWare 5.*x* and later. Displays limited NDS information.
Config	Used at the command line. Displays network adapters and configurations.
Inetcfg	Displays network adapters and protocols and allows configuration.
Install or Nwconfig	Used during installation and afterward to configure hard-drive partitions, volumes, startup configuration, and NLMs.

Table 9-2 NetWare Server Utilities

NetWare Utility	Function
Monitor	Displays performance statistics.
NDPS	Provides NDPS printer administration.
Set	Sets configuration parameters.

Much of the NetWare administration is performed through client applications. NetWare provides several applications, as listed in Table 9-3.

Table 9-3 NetWare Utilities Used on Client Workstations

Client Utility	Function
DNS/DHCP Manager	Graphical application for managing DNS and DHCP.
Filer	Character-based, menu-driven utility for managing files.
Flag	Command-line utility for changing extended file attributes.
Ncopy	Command-line utility for copying files while retaining their extended file attributes.
Ndir	Command-line utility for listing contents of directories, including their extended attributes.
NetWare Administrator	Graphical application that enables management of the entire NDS tree and resources.
NDS Manager	Graphical application that allows the management of the replication of NDS partitions.
Rconsole	Character-based, menu-driven utility that enables terminal emulation of the server's console. Requires REMOTE.NLM and RSPX.NLM to be loaded.
Rights	Command-line utility for granting, revoking, and displaying the trustee assignments on a file.

NDS Schema

Every NDS object within the NDS tree represents a resource within the Novell network. These objects represent users and groups, servers and workstations, printers, applications, and more. For each object that you can create in NDS, options are available. These options are listed in the NDS schema.

Note Both NDS and Active Directory use a schema to define the types of objects that are contained within them.

You can think of the NDS schema as a cookbook containing several different recipes. The user recipe includes the password ingredient, as well as personal information about users. The NDS schema must be identical for every NetWare server for the NDS partitions to replicate. For example, if you have a

NetWare server that has a schema with a user object that has a business telephone property, it will not be able to replicate with a server that uses an NDS schema which does not have the business telephone property on user objects. The best way to ensure that replication always takes place is to upgrade all servers to the same version of NDS at the same time.

Three types of objects exist within the schema. There is a single [root] object in every tree, which is the main container for all other objects. The second type is a container object. The third type is a leaf object that can't contain any other objects.

Every object is created using the schema. When an object is created, it has the properties that the schema provides for it. These properties might or might not have values. For example, a user object will always have a first-name property, but the administrator might not give that property a value.

Apple Macintosh

Even though Windows, NetWare, and UNIX servers are the most common network operating systems, there are others. One of these is Apple Macintosh.

Networking among Apple computers began as peer-to-peer. Each client could share its storage and printers with every other Macintosh. This network used the AppleTalk protocol suite, which was integrated so deeply into the operating system that connecting to network resources was transparent to users.

When a Macintosh computer is dedicated to sharing files, printers, or applications on the network, it is called an AppleShare Server. Every client can use its Chooser application to display network resources available to it. These resources are located in the local zone, which is similar to a workgroup. Each Mac belongs to a zone. Given the Zone Information Protocol (ZIP) and Name Binding Protocol (NBP) in AppleTalk, resources appear as members of the zone regardless of their physical location within the internetwork. For example, if a printer is shared by a computer in the Accounting zone and the printer is located in San Francisco, a person using a computer in the Accounting zone in Dallas will see that printer available in its Chooser application and can print to it.

The AppleTalk protocol provides a network address for each computer and network segment. The node addresses are available from 1 to 254 on each segment. If more than 255 computers are on a network, a second network segment address can be applied to the segment. The group of addresses applied to a single network segment is called a *cable range.*

A seed router assigns the AppleTalk addresses to the nodes. It works through a system that is completely transparent to users; they are unaware of any addressing taking place.

UNIX/Linux

Originally developed by Bell Laboratories, UNIX was created for use on minicomputers. The operating system was not intended to be proprietary and as a result, many versions of UNIX are available on the market today—along with Linux, an operating system whose source code is freely available. UNIX and Linux have slightly different implementations, but the core features of all versions are basically the same.

UNIX provides a peer-to-peer networking system and uses TCP/IP as its native protocol. UNIX is available on mainframes, minicomputers, and PCs. The largest hosts provide user access through terminal emulation.

Older UNIX versions are command-line driven. While the command-line applications are still in use, UNIX also provides a graphical interface. Commands vary somewhat among the versions of UNIX, given that they can be based on different command interfaces (called shells)—Korn, Bourne, and C, among many others. Even though the commands tend to be somewhat cryptic, they're popular with longtime UNIX users.

Multiple graphical interfaces are also available in UNIX. The two that are most popular are X Windows and Motif. Both look and function similar to Windows.

To administer a UNIX machine, you must have a root account or a super user account. This account is named "root," and it is capable of overriding permissions. Each UNIX version uses different administration utilities but usually stores configuration files in the /etc directory.

For example, the /etc/passwd file holds user information. Editing this file allows you to add new accounts and modify and delete existing ones. The file is a simple text file in which each row is a different entry representing a distinct user account. Each row includes multiple fields that are separated by colons (:). Some of the fields in this file are listed below:

- **Login-id:** The user's login account name.
- **Uid:** This is the user account number, or a numerical identifier for the account. For example, the root account is always uid 0.
- **Gid:** The group account number that specifies the group to which the user belongs.
- **User information:** The user's real name.
- **Home directory:** The path to the user's login directory.
- **Shell:** The initial shell program that the user uses. The shell field is usually blank, allowing the default shell to be used.

The /etc/group file describes the groups that are available on the UNIX machine. Like the /etc/passwd file, /etc/group is a text file in which each row represents a different group. Each row consists of several fields that are separated by colons (:). Some of these fields are listed below:

- **Group-name:** The name given to the group.
- **Password:** The password assigned to the group.
- **Gid:** The group account number that is cross-referenced to the gid field in the /etc/passwd file.
- **Users:** This is a comma-delimited (separated by commas instead of colons) list of the users that have been assigned to this group.

The hosts file is common to anyone who uses TCP/IP. On a UNIX machine, this text file is found in /etc/hosts. The file is actually a table that maps host names to IP addresses. There are only three fields in this file, as listed here:

- **IP address:** The IP address of the machine.
- **Hostname:** The name of the machine.
- **Alias:** An additional name that is available for machines but left blank when no alias is used for a machine.

The UNIX Network File System

While not all UNIX versions have the same file system, several versions of UNIX use NFS, which stands for Network File System, a system that uses long filenames and provides file attributes. NFS was created to be a networked service. When NFS is used, it is transparent to the end user because it's based on Remote Procedure Calls (RPCs), a session-layer service that makes remote applications appear to be functioning locally.

When a user opens a file through NFS, the UNIX client places an RPC call to the UNIX server. The RPC call specifies the NFS filename and the user's Uid (user account number) and Gid (group account number). The UNIX server examines this information in the RPC call to determine what access rights the user has to the file. Then the UNIX server delivers the file to the UNIX client. For access to be granted, the Uid and Gid must match on both the UNIX server and client.

Key Points

- Network operating systems are installed on servers to provide shared files, printers, and other services on the network.
- Peer-to-peer networks enable every client to also function as a server.
- Windows NT servers use a domain flat file containing users and computers.
- In Windows NT, domains must have a trust relationship for users in one domain to access resources in the other domain.
- Trust relationships among Active Directory domains are created using Kerberos and are both transitive and reciprocal.
- Active Directory domains are grouped together in a logical organization called a forest, which is made up of one or more domain trees consisting of contiguous DNS namespaces.
- Each Active Directory domain includes organizational unit containers that build a hierarchy of objects.
- NTFS is the file system used by Windows NT and Windows 2000 servers. It offers long filenames and extended file attributes.
- Novell NetWare uses a hierarchical directory service called Novell Directory Services (NDS).
- NDS uses organizational units as container objects and resources as leaf objects.
- The NDS schema acts as a recipe for objects that can exist within the NDS directory.
- Macintosh computers that are dedicated as servers are called AppleShare Servers.
- Macintosh networks use the AppleTalk protocol natively and are often configured as peer-to-peer networks.
- UNIX and Linux computers use TCP/IP natively.
- NFS is a UNIX file system that provides transparent access to network files.

Chapter Review Questions

1 Which of the following is a network in which every client can also share resources?

a) Complete trust

b) Client/server

c) Master domain

d) Peer-to-peer

Answer d is correct. A peer-to-peer network is one in which each client can also act as a server and share its resources. Answer a, complete trust, is incorrect because it is a Windows NT domain model. Answer b, client/server, is incorrect because it's the type of network in which the server is dedicated to sharing resources and clients are only consumers of resources. Answer c, master domain, is incorrect because it is a Windows NT domain model.

2 When domain A trusts domain B and domain B trusts domain C, if domain A automatically trusts domain C, what type of trust relationship is in effect?

a) Unidirectional

b) Transitive

c) Bidirectional

d) Reciprocal

Answer b is correct. When a domain trust relationship seems to pass through to other trusted domains, it is called transitive. Answer a, unidirectional, is incorrect because it wouldn't affect the nature of a third domain. Answer c, bidirectional, is incorrect because it wouldn't affect the trust relationship of a third domain. Answer d, reciprocal, is incorrect because it implies that the trust relationship of A to B works in the opposite direction.

3 Which of the following is used to authenticate users when they log on to a Windows 2000 domain?

a) Active Directory

b) SAM

c) NDS

d) /etc/passwd

Answer a is correct. The Active Directory contains the information necessary to authenticate users who log on to a Windows 2000 domain. Answer b, SAM, is incorrect because the SAM is used for Windows NT domains. Answer c, NDS, is incorrect because NDS is used for NetWare 4.*x* and later versions. Answer d, /etc/passwd, is incorrect because it is used to authenticate users on UNIX systems.

4 Which of the following provides the template for objects when they're created in the NDS tree?

a) Master domain

b) TCP/IP

c) Schema

d) NetWare Administrator

Answer c is correct. The NDS schema provides the template information to create user objects in the NDS database. Answer a, master domain, is incorrect because it is a Windows NT domain model. Answer b, TCP/IP, is incorrect because it is a protocol suite. Answer d, NetWare Administrator, is incorrect because while it is the utility used on NetWare, it doesn't provide the template of information.

5 You've been called in to troubleshoot a network employing a NetWare server and Windows XP clients. The users are complaining that they're not allowed to save files to the server unless they use very short names. What is the problem?

a) The NDS tree needs to be synchronized.

b) The volume is too large for the Windows XP clients to understand.

c) The LONG namespace hasn't been installed on the server.

d) The users should be using the Filer utility.

Answer c is correct. The LONG namespace must be installed on the server for filenames longer than 8.3 to be supported. Answer a is incorrect because NDS tree synchronization would have no effect on filenames. Answer b is incorrect because the users were able to save files, just not longer names. Answer d is incorrect because the Filer utility is not needed to create long filenames.

6 You've been called in to troubleshoot a Windows NT network. The administrator has created a new domain named NEW, and users who are members of the OLD domain are not able to access resources in the NEW domain. What can solve this problem?

a) You can create a nontransitive unidirectional trust relationship in which OLD trusts NEW.

b) You can create a nontransitive unidirectional trust relationship in which NEW trusts OLD.

c) You can create a transitive, bidirectional trust relationship in which OLD trusts NEW.

d) You can create a transitive, bidirectional trust relationship in which NEW trusts OLD.

Answer b is correct. A nontransitive unidirectional trust relationship in which NEW trusts OLD will allow the users in the OLD domain to access the resources in the NEW domain. You'll also need to grant users rights to the resources. Answer a is incorrect because OLD trusting NEW will allow only users in the NEW domain to use OLD resources. Answers c and d are both incorrect because all trust relationships within Windows NT are nontransitive and unidirectional.

7 Which of the following is used to create trust relationships in the Active Directory?

a) Kerberos

b) LDAP

c) X.500

d) NFS

Answer a is correct. Kerberos is used to create trust relationships within the Active Directory. Answer b, LDAP, is incorrect because it is a protocol used to access directory services. Answer c, X.500, is incorrect because it is a standard specification for directory services. Answer d, NFS, is incorrect because it is the file system used in UNIX systems.

8 You've been called by a network administrator to handle a problem. The administrator has recently installed a new Active Directory service in the lab and is beginning to create a test directory service. The administrator has been trying to create user objects within a group object but hasn't been able to. What's the problem?

a) The DNS service isn't functioning.

b) The administrator was trying to create the user in a domain global group but should have used a domain local group.

c) The administrator should use TCP/IP.

d) The group object is a leaf object and can't contain user objects.

Answer d is correct. Group objects in the Active Directory are considered leaf objects and can't contain other objects. The administrator should create users within the organization or organizational unit containers and then add the users to the group. Answer a is incorrect because DNS will not affect the creation of objects. Answer b is incorrect because these are both leaf objects, and the type of group will not change whether or not the users can be contained within them. Answer c is incorrect because TCP/IP is required any way you look at it and wouldn't affect only the creation of objects but rather the functionality of the entire system.

9 Which of the following NOSs is a true client/server system?

a) NetWare

b) Windows NT

c) Macintosh

d) UNIX

Answer a is correct. NetWare is considered a true client/server system because the servers are dedicated to sharing services while clients are considered consumers of services. Answers b, c, and d are all incorrect because these are peer-to-peer systems.

10 Which file system is used by UNIX?

a) NTFS

b) Novell File Services

c) NFS

d) AppleShare

Answer c is correct. NFS is the file system used by UNIX. Answer a, NTFS, is incorrect because this is the file system used by Windows Servers. Answer b, Novell File Services, is incorrect because it is used by NetWare servers. Answer d, AppleShare, is incorrect because it is the name applied to a dedicated Apple Macintosh server.

Chapter 10

Establishing Network Connectivity

Networking enables two disparate types of machines to communicate and share information across a common link. A lot of background work goes into implementing that common link. The network infrastructure has to be put into place; at least one protocol must be installed and configured to function on every network server and workstation; and behind-the-scenes automation should be put in place to reduce administrative overhead, such as dynamic addressing and naming. When you think of network connectivity, remember that you must configure both the client and the server before the two will be able to communicate.

In this chapter, we'll review the configuration steps needed for Transmission Control Protocol/Internet Protocol (TCP/IP) and Internetwork Packet Exchange/Sequenced Packet Exchange (IPX/SPX). We'll discuss how Network Beginning Input Output System (NetBIOS) names and TCP/IP host names apply to networking. Then we'll go over the protocols that help automate and simplify both addressing and naming in a network: Dynamic Host Configuration Protocol (DHCP), Domain Name System (DNS), and Windows Internet Name System (WINS).

Basic Connectivity

Protocols are the language of the network. For two computers to communicate, they must both speak the same language. To illustrate, imagine two identical computers sitting side by side that are connected to the same hub, but one of the two uses TCP/IP and the other uses IPX/SPX. Neither of these machines knows that the other even exists because they operate according to two different protocols—they speak different languages. Therefore, before you ever begin configuring a computer, you need to know something about the network it is on and what other computers it needs to communicate with.

Connecting with TCP/IP

It's rare to find a network that is not connected to the Internet. Most organizations use the Internet to exchange e-mail messages, and many use it for research purposes. Businesses can create a Web site as a marketing presence, a revenue generator, or even an extension of business applications. The Internet has become pervasive, and as a result, so has its native protocol suite, TCP/IP.

The protocols within the TCP/IP stack function as both a mode of communications as well as a set of utilities that help a user interface with the network. When TCP/IP is properly configured, it enables every host on the network to communicate with every other host. In this context, a host is not just a workstation or a server, it also can be a printer, router, fax server, or just about anything with an IP address assigned to it that is directly connected to the network.

On Windows NT-based operating systems, including Windows NT 4.0, Windows 2000, and Windows XP, you can use the ipconfig command to view your TCP/IP configuration.

1. Click the Start menu and then Run.
2. Type **cmd** in the Run box. Depending on the version of Windows you are using, you might need to type **command** instead of **cmd**. This will open up a command prompt window.
3. At the prompt, type **ipconfig/all**, and you will see results similar to those shown in Figure 10-1.

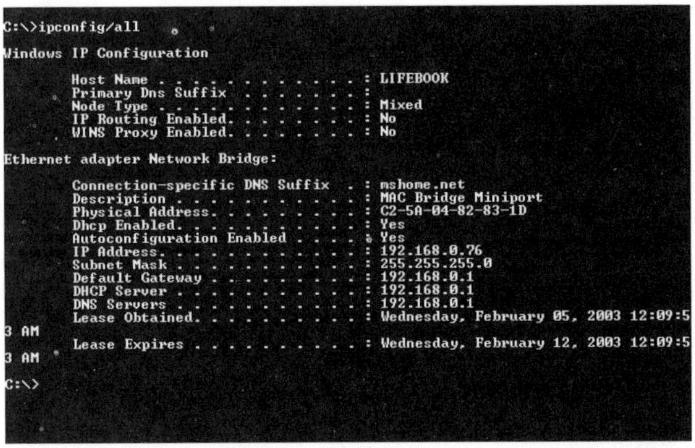

```
C:\>ipconfig/all

Windows IP Configuration

        Host Name . . . . . . . . . . . . : LIFEBOOK
        Primary Dns Suffix  . . . . . . . :
        Node Type . . . . . . . . . . . . : Mixed
        IP Routing Enabled. . . . . . . . : No
        WINS Proxy Enabled. . . . . . . . : No

Ethernet adapter Network Bridge:

        Connection-specific DNS Suffix  . : mshome.net
        Description . . . . . . . . . . . : MAC Bridge Miniport
        Physical Address. . . . . . . . . : C2-5A-04-82-83-1D
        Dhcp Enabled. . . . . . . . . . . : Yes
        Autoconfiguration Enabled . . . . : Yes
        IP Address. . . . . . . . . . . . : 192.168.0.76
        Subnet Mask . . . . . . . . . . . : 255.255.255.0
        Default Gateway . . . . . . . . . : 192.168.0.1
        DHCP Server . . . . . . . . . . . : 192.168.0.1
        DNS Servers . . . . . . . . . . . : 192.168.0.1
        Lease Obtained. . . . . . . . . . : Wednesday, February 05, 2003 12:09:5
3 AM
        Lease Expires . . . . . . . . . . : Wednesday, February 12, 2003 12:09:5
3 AM

C:\>
```

Figure 10-1 Ipconfig provides a view to the computer's TCP/IP configuration.

4. From this information, you'll be able to identify the following:
   - The IP hostname; in this case, it's lifebook.
   - The MAC address of the Ethernet card; in this case C2-5A-04-82-83-1D.
   - The IP address of the computer; in this case 192.168.0.76.
   - The subnet mask of the computer; in this case, it's 255.255.255.0.
   - The IP address of the DNS server; in this case, it's 192.168.0.1.
   - The IP address of the DHCP server; in this case 192.168.0.1.
   - The IP address of the default gateway; in this case, it's 192.168.0.1.
   - The DNS suffix; in this case, it's mshome.net.
   - The WINS address, if any; in this case, WINS is not used.
   - The DHCP lease expiration; in this case, it expires on February 12, 2003.
5. Type **cls** at the prompt, press Enter to clear the screen, and then type **ping127.0.0.1**. Ping will send packets to the IP address you specify to determine whether that other host is reachable. You'll be pinging your own machine, even though this isn't the same IP address you saw in the ipconfig command screen. (This is because 127.0.0.1 is the loop-back address and represents the local host, no matter what host you happen to be using.)

Testing the IP Network

When you migrate to a TCP/IP network from another protocol stack, you're bound to encounter a few problems. During the migration, two different types of protocols will be running on the network simultaneously. These two protocol stacks will probably consume more network bandwidth than if the network were dedicated to a single protocol. Performance issues might even result in a torrent of user complaints and frustrations.

When you start testing the network to determine where the performance issues might stem from, you first need to log on to the network during peak usage. Typically, this is when the most users are doing the most online work at the same time—usually between 10:00 A.M. and 2:00 P.M. if an organization holds typical business hours. While you're online, run the applications that users normally run and experience the network's performance personally. If you confirm that network performance is unacceptably slow, you need to review the properties of the workstation's network connections.

You review the workstation's network properties first because the most recent change on the network is the migration from one protocol stack to TCP/IP. This is a troubleshooting rule: the last change is the most suspected reason for current problems.

If you are using Windows 2000, to check the Network properties open the Control Panel and select the Network and Dial-Up Connections icon. From there, select Local Area Connections and then Properties. This will reveal the network services that the computer is configured to use, as shown in Figure 10-2.

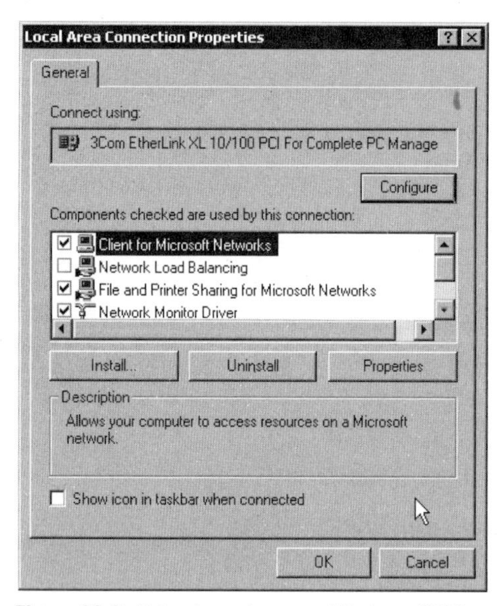

Figure 10-2 Network services on a Windows 2000 machine can be viewed in the Network and Dial-Up Connections applet.

As you can see in the figure, the computer is configured with more than one network service, and those in turn, are using more than one protocol (which are not shown). Multiple protocols cause some overhead. There might be duplicate data transmissions and multiple routing information packet transmissions, both caused by running more than one protocol stack. There are two ways to handle this problem, both of which depend on the goals for the network.

- If the network goal is to run TCP/IP as a single protocol, and all other systems that the workstation must communicate directly with are already running TCP/IP, the solution is to remove all other protocols from the workstation *except* for TCP/IP.
- If the network will need another protocol stack in addition to TCP/IP, or if the other hosts on the network are not yet migrated and this workstation must communicate directly with them, the solution is to change the binding order of the protocols so that the most common one is used first. This action will result in the first protocol being used over any others, and if you make TCP/IP the first, the workstation will be optimized. You can do this in the Advanced Settings on the Advanced menu by selecting the adapter and then using the arrows to move the protocols in the correct order.

Situations in which multiple protocols are used on the network are commonplace. The fewer protocols in use, the better the network's performance, not to mention the higher its level of security. Unfortunately, because so little effort is needed to implement NetBEUI or IPX/SPX, all too often these protocols will be installed on a network with the thought that they'll provide backup in case of a TCP/IP failure. But this approach signals a lack of understanding. Once configured correctly, TCP/IP will function without a problem as long as the underlying network is functioning.

When you configure TCP/IP across the network, consider the following issues:

- Use of NetBIOS names and Windows Internet Name System (WINS)
- Use of IP host names and Domain Name System (DNS)
- Implementation of dynamic IP addressing through DHCP

Installing TCP/IP on a Windows 2000 Workstation

To install TCP/IP on a Windows 2000 workstation, follow these instructions:

1. Open Control Panel and double-click the Network And Dial-Up Connections applet.

2. Click Local Area Connection. You'll see the Local Area Connection Status property sheet.
3. Click the General tab if it isn't selected, and then click Properties.
4. Click Install and then fill in the appropriate information for the following, unless you'll be configuring the computer to use DHCP:
   - IP address
   - Subnet mask
   - Default gateway IP address (only if applicable)
   - DNS server IP address (only if applicable)
   - WINS server IP address (only if applicable)

Reviewing TCP/IP Configuration Settings

It's a good idea to familiarize yourself with TCP/IP configuration settings. Let's look at a Windows XP computer. To start, right-click My Network Places and select Properties. Under Windows XP, each network and dialup interface is shown as a separate icon. Select one and right-click on it. You'll see a screen similar to that shown in Figure 10-3.

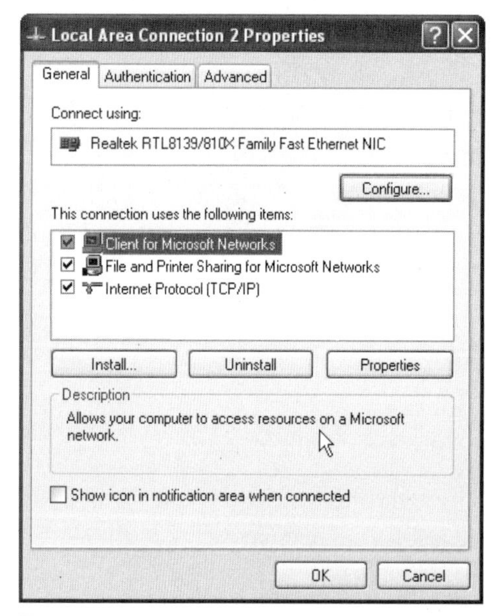

Figure 10-3 Windows XP displays its network connection properties on an interface-by-interface basis.

The General tab displays all the protocols and services that are bound to this interface. Click on Internet Protocol to select it, and then click the Properties button to display the TCP/IP properties that are specific to this interface. You'll see a screen similar to that shown in Figure 10-4.

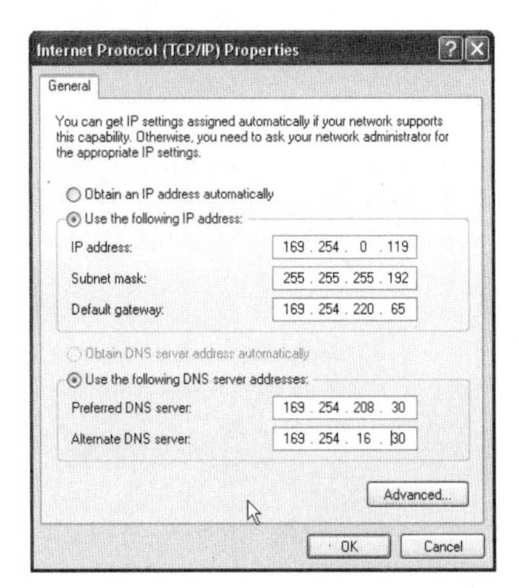

Figure 10-4 The properties for TCP/IP are different for each interface within a computer.

The status of the DHCP client is indicated when the option at the top, Obtain an IP Address Automatically, is selected. The IP address and subnet mask should be supplied by a DHCP server. A DHCP server can also supply the default gateway DNS information, in which case you would select the option to Obtain DNS Server Address Automatically, and just about any other TCP/IP-related information. Click the Advanced button to see additional information. (See Figure 10-5.)

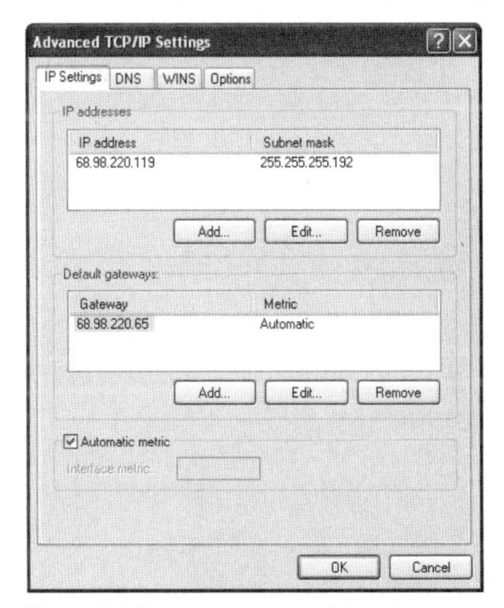

Figure 10-5 The Advanced TCP/IP information dialog box provides additional options for configuring IP on a Windows XP computer, as well as Windows NT and Windows 2000.

On the IP Settings tab, you can add another IP address to the network adapter. This is helpful for unique situations, such as a Web server that hosts multiple Web sites with different IP addresses. Click the DNS tab to see the advanced DNS properties dialog, shown in Figure 10-6.

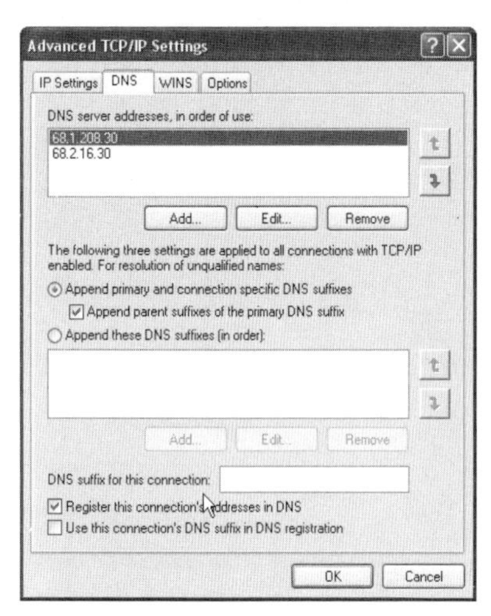

Figure 10-6 DNS properties are expanded in the Advanced DNS Properties dialog box.

In the Advanced DNS Properties dialog box, you can add multiple DNS server addresses and place them in order of usage. The workstation will check each DNS in succession. You need only a single DNS server to provide name services, but it's nice to have a backup DNS server just in case of a failure. For example, if your Internet Service Provider (ISP) supplies you with a DNS server and the Internet link goes down, a backup DNS server can provide internal services. In addition, if your ISP's DNS server goes down but the Internet connection stays up, a backup DNS server on the Internet will ensure that users don't experience an outage of services.

At the bottom of the Advanced DNS Properties dialog box, you have the option of entering domain names that can be used to complete the unqualified name of a computer. An *unqualified name* is simply the name of the host, such as Fred or Wilma. A *fully qualified domain name (FQDN)* is the name of the host plus the name of the domain, such as Fred.contoso.com or Wilma.flintstone.contoso.com. You add the domain suffixes—contoso.com and flintstone.contoso.com, in that order—in the Append these DNS Suffixes box. Then when the user uses a command, such as ftp fred, this will automatically add contoso.com to the name fred and will try to ftp fred.contoso.com. If that doesn't work, it will add flintstone.contoso.com to fred and try to ftp fred.flintstone.contoso.com. In a network supporting heavy command-line users, this is a very helpful device. In a world in which the graphical interface covers up the background networking, it's less necessary.

Many Windows networks use Windows Internet Name Service (WINS) to map NetBIOS names to IP addresses. The NetBIOS name is used to identify the computer in "human-friendly" terms and is built into Windows NT. With the advent of Windows 2000's native usage of TCP/IP, WINS is no longer necessary in a network that doesn't use any Windows operating system older than Windows 2000. If you click on the WINS tab, you'll see the settings for WINS and NetBIOS, as shown in Figure 10-7. Here you can add the IP addresses of the WINS servers on the network and import an LMHosts file, which lists NetBIOS names that are mapped to IP addresses and used by local hosts to look up names.

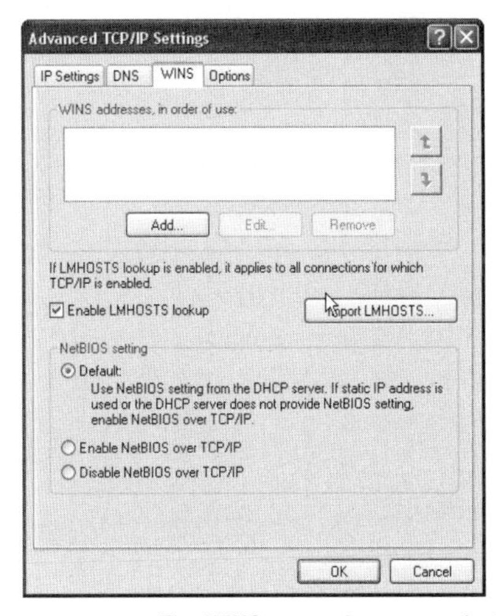

Figure 10-7 The WINS properties appear in the Advanced TCP/IP Settings.

LMHosts can be used in place of WINS or as a supplement to WINS. I've actually used LMHosts files in the past as a troubleshooting device. If I can access a computer with the LMHosts file but not when I remark that listing out of the file, I know I have a WINS problem to resolve. The NetBIOS settings in this dialog are either to enable NetBIOS over TCP/IP, disable NetBIOS over TCP/IP, or let the DHCP server provide the correct settings. If you're using DHCP, the default option is the best because you can manage all WINS settings from the DHCP server. This means that if you're migrating away from WINS, you can easily do so at one machine. If you're not using WINS, you should opt to disable NetBIOS over TCP/IP.

The Options tab provides you with the ability to perform TCP/IP filtering, as shown in Figure 10-8.

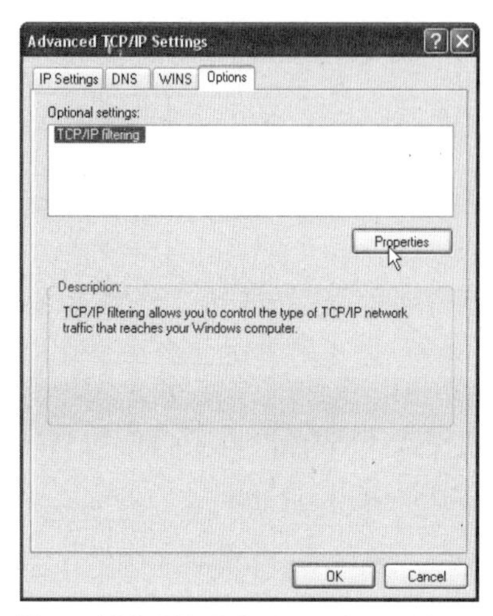

Figure 10-8 TCP/IP Options refers to "filtering" TCP/IP packets.

There is only one button to click, the Properties button. This will lead you to the dialog box shown in Figure 10-9.

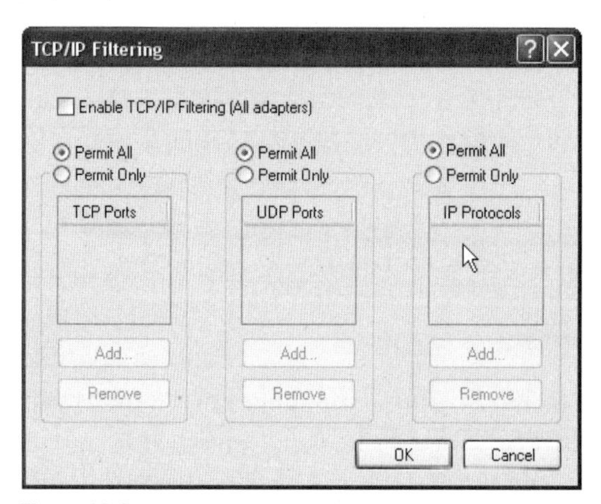

Figure 10-9 Windows XP allows you to filter incoming TCP/IP packets.

In the Windows XP dialog box, you can filter incoming packets based on the TCP port, UDP port, or protocol type. If you decide to filter packets, you need to add the ports that you want to allow through this interface. If you forget a port or protocol, you'll accidentally filter that port or protocol out. You may notice that filtering incoming packets and ports is the same function that is performed by a firewall.

Naming Computers on a TCP/IP Network

Computers depend on addresses that are unique on a network. In IP, these addresses are numerical in nature. Humans don't remember numbers easily. Just imagine trying to remember 188.96.52.183 and differentiate it from 188.96.52.185, 188.96.52.193, 188.96.52.138, and more. Given hundreds of these IP addresses to remember, things can become quite confusing. That's why names are assigned to computers. It's much easier to remember a name like "Fred" than it is a string of numbers. Unfortunately, the computer doesn't understand what "Fred" means—you have to provide the IP address somehow. The computer uses a method called name resolution. You can map a NetBIOS name to an IP address on a Windows network that uses TCP/IP. On any TCP/IP network, you map the host name to the IP address.

NetBIOS Names on Windows Computers

Since Windows for Workgroups first was introduced, Windows networking has been dependent on NetBIOS names and the NetBEUI protocol. It has been built into the operating system of every version of Windows up to Windows 2000. (Windows 2000 and later computers use TCP/IP as the native protocol, removing the need for NetBIOS names. Windows 2000 and later can still use NetBEUI, and still use NetBIOS names. The NetBIOS name, however, is no longer required.)

Note A NetBIOS name on a Windows computer is 16 bytes long and can be up to 15 characters in length. The sixteenth byte is reserved for the NetBIOS suffix, which is used to identify the functionality of the device.

But even though legacy Windows computers use NetBIOS, they also can function on a TCP/IP network and still keep the NetBIOS name. There is a string of logic that leads a Windows machine from its human-friendly NetBIOS name to an IP address. All applications interact with the network through the application layer. The NetBIOS application programming interface (API) handles networking via the NetBEUI protocol on a Windows network. NetBEUI, in turn, requires NetBIOS host names to communicate across the network. On a Windows network that uses NetBIOS names over TCP/IP, the NetBIOS name must be resolved to the IP address.

When a source Windows computer must send data to another host, it must first know the destination host's name. The source computer broadcasts a NetBIOS Namequeryrequest on the segment. If there is a destination host with that particular name, that host will reply with its own Media Access Control (MAC) address. This is all that a Windows computer needs to know to communicate.

However, on a TCP/IP network, a Windows computer that depends on a NetBIOS name will not be able to communicate. A TCP/IP network uses only IP addresses for communication. The only way to get the two to function together is to implement NetBIOS over TCP/IP and then resolve the NetBIOS name to an IP address.

With NetBIOS over TCP/IP, an application will interact with the NetBIOS API, which will then invoke NetBIOS over TCP/IP. This in turn will query the network for the destination computer's NetBIOS name resolved to an IP address. The final step is to resolve the IP address to the MAC address, which is performed using the Address Resolution Protocol (ARP). When the MAC address is returned to the source machine, it can begin communicating with the destination computer.

One of the ways to resolve a NetBIOS name to an IP address is through the LMHosts file. You can find this file in the c:\windows\system32\drivers\etc directory in Windows NT, Windows 2000, and Windows XP. In Windows 95 and Windows 98, it is found in the c:\windows directory. Using LMHosts can be quite time consuming. Imagine that you're administering a small network of 21 computers, and you decide to change the names of two of the computers. You would need to change the LMHosts file on each of the 21 computers. In a network with hundreds of computers, well, adding even a single computer to the network would be a burden.

WINS was developed to overcome the problems with LMHosts files. A WINS server maintains a database that resolves NetBIOS names to IP addresses of computers on the network. When a Windows computer needs to find a destination computer, it can query the WINS server.

Host Name

Unlike NetBIOS names on a NetBEUI network, host names on a TCP/IP network aren't necessary for establishing communications. A user can type in the IP address of a destination computer and access that computer without a problem. The host name in a TCP/IP network, therefore, is provided solely for the user's benefit.

A fully qualified domain name (FQDN) is in the format of a series of host and domain names separated by periods. The first name is the name of the host; the second is the name of the domain or subdomain that the host is a member of; the third is the name of the domain (if the host is within a subdomain) that houses the subdomain; and the last is the top-level well-known domain. All are listed in Table 10-1. The table shows a few examples of country domain names but doesn't define all of them because there are so many. The final FQDN is computername.subdomain.domain.com. Note that there can be subdomains of the subdomains, too, but a computer that has more than a couple of subdomains is rare.

Table 10-1 Top-Level Domains

Domain Name	Intended For
.com	Commercial networks
.mil	U.S. military
.gov	U.S. government
.edu	Educational facilities
.us	The two-letter name for United States networks
.net	Network provider
.org	Nonprofit organizations
.biz	Businesses
.museum	Museums
.de	The two-letter name for German networks
.au	The two-letter name for Australian networks
.uk	The two-letter name for networks in the United Kingdom

In a TCP/IP network, there are two ways to resolve a host name. One is to use a hosts file; the other is to use DNS. The hosts file was the precursor to DNS. It provided a simple name to IP address resolution on a local computer. You can always find the hosts file in a directory named /etc, which holds true for NetWare servers, Windows NT, Windows 2000, and Windows XP computers, and UNIX hosts. This is because UNIX computers originally placed the hosts file in the /etc directory, and as TCP/IP was added to other operating systems, the same directory name was used to house the hosts file.

The format of a hosts file is simple. Each line in the file represents a single name to IP address mapping. Each line begins with the IP address followed by the host name. Any comments in the file begin with a pound sign (#) and are not read by the computer. For example, the following might be lines in a hosts file:

```
192.68.33.52     myserver.domain.com
202.5.189.7      joeserver.jzzsyss.net
179.8.228.25     fred.contosocity.org #Mail server
```

Managing hosts files causes a tremendous amount of overhead, making centralizing and automating name to address mappings desirable. Imagine using

DHCP—a system in which a pool of IP addresses can be shared by multiple computers—and having to change hosts files on hundreds of machines every time an IP address was passed on to a different computer. It would be impossible to keep up. That's where DNS comes into play.

The other way to resolve an IP address to a host name is to use DNS. The DNS client contacts a DNS server with a query for an IP address matched to the host name. The DNS server manages and maintains IP address to host name mappings for a network. It works with other DNS servers to resolve names and addresses for an entire global network. Using a hierarchical structure, the DNS system can work efficiently on any size network. We'll go over DNS in more detail later in this chapter.

Configuring the Server Side

To create a connection on the network, services must be installed and configured on the server as well as on a client. For example, if you configured all of your client workstations to receive IP addresses via DHCP and to use DNS and WINS for name resolution, but you neglected to install and configure DHCP, DNS, and WINS servers, none of the workstations would be able to connect to the network.

DHCP

DHCP is the service that delivers IP addresses to clients, but DHCP servers can provide far more than just IP addresses to clients; they can provide WINS and DNS server addresses, default gateway addresses, and subnet masks. I recommend that when you configure a DHCP server you include this extended information. In doing so, you'll save yourself a lot of work whenever you need to change a DNS server or remove WINS from the network altogether.

DHCP is a known time-saver. Before DHCP, each IP address was assigned to each workstation and server individually, along with all of the extended information. If a workstation was moved to a different segment of the network, its IP address needed to be changed. A single change is not such a big problem, but hundreds or thousands of installs, moves, changes, and removals on an enterprise network creates an IP address–management headache. When you manually administer IP addresses, you run the risk of conflicts among them.

DHCP allows you to manage only the IP addresses for a small set of servers that require static IP addresses. These include the DHCP server, DNS and WINS servers, and Web servers.

DHCP grew out of the Bootstrap Protocol (BOOTP), a protocol within the TCP/IP stack that enabled diskless workstations to boot up to the network.

While DHCP doesn't perform the same functions as BOOTP, it does enable a DHCP client to request and receive the following common items from a DHCP server (and often receives more information along with this):

- The IP address to be assigned to the client
- The subnet mask for that IP address
- The IP address of the default gateway
- The IP address of the DNS server
- The IP address of the WINS server
- NetBIOS options

How DHCP Works

DHCP clients initiate the process by sending a request for an IP address. The DHCP server receives the request and responds with an IP address. Depending on how the server is configured, the IP address can be reserved for that specific machine or it can be a member of a pool of available IP addresses. The term given to a group of IP addresses in the DHCP database is *scope*. Each scope is given its own settings. You use different scopes when you have different settings to pass down to workstations. For example, a DHCP server can provide its services to several different subnets on the network. Each different subnet can be given its own scope of IP addresses. Keep in mind that you can have only one scope per IP subnet.

DHCP clients *lease* their IP addresses from the DHCP server via a four-step procedure:

1. The client initiates the process with DHCPDISCOVER. The client broadcasts the network looking for a DHCP server. Because the client doesn't know the DHCP server's address, it simply sends out its own MAC address with a 0.0.0.0 address, representing itself, and a destination address of 255.255.255.255. Note that the DHCP server must be on the same physical segment or the routers sitting between the client and the DHCP server must be able to forward this broadcast to the segment on which the DHCP server does reside. This is performed through a DHCP relay agent.
2. The server responds in the DHCPOFFER process. The DHCP server receives the client's request and broadcasts the client's IP address, subnet mask, and IP address of the DHCP server. The DHCP server also communicates the lease length, which is the period of time that the DHCP client rents the IP address. The DHCP server also places the IP

address in reserve so that it won't be offered to another DHCP client and cause a conflict.

3 The client accepts the IP information in the DHCPREQUEST process. It is possible that more than one DHCP server will offer an IP address to the client because the original client DHCPDISCOVER broadcast is transmitted to all DHCP servers. When the client does not accept the IP information, the DHCP server releases the IP address back into the pool of available addresses.

4 The next step is the DHCPACK process. The DHCP server sends a response to the client's acceptance of the IP information and includes the remaining IP configuration, including the DNS and WINS server IP addresses.

   If the client is trying to lease an IP address that is not available, the server will respond with a DHCPNACK, which is a negative acknowledgment. When the client receives this, it will begin a new DHCP lease process.

An easy way to remember the DHCP process is as Dora N.—D-discover; O-offer; R-request; A-acknowledgment; and if it doesn't work, N-negative acknowledgment.

Remember that IP addresses are leased, so the assignment is not permanent. At the half-point of the lease, the DHCP client sends a DHCPREQUEST packet to the DHCP server to let it know that it wants to keep the IP address. The DHCP server will usually respond with a DHCPACK and update the lease time. In the event that the DHCP server does not respond at the half-point, the client will try to send a DHCPREQUEST again at the three-quarter (¾) point, and then at the seven-eighths (⅞) point, the client will send a DHCPDISCOVER packet to any available DHCP server on the network.

If you want to force a release or renewal of an IP address on a Windows NT, Windows 2000, or Windows XP computer, you can use the ipconfig command. (On Windows 95 and 98, you can use the winipcfg graphical utility.) Open a command prompt, type **ipconfig /renew**, and the renewal process with the DHCP server will be forced. If you type **ipconfig /release**, the IP address will be released.

Installing DHCP Services on Windows 2000 Server

Installing DHCP is similar to installing any other Windows component. You start in Control Panel and use the Add/Remove Programs applet. Once you've started that, you follow these instructions:

1. Click Add/Remove Windows Components, and then click Components.
2. The Windows Component Wizard appears. Click Next in the first dialog.
3. Select Networking Services from the list, and then click Details.
4. Select the Subcomponents of Networking Options list, and then click Dynamic Host Configuration Protocol (DHCP). Click OK.
5. Click Next. You'll be prompted for the location of the Windows 2000 setup files. Type in the path to the Windows 2000 CD-ROM, or another location that contains these files. Click OK.
6. Windows 2000 will install the files. Click Finish.

Installing DHCP Services doesn't make a DHCP server. You must also create a scope and populate it with IP addresses.

Note A DHCP scope contains IP addresses. An IP address can belong only to one scope, and a scope can only contain addresses for a single IP subnet.

To create a scope, you need to have a strategy for the IP addresses to start with. You must know what you will call the scope. This will be the value you place in the scope's Name. Then you should know the starting IP address for the scope, which is called the IP Address Range From Address parameter, as well as the ending IP address for the scope, which is called the IP Address Range To Address. You must know what the subnet mask is for this set of IP addresses, because you'll be using that in the Mask parameter. If you want to exclude any addresses from the address range, put the first excluded IP address in the Exclusion Range Start Address and the last one in the Exclusion Range End Address. (Excluded IP addresses are ignored by the DHCP server. You would include IP addresses in this range only if you have manually assigned static IP addresses that are a part of the scope's IP address range.) You should also know how long you'll be leasing the IP addresses. You can use the Lease Duration Unlimited parameter to provide a permanent lease on an IP address, or you can provide a time limit in the Lease Duration Limited To parameter.

It's a good idea to place a limit on the lease, because that will make more IP addresses available on the network. The actual time limit for a lease depends on your network environment. If you have a busy network with a lot of moves and changes, you should use a short duration lease. I've found that a duration of about a half-week works well in most networks.

Note The default duration for a DHCP lease in Windows 2000 is eight days.

To create the scope, follow these steps:

1. Open the DHCP manager window by clicking the Start menu and pointing to Program Files and then Administrative Tools. Then click DHCP.
2. Once the application starts, right-click your DHCP server and click New and then Scope.
3. The Create Scope Wizard starts. Click Next.
4. Provide the name of the scope, and click Next again.
5. The next screen is important because this is where you specify the IP addresses that are held within the scope. Specify the IP address range and the subnet mask, and then click Next.
6. In the following dialog box, you have the opportunity to specify excluded IP addresses. You don't have to have an excluded range, but if you do have it type the address range in. Click Next.
7. You are taken to the final screen where you specify the lease duration. Fill in the information. Click Next, and then click Finish.
8. You'll see the scope, but it won't be able to distribute IP addresses until you activate it. Right-click the scope in the DHCP window, click Task on the popup menu, and then click Activate. (If you are running DHCP on a network using Active Directory, you must also authorize the DHCP server in the Active Directory.)

DNS

As we discussed earlier, the Domain Name System (DNS) resolves IP addresses to host names. The DNS system is universally used throughout the Internet. You can think of DNS as the yellow pages of the phone book. It contains names matched to numbers. The phone book is organized in such a way that you can look up a subject such as automobiles, and then a specific subject such as automobile sales, and then the name and phone number of a company that sells automobiles. The phone book is distributed so that everyone in Phoenix, Arizona, has a copy of the Phoenix, Arizona, phone book; while the people in Detroit, Michigan, have a copy of the Detroit, Michigan, phone book. A person in Phoenix who wants to look up a number in Detroit has to call up to a higher level to get the number for information in Detroit, and then call Detroit to get the information for automobile sales. DNS is similar.

First, like the yellow pages, DNS is distributed. DNS servers in the microsoft.com domain have microsoft.com IP address mapping information, while DNS servers in the comptia.org domain contain comptia.org IP address mapping information. If you were sitting at a microsoft.com computer and needed to access data on a comptia.org computer, the information wouldn't be available locally. However, DNS calls up the DNS hierarchy to a higher level (the DNS root servers) to get the location of the comptia.org DNS server, and then refers your request to that comptia.org DNS server. If the comptia.org DNS server has the information, it will transmit the correct IP address and your request will go through. This transaction is entirely transparent to you, the user.

You can subdivide a domain to make it easier to maintain DNS information. Let's assume you're at microsoft.com, and you have several domains to manage with thousands of computers located around the world. You would probably want administrators to handle problems locally, because that tends to have economic advantages. You could create a DNS server to manage only the pacrim.microsoft.com domain, another to manage the euro.microsoft.com domain, and so on (somewhat similar to separate phone books for a church or homeowners association, except that an IP address in one does not appear in any of the others).

How DNS Domains Are Organized

DNS domains are organized hierarchically. The arrangement is similar to a directory structure on your hard drive, given that each domain is a container that can contain either hosts or other containers (subdomains).

The very top of the DNS hierarchy is the root, which is usually represented by a period (.). Below the root lie top-level domains, such as .com, .net, .mil, .au, and so on. The second-level domains are the ones you are probably most familiar with. These are the microsoft.com, comptia.org, and contoso.net names that represent individual organizations.

Below the second-level domains are any subdomains created by the organization that registers that domain name. There is no hard and fast rule regarding whether an organization should create subdomains. The DNS hierarchy is displayed in Figure 10-10.

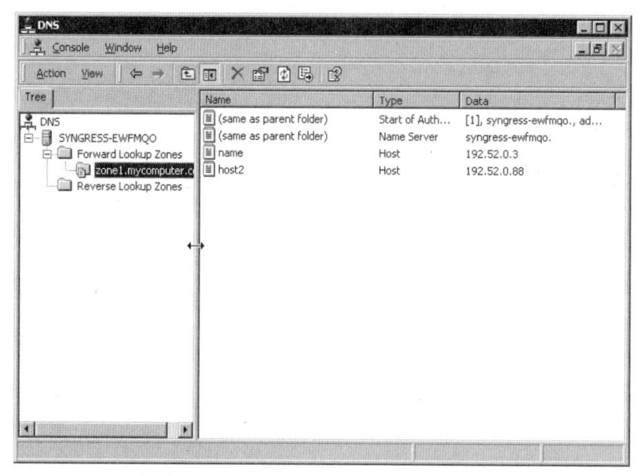

Figure 10-10 The DNS hierarchy has an inverted tree structure.

How DNS Resolves Names

Let's walk through what happens when you resolve a name using DNS.

1. The first step is to attempt to connect to a server. This happens whenever you type a URL (Uniform Resource Locator) into a Web browser address window, such as *www.microsoft.com*. This will attempt to connect to a server providing Web service at microsoft.com.
2. When you type the name into the Web browser and press Enter, a recursive query is transmitted to the primary DNS server on your computer.
3. The primary DNS server checks its database for an address record for *www.microsoft.com*. It looks in its cache to see whether it has already successfully resolved that same address. If the DNS server is not authoritative for the domain (meaning that it doesn't contain records for microsoft.com), it sends an iterative query to the root DNS server.
4. The root name server will not have the address record for *www.microsoft.com*. However, it will have the IP address for the top-level domains, such as .com, .net, .org, and so on. So, the .com DNS server's IP address is sent to your DNS server.
5. Now, your DNS server will query the .com DNS server for *www.microsoft.com*. The .com DNS server won't have the record, but it will have the record for a DNS server at microsoft.com, and it will send your DNS server that IP address in a best-effort attempt.
6. Now, your DNS server connects to the DNS server at microsoft.com and queries it. The microsoft.com DNS server checks its database for the *www.microsoft.com* record and successfully finds it. The

microsoft.com DNS server then responds to your DNS server with the IP address associated with *www.microsoft.com*.

7. Your preferred DNS server completes the recursive query by sending your Web browser the IP address of *www.microsoft.com*, and your Web browser connects and displays the Microsoft home page. If the DNS server at microsoft.com did not have an address record, the DNS server would have returned an error message.

DNS Zones

A DNS server manages domains through the use of DNS zones. Each DNS zone is a database file containing the resource records for one or more domains. There are two types of zones:

- **Forward lookup.** This is the type of zone that is used when the query knows the host name but not the IP address.
- **Reverse lookup.** This is the type of zone used when the query knows a host's IP address but not the host name.

As you can probably guess, forward lookup zones are the most common. In the case of a company with three domains named lab.mycompany.com, corp.mycompany.com, and mycompany.com, you'd probably find that queries are best handled using a separate zone for the domain that is geographically different from the others. For example, most people will be using the corp.mycompany.com and mycompany.com domains, but mainly the people who work in the lab would need to use IP address lookups in lab.mycompany.com. You can separate the zones and make it easier to manage the IP addresses by creating a separate zone for lab.mycompany.com.

Note DNS zones are not the same as domains. The zone is a file that contains records for hosts that reside on one or more domains. A zone file can contain several domains.

The zone file is populated with resource records, of which there are several types. The following are some samples of resource records:

- **Start of Authority (SOA):** Identifies the DNS server that is authoritative for the data in the domain.
- **Name server (NS):** A record for DNS servers in the domain.
- **Address (A):** The standard name to IP address mapping record for a host.

- **Service (SRV):** Name to IP address mappings for particular services in the domain.
- **Canonical name (CNAME):** Used to create alias names for hosts that already have an A record. For a server named myserver.mycompany.com that is also a Web server, you can create a CNAME record for www.mycompany.com to point to that server.
- **Mail Exchanger (MX):** Identifies the mail servers.
- **Pointer (PTR):** Used for reverse lookups.

Just as you wouldn't present a single physical phone book for everyone in a large organization to use, you wouldn't use a single DNS server to provide lookups. You'd need to plan for some degree of fault tolerance, so that if a DNS server fails, it doesn't take down the entire network.

Zone files in the DNS database can have copies. You can create a secondary zone file on a DNS server. The secondary zone file is read-only, meaning that it will resolve names but it won't allow you to make changes to the zone file. The only zone file that you can make changes to is the primary one, which is the first one that you create for a zone.

A zone transfer is the process that copies the primary zone file to a DNS server that contains a secondary zone file. When changes are made to the primary zone file, the entire file is copied to servers with secondary zone files. The zone transfer usually takes place on a periodic basis. The larger the zone file, the longer the transfer. Newer DNS servers support incremental zone transfers (IXFR), which transfers only the changed information between primary and secondary zone files, so this problem becomes less of an issue.

Installing DNS on a Windows 2000 Server

DNS can be installed while you're installing a Windows 2000 domain controller, or you can add it later on any Windows 2000 server. If you add it later, this is the process to use:

1. Go to Control Panel and open the Add/Remove Programs applet. Select Windows Components. The Windows Components Wizard will start. Click Next.
2. Select Components, scroll down to Networking Services and select it, and then click Details.
3. Click Domain Name System (DNS), and then click OK.

4. You'll be prompted for the path to the Windows 2000 installation files. Type it in, and then click Continue.
5. The server installs the service.

WINS

Windows Internet Name Service (WINS) is used for only one task: to resolve IP addresses from NetBIOS names. Because Windows 2000 and later versions no longer require WINS, you might never need this service, even on a Windows network.

WINS works in a similar fashion to DNS in that it maintains a database of NetBIOS names and maps them to IP addresses. The difference between the two is that WINS is self-contained. It doesn't refer a name query to any other authority. In fact, the WINS database is similar to using a single phonebook for everyone in a large organization.

When a WINS client gets online, it registers its own name and IP address with a WINS server. Whenever the client needs to connect to another client, it will query its WINS server to find the IP address, and then continue with establishing a connection.

Every WINS client must be configured with the IP address of at least one WINS server. This is the server that the client registers with as well as queries.

Multiple WINS servers can coexist on the network as replication partners that either push (transmit) updates to other WINS servers, or pull (request) updates from other WINS servers. You'd want at least one additional WINS server on the network for fault tolerance, and generally more because additional WINS servers increase the performance of queries and responses by looking first at local WINS servers.

Windows 9*x* and Windows NT clients are allowed to be configured with a primary WINS server and a secondary WINS server. If the primary WINS server can't be reached, the client will try the secondary WINS server. (Windows 2000 and later clients can have up to 12 additional secondary WINS servers.)

Key Points

- To establish connectivity with TCP/IP, you must install the protocol on both servers and clients.
- To connect, clients must be configured with the correct IP address, subnet mask, and may also have the default gateway, and DNS server information.

- The client's IP address information can be manually configured, or it can be obtained from a DHCP server.
- NetBIOS names are used on Windows networks.
- Host names are used to identify computers on TCP/IP networks.
- DHCP reduces the administrative overhead of monitoring IP addresses.
- The DHCP server manages DHCP scopes, or pools of addresses that are assigned to a single subnet. There can only be one scope per subnet.
- DNS translates between IP addresses and host names.
- DNS clients query a DNS server to find the IP address of a computer.
- DNS servers maintain zone files that contain the address records for one or more domains.
- A primary zone file can be updated. A secondary zone file is read-only. Zone transfers send new information from a primary zone to a secondary zone.
- WINS is used by clients to resolve NetBIOS names to IP addresses.

Chapter Review Questions

1. True or False? To communicate across a TCP/IP network, you must have a DNS server and DNS clients installed.

    a) True

    b) False

    Answer b, false, is correct. A TCP/IP network in which users either know the IP addresses of all hosts, or a hosts file contains the IP address to name mappings, can exist. DNS is not necessary, but it certainly makes life easier.

2 You're called in to resolve a problem on a network. Users can't communicate with a certain server named fred.flintstone.com, or FRED which is its NetBIOS name, but they can communicate with all other servers, and they can connect to the Internet. The network clients are all Windows 98 and Windows NT 4.0. The network servers all use Windows NT 4.0. TCP/IP is the protocol used on the network. What could be the problem?

a) DNS servers are not installed correctly.

b) The A record for fred.flintstone.com is not in the DNS zone file.

c) WINS is not installed on the network.

d) The WINS record for fred.flintstone.com is not listed in the WINS server.

Answer d is correct. No listing of the WINS record for FRED in the WINS server is most likely the cause. The servers and clients are all Windows versions that require WINS to resolve the NetBIOS name to an IP address. Answer a is incorrect because clients can connect to the Internet. Answer b is incorrect because DNS is not the method that legacy Windows clients use to resolve IP names for local servers. Answer c is incorrect because the clients can communicate to all other servers.

3 Put the following DHCP process in the correct order.

I. DHCP client requests the IP configuration information.
II. DHCP client broadcasts to discover a DHCP server.
III. DHCP server acknowledges the IP acceptance and transmits additional IP information.
IV. DHCP server offers the client an IP address.

a) II., IV., I., III.

b) II., I., IV., III.

c) III., IV., I., II.

d) IV., II., I., III.

Answer a is correct. The order of execution is as follows: II. The DHCP client broadcasts a DHCPDISCOVER packet; IV. The DHCP server offers an IP address with DHCPOFFER; I. The DHCP client accepts the IP address with DHCPREQUEST; and III. The DHCP server acknowledges the acceptance with DHCPACK. Answers b, c, and d are all in the incorrect order.

4 If you have a computer that is named JOE, and you want to automatically add a domain name, microsoft.com, to create a FQDN of joe.microsoft.com, what should you configure?

a) Default gateway

b) DHCP

c) DNS suffix

d) WINS

Answer c is correct. The FQDN is created by concatenating the DNS suffix with the host name. Answers a, b, and d do not affect the FQDN structure.

5 What does a DHCP server hold IP addresses in?

a) A pool

b) A scope

c) A zone

d) A subnet

Answer b is correct. A DHCP server holds a group of IP addresses in a scope. Answer a, a pool, is incorrect, because that term is used only as a descriptor. Answer c, a zone, is incorrect; a zone is used to hold records in a DNS database. Answer d, a subnet, is incorrect because that is the logical grouping of IP addresses assigned to a physical segment.

6 You've been called in to troubleshoot a network in which the administrator is unable to update DNS information. What might be the problem?

a) The administrator is trying to update a secondary zone file.

b) The administrator is trying to update a primary zone file.

c) The administrator is using hosts files.

d) The administrator has accidentally copied a scope.

Answer a is correct. If the administrator is trying to update a secondary zone file, it won't work because a secondary zone file is read-only. Answer b is incorrect because the administrator can make changes to DNS information in a primary zone file. Answer c is incorrect because there is no need to use hosts files with DNS. Answer d is incorrect because scopes apply to DHCP services, not DNS.

7 If you were configuring a DHCP client, which of the following should you do?

a) Fill in the IP address.

b) Fill in the DHCP address of the DHCP server.

c) Obtain an IP address automatically.

d) Fill in the subnet mask.

Answer c is correct. When configuring a DHCP client, you simply need to configure it to obtain an IP address automatically. Answer a is incorrect because you don't need to assign an IP address to DHCP clients. Answer b is incorrect because DHCP clients broadcast to find DHCP servers and don't need their addresses to function. Answer d is incorrect because the DHCP server will provide the subnet mask to the client.

8 When you configure DHCP on a server, which of the following must you do?

a) Configure a scope that includes a range of IP addresses, the subnet mask, and information such as IP addresses of DNS servers and WINS servers.

b) Configure a zone that includes the IP address to name mappings for the computers within the domain.

c) Configure the server to obtain an IP address automatically.

d) Add the IP address to NetBIOS name mappings.

Answer a is correct. When you configure a DHCP server, you must add a scope that includes a range of IP addresses, the subnet mask, and other information such as the IP addresses of DNS and WINS servers. Answer b is incorrect because you do this when you configure DNS servers. Answer c is incorrect because a DHCP server must have a static IP address. Answer d is incorrect because this is what you do with a WINS server.

9 Which of the following is read-only?

a) Primary WINS server data

b) Primary DNS zone file

c) Secondary DNS zone file

d) DHCP scope

Answer c is correct. A secondary DNS zone file is read-only. Answers a, b, and d are incorrect because they can be changed.

10 If you want to communicate with COMPUTER8, which is the NetBIOS name of a server, which of the following will be used to resolve the name to an IP address?

a) A WINS server

b) A DNS server

c) A DHCP server

d) A hosts file

Answer a is correct. A WINS server is used to resolve NetBIOS names to IP addresses. Answer b is incorrect because DNS resolves host names to IP addresses. Answer c is incorrect because DHCP provides an IP address to a DHCP client. Answer d is incorrect because hosts files will resolve host names to IP addresses. An LMHosts file could resolve a NetBIOS name to an IP address, but that isn't an option for this question.

Chapter 11

Network Configuration

Most networks operate behind the scenes. Users don't interact with a virtual local area network (VLAN); in fact, they rarely even know that there might be one on the network. A VLAN is simply something that the administrator has implemented to increase performance. This stands true for storage solutions, fault-tolerant systems, and most remote access and security systems.

The small office home office (SOHO) network is a bit different from the network in a typical organization. A SOHO accommodates few users, and most support themselves except in dire emergencies. SOHO users are forced to be aware of how to care for their equipment, connect to other computers, and troubleshoot on the fly.

In this chapter, we'll review VLANs, storage, disaster recovery, remote connectivity, and security. In addition, we'll look at the SOHO network and its own connectivity options.

Using Switches to Create VLANs

A local area network (LAN) is often considered a broadcast domain (however, many consider a LAN to be any type of network that's confined to a small geographical area, such as a building). As a broadcast domain, this means that every host located on that LAN is able to broadcast to every other host and receive broadcasts from every other host. (A broadcast is a message that is directed to every host.) Broadcast domains are bounded by any device that translates above the physical layer. This means that a bridge, switch, router, and gateway are each considered a boundary for the broadcast domain.

VLANs are created using switches. A switch is essentially a bridge with multiple ports that make the switch appear physically the same as a hub. With a bridge, the only option is to connect two LANs. Using a bridge to connect an individual workstation or server to a network simply doesn't make sense. However, a switch gives you the option of connecting a single host to each port, if you want to. (Connecting individual computers to each port of a switch is sometimes called *micro segmentation.*) Plus, you can connect entire LANs to each port. These options are depicted in Figure 11-1.

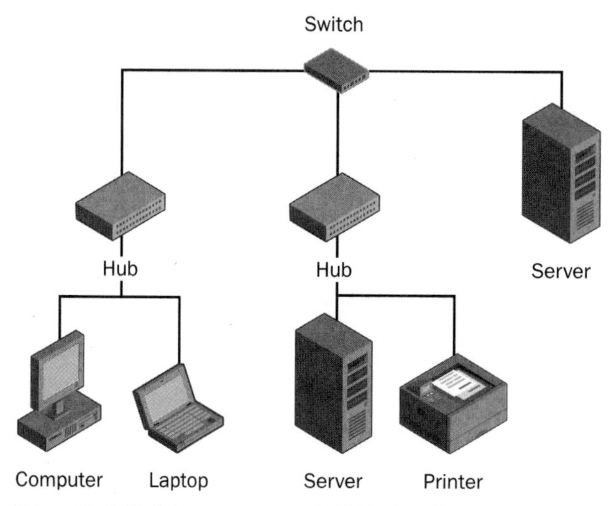

Figure 11-1 Switches can connect individual stations or entire LANs on each port.

If you micro segment the network, no broadcasts can get through—this is where a VLAN comes in. VLANs are able to group ports into a broadcast domain, which is important as applications become more collaborative.

The technology behind VLANs began as proprietary systems implemented by vendors. Most of those systems couldn't be mixed or matched. The Institute of Electrical and Electronics Engineers (IEEE) developed standards for VLANs so that different vendors' products could interoperate. The IEEE 802.1Q specification is now implemented by a majority of VLAN vendors.

You can create a VLAN in three ways:

- **Port-based.** Early on, VLANs were constructed through the simple assignment of ports on a switch to a VLAN identified by a number. The disadvantage to port-based VLANs is that a growing mobile workforce often ruins the best intentions of the network administrator. Each person who moves to a different location will no longer be in the VLAN as intended.

- **Layer 2 grouping.** The layer 2 grouping method is also called the MAC Membership VLAN. As you might guess, it's based on the MAC address of the computer. This method has a great advantage in that the VLAN works regardless of the physical location of the computer or the address it has been given.
- **Layer 3 grouping.** This method uses the network layer address to identify the computers that should participate in a particular VLAN. If TCP/IP is in use, this would be the IP address; if IPX/SPX is in use, it would be the IPX address. The switch looks at the address in the network layer header to determine which VLAN a packet belongs to. A disadvantage to this method is that a computer can change its IP address with DHCP, or its entire network layer address will change when the computer moves to a different subnet.

No matter which VLAN method is used, there is a high administrative component. First the VLAN must be managed by VLAN management software. Second, each VLAN must be configured by the administrator, and in the cases of port mapping and layer 3 grouping, it must be watched carefully.

There are benefits to VLANs, as well. VLANs are based on the use of switches, and switches offer increased network performance. The need to use routers is decreased, which further increases performance given that routers tend to introduce latency to network traffic. (Latency is not necessarily caused by the router itself. Think of it this way: if a small, fast packet gets stuck behind a large, slow packet, there is a higher latency on the network.) VLANs function well in large organizations in which groups that work closely together are physically far away because of their seating arrangements. A VLAN enables a group to keep its traffic to itself, increasing security.

Network Attached Storage

Most people are familiar with server attached storage (SAS), which is also referred to as directly attached storage, because it's directly attached to a server. These are usually large arrays of hard disks that are kept in external boxes connected directly to a high-performance interface in the server. Network attached storage (NAS) is something different.

A NAS device is dedicated to nothing more than serving the network with storage. A NAS device is a server of sorts because it provides file services, but it doesn't have the capacity to do anything else.

A workstation in a network using NAS does not authenticate to the NAS device, nor does it run a client/server application utilizing the NAS's processor. Instead, the workstation communicates directly to a standard server; but when data is delivered out of storage, it is done by the NAS device. One common NAS device is a CD-ROM tower connected directly to the network. A typical NAS configuration is shown in Figure 11-2.

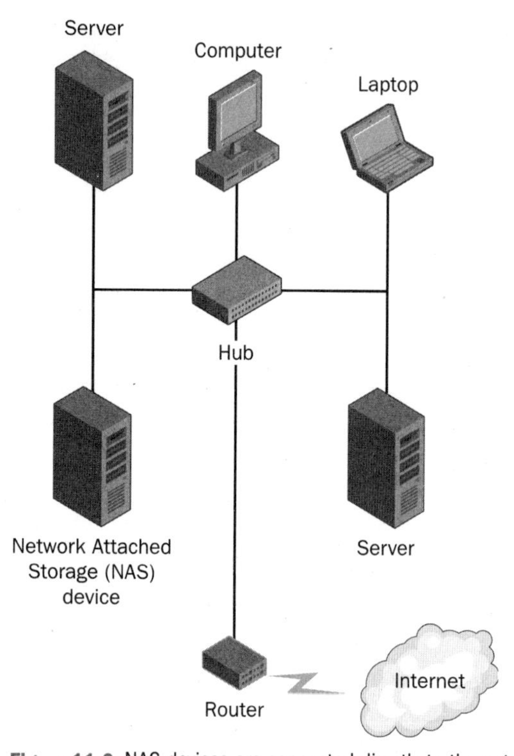

Figure 11-2 NAS devices are connected directly to the network.

NAS communicates using common protocols, including:

- Network File System (NFS) typically found in UNIX environments
- Common Internet File System (CIFS) in Windows networks
- File Transfer Protocol (FTP) and Hypertext Transfer Protocol (HTTP) in IP networks

Storage Area Networks

A Storage Area Network, also called a System Area Network (SAN), is quite different from NAS, even though the acronyms are similar and they're both storage solutions. A SAN is a network of one or more storage devices that communicate outside the regular network with one or more servers.

In the SAN, storage devices are placed on a separate network using fibre channel technology to connect to a hub, as depicted in Figure 11-3. Servers are also equipped with special adapters to connect to the hub and access the stored data on any of the connected storage devices. When using fiber optics, the SAN can communicate across fairly long distances, making it easier for fast access to stored data to be shared among buildings on a campus.

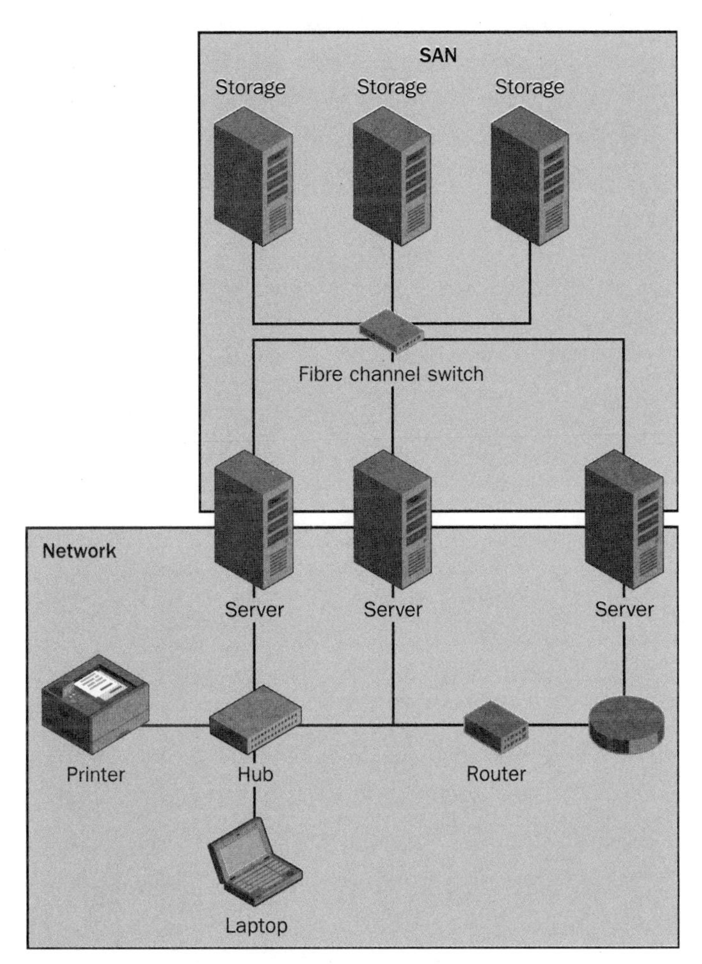

Figure 11-3 A SAN is separate from the LAN.

SANs make it possible to separate large transfers of data from the LAN. Because the SAN can communicate across a distance, it can bypass routers, bridges, and switches that might be necessary because of the large number of users on an enterprise network.

Central management of data is possible via a SAN, yet the data is still available to many different systems and applications. A SAN can scale up to terabytes (TB) of data. SANs are still proprietary in nature. Most vendors' SAN products do not interoperate with others', and the actual installation and configuration procedures must be performed according to the manufacturer's specifications.

What Is Fibre Channel? Fibre channel is a technology specification for one gigabit per second (Gbps) data transfer across fiber optic cabling, and in some configurations, over copper wires. Fiber optic cables can be used up to a length of 10 kilometers, which is more or less four miles. Copper cable can be used up to 30 meters in length, or about 100 feet.

Fibre channel includes its own model for networking. FC 0 is similar to the physical layer of the OSI reference model in that it specifies the physical characteristics of the interfaces, media, and data transfer rate. FC 1 defines the encoding scheme for the data that is transferred, which is also a physical layer type of specification. FC 2 offers a description for data transfer sequence and data framing, much like the data-link layer of the OSI model. FC 3 defines bandwidth management, and FC 4 describes the application and protocol management.

Fibre channel can be a star topology, physically, because it uses a hub with point-to-point connections between the hub and each of its nodes. Fibre channel is also found in a physical ring topology, with point-to-point connections between each node until the last node connects to the first. The manufacturer's specification and the media you use will determine which physical topology will be applied.

Disaster Recovery

Disaster recovery is the one thing that you should always plan for and hope to never have to do. If your hopes come true, your preparation can seem to have been a waste of time. But if a disaster does occur, your plans will save you time and money.

Part of disaster recovery involves budgeting. You need to know how much money you'll be losing before you can determine how much to spend to prevent its loss. For example, say you're administering two servers: one houses a mission-critical application that is used in production transactions at a rate of $50,000 per hour; and the other houses noncritical file storage. Clearly, the money you spend to protect the server housing the mission-critical application should be far more than that spent to protect the server housing noncritical file storage.

Keep in mind that a disaster doesn't necessarily mean a total loss of everything. Usually it will affect only part of a network, whether it is a single building, a room, or even a server. When you plan for disaster recovery, you need to determine what threats can occur and how likely they are to take place. Whatever vulnerabilities exist in the environment, it's best to plan ways of overcoming them. You should monitor and measure the network for any events that could trigger a disaster, even if they are simple, such as a utility that monitors the disk status on the storage systems. You should deter any possible disaster, using uninterruptible power supplies (UPS) and firewalls. But what is most often considered disaster recovery are the corrective measures planned for and then implemented in the case of a disaster.

These are the types of disasters you might encounter:

- **Mission-critical server failures.** Any component within a server can cause it to fail to provide services, including hard disk, power supply, processor, memory, and even the network interface.
- **Network component failures.** When network components such as hubs, switches, and routers fail, entire sections of the network become unusable.
- **Building problems.** If the building power goes out, the wiring is damaged, or the wide area network (WAN) link is cut, the network fails.
- **Mother Nature.** Floods, tornadoes, fires, and the occasional monsoon are the types of disasters that can completely devastate the entire network and all the data stored within it.

For each type of disaster, consider the likelihood that it will take place. Then weigh the value of each network component (and its data) that could be affected. Next look at the methods that you can use to prevent the disasters. And finally, consider the ways to recover from a disaster if it does take place. You'll always need to have recent network data available at some location outside the building (and preferably the city) that can be restored. In the event of a total disaster, these are your options:

- Wait to resume operations after a new site can be constructed.
- Temporarily move operations to another site in a different location.

- Use a "cold" site, which is another location with equipment available to begin operations as soon as you can move personnel to that location and restore data to that equipment.
- Use a "hot" site, which is another location that mirrors your location's data operations, usually is directly connected to your network, and is able to begin operations the moment your location is affected by the disaster. (Obviously, this is the most expensive option.)

The disaster recovery plan should specify how personnel safety will be handled (evacuation procedures), what systems and applications are required for ongoing operations, the data that is needed, and the recency of that data. Disaster recovery plans usually define manual procedures to use if computing is unavailable and the restoration procedures for data.

Note Backing up data to tape or optical storage media is always part of a disaster recovery plan.

Fault Tolerance

Most people think about disaster recovery in terms of restoration of the damaged network, but it's actually less expensive to prevent a disaster than to restore one. I remember a client who used a server to log minutes for telephone calls. This was a very expensive server, but the client didn't want to spend money on a fault-tolerant disk array that would make certain the server could continue working even if a disk failed. Well, you guessed it. The disk failed, and the server stopped logging minutes for telephone calls. The distressed client called, desperate for restoration of the server, willing to pay anything because, as it turned out, the money lost for every hour that the server was not logging minutes was in the hundreds of thousands of dollars. No less than a week later, the client had a fully fault-tolerant solution in place.

Fault tolerance is another term for redundancy. You can have redundant components within a server, redundant servers, and even redundant networks, in the case of a hot site. A fault-tolerant system simply has a spare part that takes over if another part fails. Fault tolerance can work for the following:

- **Memory.** Some servers support error-correcting memory with a spare memory module to use in case of memory failure.
- **Network interface cards (NICs).** NICs can be redundant in two ways. They can share the network traffic, or one of the NICs can wait until the first fails before it kicks in.

- **Redundant Array of Inexpensive Disks (RAID).** Data is mirrored, shared, or striped across multiple disks. Pay attention to these versions of RAID:
    - RAID 1: Mirroring disks connected to a single hard disk controller, or duplexing disks connected to two different hard disk controllers.
    - RAID 5: A group of three or more disks is combined into a volume with the disk striped across the disks, and parity is used to ensure that if any one of the disks fails, the remaining disks will still have all data available.
- **Power supplies.** One power supply takes over if the original fails.
- **Clusters.** Two or more servers are grouped to provide services as if the group were a single server. A cluster is transparent to end users. Usually, a server member of a cluster can take over for a failed partner with no impact on the network.

Working with a Small Network—The SOHO

The goals of a SOHO network are different from those of an enterprise network. Most, listed below, are fairly basic:

- Share an Internet connection.
- Share printers and scanners.
- Share files, access e-mail messages, and back up the data.
- Access a corporate network via virtual private network (VPN).
- Send or receive faxes without a fax machine.

The primary goal is usually to establish a connection to the Internet and share it with every other computer on the network. The remaining services usually depend on that connection being in place. We'll now look at several methods of Internet connectivity common to SOHO networks: DSL (digital subscriber line), cable, home satellite, and the plain old telephone system (POTS). We'll also look at wireless SOHO networks.

Using DSL to Access the Internet

DSL uses standard telephone wires to carry high-speed data—which makes it just right for a home office—but doesn't interrupt standard telephone calls. Asynchronous DSL, or ADSL, is the most common form of DSL used for broadband Internet access. The term *asynchronous* refers to the fact that download speeds are faster than upload speeds. Because Web browsing tends to demand a faster speed, and because uploading (such as sending e-mail messages) takes place behind the scenes, ADSL is ideal for SOHO networks. The telephone company benefits because DSL removes long-duration data calls from the public switched telephone network while providing another source of revenue. DSL usually requires a splitter, as shown in Figure 11-4.

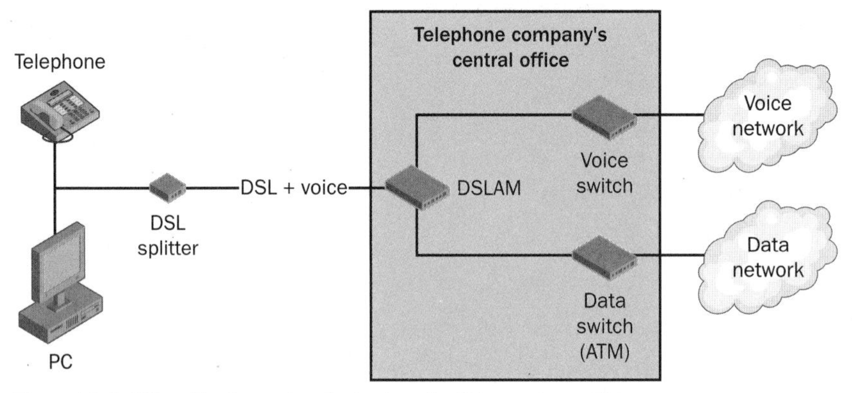

Figure 11-4 DSL traffic shares the wire but is split off from voice traffic.

The data travels along the same wire as voice traffic to the telephone company's central office. Once there, it passes through the Digital Subscriber Line Access Multiplexer (DSLAM). Voice traffic is filtered into the public switched telephone network. The data is usually transmitted using the Asynchronous Transfer Mode (ATM) protocol to access the Internet.

At the SOHO, the DSL modem connects via an Ethernet interface to the computer. To share the DSL connection, the computer connected to DSL must also have a second interface (whether wireless, Ethernet, or other) with which it will network with the other computers. That computer will then share the DSL connection, and all traffic will pass through it to the Internet, as shown in Figure 11-5. Another common scenario is to use a DSL modem connected directly to a router, which then routes data to the remaining network. This eliminates the need for a multi-homed computer (a PC with two or more NICs) to share the connection.

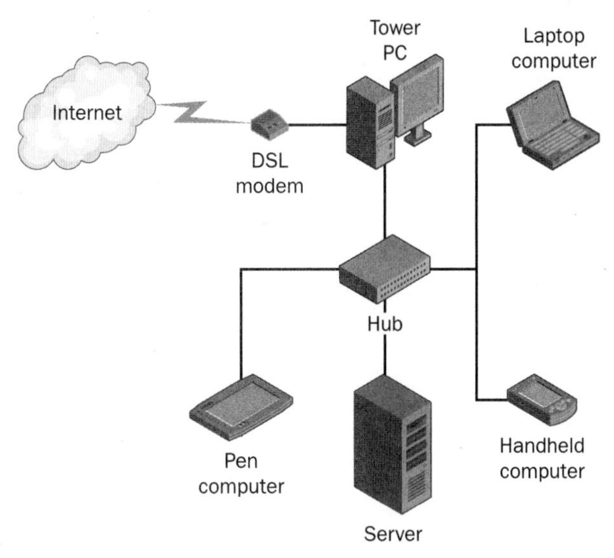

Figure 11-5 A DSL connection can be shared.

Using Cable Modems for Internet Sharing

The Data Over Cable Service Interface Specification (DOCSIS) is the cable TV industry's response to the telephone companies' domination over Internet connectivity. Cable modems offer high-speed Internet service via the existing cable network. Because cable TV was originally a unidirectional medium, they had to find a way to make it work with bidirectional data. Cable TV companies have since upgraded their backend networks with hybrid fiber coax (HFC) to bring fiber optics into the cable TV network, which does the trick. In this configuration, fiber optics is deployed within much of the network, and traditional copper coax exists only in short runs.

Because Internet access requires bidirectional data transmissions, a TV channel is reserved for data service. In DOCSIS 1, a download bandwidth of 40 Mbps is available, with a 10-Mbps upload speed. DOCSIS 2 increases the upload speed to 30 Mbps. The media is shared by all the cable subscribers within a local area. To maintain a shared network in which no subscriber can dominate the bandwidth, the cable provider caps the upload and download rates for each subscriber.

A cable modem is shared in the same way that a DSL connection can be shared. A computer will connect to the cable modem and through it to the cable network. That computer will have another interface to network with the remaining computers at the SOHO. All data traveling to and from the Internet will be routed through that single computer. In addition, a router can be used in place of the multi-homed PC and be able to deliver the same connection.

Home Satellite

Direct broadcast satellite TV is another medium under exploration for delivering high-speed Internet services. The problem with it so far is that latency from geosynchronous orbit makes interactive browsing activities difficult. It is also not a cost-effective solution.

The other type of home satellite is a point-to-point wireless service using line of sight connections between the Internet Service Provider (ISP) and the subscriber. The system is similar to that shown in Figure 11-6. A wireless transceiver is placed at the SOHO and pointed at the large satellite hosting the ISP, which can exist up to 35 miles away. This system offers high-speed Internet connectivity, but it also has some connectivity issues. Birds, storms, trees, buildings—in fact, any object—can block the line of sight between the SOHO and the ISP.

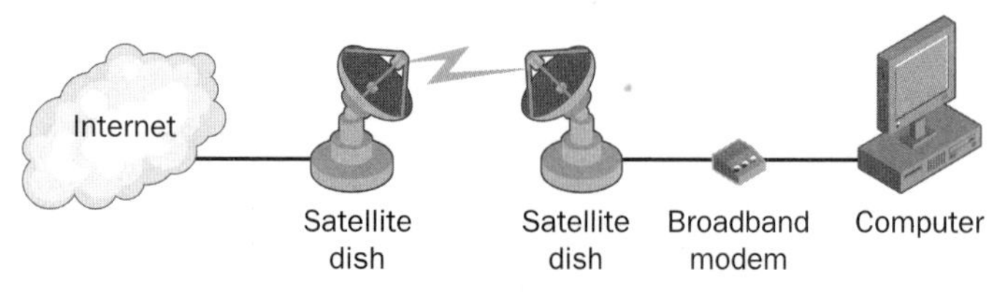

Figure 11-6 A satellite connection uses line of sight connectivity for transmission.

With home satellite, a broadband modem is used to connect to the computer. It also connects to the satellite using copper coax cabling. The data is transmitted between the two. Again, the way to share this type of connection is to add a second network interface to the computer connected to the broadband modem and route Internet data through that computer. Satellite modems can also be connected to a router for the connection to be shared.

Creating a Wireless Network

SOHOs no longer must depend on cabling, NICs, hubs, and routers to communicate. A SOHO can use the Ethernet 802.11(b) wireless technology, and the computers will connect with little installation effort. Using 802.11b allows a SOHO to become mobile. A person who creates a home office can easily move it to any location within the house without causing major wiring requirements.

802.11(b) delivers 11-Mbps connectivity, which is a little faster than Ethernet 10BaseT. It uses the 2.4-GHz band, which is popular with wireless telephones. When you implement 802.11(b), you should avoid using wireless phones in the same range.

Wireless networks can be created among individual devices, which is the ad-hoc peer-to-peer mode. Or they can be created using a wireless access point (WAP) in the managed mode. WAPs can connect to wired networks, and they can bridge between one another so that there is seamless connectivity as a mobile device moves throughout the wireless network.

Using Dialup

Dialup connections are the mainstay for many a SOHO. Often, a SOHO will access the Internet only through individual dialup connections on each of the network's PCs. However, it is possible to share a dialup connection.

Let's take an example. Joe has a home-based property-management business. He manages three properties, and he uses the Internet to communicate with homeowners association managers. He also uses the Internet to research competitive pricing and to submit bids. Joe spends a total of four hours a week online. For him, a dialup connection is sufficient for these activities. Now, he uses a laptop in the field and transfers data to his PC using an 802.11(b) ad hoc peer-to-peer network. Joe can transmit data from the laptop directly to the Internet if he shares his dialup connection. When Joe's PC discovers data packets that are intended for the Internet, his PC will dial up its default Internet connection and then transmit the data.

Security Impact on the Network

In security systems, you must be able to restrict access for elements that can damage your network. In other words, you must determine whether data is "good" or "bad," and to do so you must examine that data. The time that it takes to look at and qualify data will usually slow the network. But bottlenecking the data stream is not the only impact that a security system can have on the network. We'll now look at how various security methods will affect the network.

Encryption

You can encrypt data that you store on a hard disk using a file encryption technology, just as you can encrypt the data transmissions sent across the network. When you allow users to encrypt data on a hard disk for later retrieval, you can encounter two different types of problems: files whose owners have lost or forgotten them and forgotten passwords. Both issues require an administrator to manage encrypted files.

Encrypting data transmissions has more of an impact on the network. Data that is encrypted usually has a larger footprint; not only is it scrambled before it is sent, but a key is also usually agreed upon through a series of transactions before the data is transmitted. Both the larger data transmission and the extra traffic slow the network down.

Another issue with encryption is the need to manage certificates. Certificates validate the resource that is sending or receiving encrypted data. A certificate is granted by a certificate authority (CA) server. CA servers can be organized into an enterprise hierarchy. The design, implementation, and management of a CA server add to the administrative work on a network.

Firewall and Blocking Ports

Typically, a firewall filters out traffic to prevent unauthorized access to a network. Most often it is used to thwart intruders who attempt to access private networks from the Internet. A firewall is simply a form of router that passes data from one network to another. But first it reads the type of data and filters out whatever matches the filters that are within the rules, or access control list, of the firewall.

You can prevent packets from coming from a specific IP address. For example, you can prevent all traffic from a known address from which you receive a lot of spam e-mail messages. The result of this is that you won't be able to receive any other type of traffic from that address. Let's take for example that a company vice president's home PC has a virus that e-mails everyone in his address book over and over. If you block that address, the VP will have no access to the network whatsoever, resulting in a serious problem. The best thing to do is examine any suspicious traffic before denying it.

You can also block ports using a firewall. When a firewall blocks ports, the router looks at the packet's TCP and UDP ports. Ports can be blocked in two ways:

- You explicitly deny traffic sent to a certain port, and allow all other traffic.
- You explicitly permit traffic sent to certain ports, and deny all remaining traffic.

These two methods might seem equivalent, but they have different effects. If, for example, you deny all traffic that is sent to TCP and UDP ports 23, this should stop Telnet traffic. You might want to do this because some denial of service attacks use Telnet. However, this method won't stop all Telnet traffic. Anyone can Telnet to a different port on your server given that those other ports are open.

The second method is more secure, but it can cause more problems if you don't know what ports are used on your network. If, for example, you have implemented a Citrix MetaFrame XP server on your network, the server uses TCP port 1494 by default. If you aren't aware that the server requires this port and you establish an access control list on the firewall that explicitly allows e-mail (Simple Mail Transfer Protocol, or SMTP) traffic via TCP port 25 and denies all other traffic, the Citrix clients won't be able to connect to the server across the Internet. However, Citrix clients will be able to connect inside the network and via direct dialup.

Using a Proxy Server

A proxy server sits between a client on a private network and the Internet to provide a layer of security between the two. When an application on the client makes a request, it directs that request to the proxy server, which then translates the request and passes it to the appropriate Internet server. The Internet server responds to the proxy server, which then passes the response to the client application. A proxy server has two interfaces, one that connects to the private network and the other that connects to the Internet.

Using a proxy server has the value of blocking inbound traffic from the Internet. A proxy server allows the network's clients to browse the Web and prevents Internet clients from accessing private servers. In addition, because the proxy server translates every client request, it has the opportunity to restrict what the internal clients are connecting to on the Internet.

The impact of using a proxy server on the Internet is that it slows connection time through the proxy for client applications. It must be manually configured in the Web browser, as shown in Figure 11-7, adding to administrative overhead. One value of a proxy server is that when the proxy server offers the ability to cache Internet content, it can speed up client access to Internet websites that are accessed frequently. Finally, the proxy server might cause some problems for those in the organization who don't want their Internet traffic viewed.

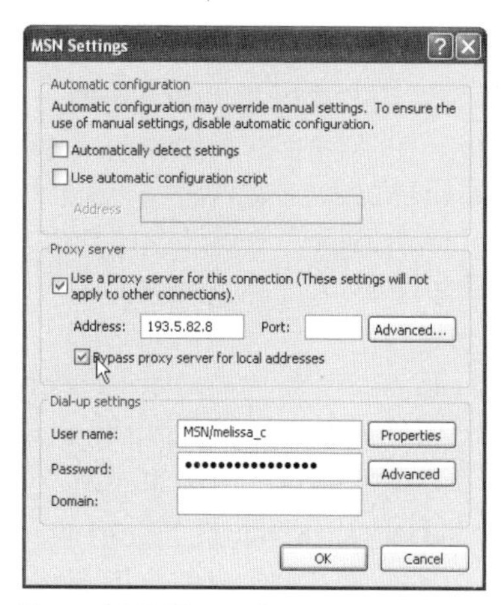

Figure 11-7 You configure the proxy server within the Web browser properties.

Tunneling

Tunneling, also known as virtual private networking, provides security for the data that is transmitted between a remote client and a network. Data is encapsulated in IP packets so that it can be transmitted without being eavesdropped upon. A corporate network that is connected to the Internet can provide VPN connections to company employees. The employees connect to the Internet anywhere in the world, then using the VPN protocol, they can create a private connection directly to the corporate network.

A VPN allows a company to reduce its costs for long distance telephone calls. For example, a businessperson traveling anywhere can still dial up a local ISP and create a connection to the network. Because the call is local, no long distance or 800 number costs are incurred.

The impact on the network is low—VPN connections are notoriously slow. The data must be processed at the client end before it is transmitted across the Internet to the remote access server (RAS). It must then be processed again at the RAS before it can be sent to the application.

Key Points

- Switches enable users to share a VLAN even if they are not physically close to each other.
- Network attached storage (NAS) is a storage unit directly connected to the network. Clients request data from other servers, and the data is delivered from the NAS storage.

- Storage Area Networks (SANs) are storage systems that are connected outside the local area network (LAN) and are accessible from one or more servers.
- Disaster recovery is the process of returning a network to working order after a disaster has taken place.
- Hot sites are locations that mirror a network and can immediately take over if a disaster takes place.
- Fault tolerance is the use of redundant components to ensure that a failure will not affect performance.
- SOHO networks can use a single DSL, cable modem, home satellite, or dialup connection to the Internet, and share it with other computers on the network.
- Security solutions often have the unintended impact on the network of slowing down data transmissions or denying access.

Chapter Review Questions

1 Which of the following pieces of equipment are necessary to create a VLAN?

a) Proxy server

b) Gateway

c) Switch

d) Firewall

Answer c is correct. A switch is used in VLANs. Answers a, b, and d are all incorrect because a proxy server, a gateway, and a firewall do not have ports that can be dedicated to a virtual LAN.

2 A CD-ROM tower would be considered which of the following types of storage?

a) RAID

b) NAS

c) SAN

d) DSL

Answer b is correct. CD-ROM towers are considered a form of network attached storage, or NAS. Answer a, RAID, is incorrect because it is a set of multiple hard disks configured to work as a single hard disk. Answer c, SAN, is incorrect because it is a storage system that connects to a network outside the LAN. Answer d, DSL, is incorrect because it is a method of Internet connectivity, not storage.

3 True or False? 802.11(b) is a method used by telephone companies to deliver DSL to SOHO networks.

a) True

b) False

Answer b, false, is correct. 802.11(b) is a wireless network specification.

4 You're troubleshooting an IP network. The administrator has added a firewall and denied all traffic to TCP port 23 to stop Telnet traffic from causing denial of service problems. However, the administrator is worried that this won't stop the problems of Telnet to other ports. What should you look for when you examine the access control list on the firewall that will alleviate the administrator's fears?

a) If there's a final Deny All statement

b) If there's a final Permit All statement

c) If there's an initial Deny All statement

d) If there's an initial Permit All statement

Answer a is correct. If there's a final Deny All statement, Telnet traffic will not be able to access any port except those that are explicitly permitted. Answer b is incorrect because this will allow Telnet traffic to access any port that is not explicitly denied. Answer c is incorrect because this will deny all traffic to all ports. As packets are examined by the router, they will automatically be dropped before any other access control list statement is processed. Answer d is incorrect because it will have no effect on any following statements.

5 A network administrator is reviewing the disaster recovery plans for the network. Which of the following is an appropriate disaster recovery activity for a network server that stores noncritical data?

a) Hot site implementation

b) Cold site implementation

c) RAID implementation

d) Data backup to tape with offsite storage

Answer d is correct. An appropriate disaster recovery activity is to back up data to tape and store the tape offsite. Given that the files are noncritical, they won't be needed until a later time. Answers a, b, and c are all incorrect because they're more appropriate solutions for mission-critical sites and servers.

6 Which of the following security methods is configured in the Web browser?

a) Proxy server settings

b) Firewall settings

c) Ports that will be blocked

d) 802.11(b) settings

Answer a is correct. Proxy server settings are configured in a Web browser's properties. Answers b and c are incorrect because these are configured on the network. Answer d is incorrect because it is configured as a network protocol.

7 Which of the following should be avoided when using 802.11(b)?

a) DSL

b) Home satellite

c) Dialup connections

d) Wireless phones

Answer d is correct. A wireless phone using the 2.4-GHz radio frequency will interrupt 802.11(b) wireless networks and should be avoided. Answers a, b, and c are all incorrect because they have no impact on 802.11(b).

8 A network administrator has asked your opinion about what type of security solution to install on a network. The administrator wants to restrict all incoming traffic and watch the Internet browsing activities of some users. Which of the following can provide these features?

a) VPN

b) Proxy servers

c) Firewall

d) Encryption

Answer b is correct. Proxy servers can restrict all incoming traffic while enabling an administrator to keep tabs on the Web browsing activities of users. Answers a, c, and d are all incorrect because none provide these services.

9 Which of the following can be an Internet connection that is shared across an 802.11(b) wireless network? Choose all that apply.

a) DSL

b) DOCSIS

c) Home satellite

d) Dialup

Answers a, b, c, and d are all correct. Any type of Internet connection can be shared across an 802.11(b) network.

10 True or False? Encryption slows the data transmission.

a) True

b) False

Answer a, true, is correct. When encrypting data, it must be processed both at the sending and receiving nodes.

Part 4

Network Support

All the tasks you'll undertake in designing, planning, and implementing a network are also used to support it. However, these tasks are dominated by the administration of networked systems, management of user issues, and troubleshooting network problems.

Network support is the final focus of the four test domains in the CompTIA Network+ examination. Nearly one-third of the exam—actually, 32 percent of it—is based on network-support concepts. This section will cover TCP/IP utilities, problem solving, configuration, and using a troubleshooting strategy.

Chapter 12

TCP/IP Utilities

Of all the important tools you'll need to have, Transmission Control Protocol/Internet Protocol (TCP/IP) utilities are at the top of the list. While an ample number of network management applications and equipment can identify and resolve problems, few are as necessary as basic TCP/IP utilities on a TCP/IP network.

We'll look at some essential utilities that help you examine and diagnose network conditions. Pay close attention to how they're used, because you'll be tested on them. Not only should you know how they function, you also need to know when to use them in a network support situation.

Tracing the Route with Tracert

Data transmitted across a TCP/IP network will often cross multiple routers on the way to its destination. This is especially true in Web browsing, sending e-mail messages to someone outside an organization, or otherwise using the Internet. If a problem occurs with the transmission of data that causes it to slow down, you'll need to locate the source of the problem before you can resolve it.

For example, if as the network administrator your e-mail server was taking several hours to transmit to your ISP's e-mail server, you'd need to know what was causing the slowdown—especially when people outside your organization start calling your users to see why they haven't responded to e-mail messages. Your first step will be to use the Tracert utility to trace the path from your e-mail server to the ISP's e-mail server, which will tell you whether a backup dialup link between your network and the ISP is in use instead of the standard route. But this might not be obvious if your backup link connects to the same ISP router as your standard wide area network (WAN) link. What you'll see instead is the time it takes for data to transmit between these two routers. From there, you can look at the router on your end and ask your ISP to look at the router at its end to identify the problem.

Tracert runs at the command prompt. It uses the Internet Control Message Protocol (ICMP) to echo each host in the path and determine the time it takes for data to travel and which hosts are crossed in the path. As you can see in Figure 12-1, the data you receive using Tracert shows both the host name and the IP address of the host, as well as the number of milliseconds (ms) that it takes data to move from the previous host to that host.

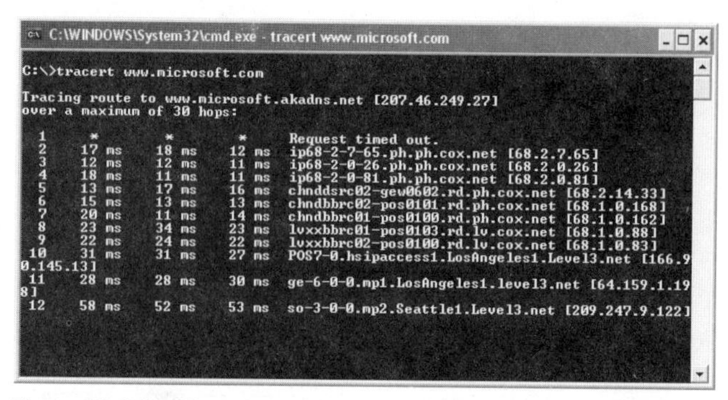

Figure 12-1 Use Tracert's output to see how data transmits across the network.

Tracert's syntax is tracert [–d] [–h maximum_hops] [–j host_list] [–w timeout] hostname_or_ip_address. The minimum that you must type after the command prompt is this: **tracert *hostname_or_ip_address***. The other items are switches, shown in square brackets, which are optional. The options you can use with Tracert are described in Table 12-1.

Table 12-1 Tracert Syntax

Switch	Required Information	Function
-d	Nothing required	Tells Tracert not to resolve addresses to hostnames.
-h	The number of routers, or hops, that you want Tracert to stop at	Tells Tracert to search through that number of routers and then stop, even if you don't reach the target host in the path.
-j	A name or IP address of one or more hosts that you want Tracert to travel across	Tells Tracert to trace a route along certain hosts to the final destination.
-2	A number representing the timeout value	Tells Tracert to wait for the timeout value in milliseconds for a response.

Pinging Hosts on the Network

Sometimes a computer doesn't seem to connect to the network at all. The user might be trying to browse the Web, or performing some other network function, and there's no apparent response from the network. The problem between the computer and the other host could be an application error, such as when the Web server's Hypertext Transfer Protocol (HTTP) service fails. In this case, it appears to the Web browser application that the client computer can't connect. To determine whether the host is reachable, you need to test end-to-end connectivity at the network layer. This can be done using the Packet InterNet Groper (ping) utility.

Ping is a command that forwards a packet of data to another host across a TCP/IP network. If a response is returned, you can assume that TCP/IP is configured correctly and that it's possible to make the connection. The packet that ping sends is an ICMP echo command.

The basic syntax of ping is "ping *hostname*." You can also ping an IP address. When you run ping 127.0.0.1 (which is the loopback IP address of your adapter), you'll essentially verify that the computer can communicate via TCP/IP with its own network interface card (NIC). You can use ping to test the maximum amount of data, or maximum transmission unit (MTU), that can be carried in a packet from one end of the link to the other. You can see the switches available with ping by typing **ping /?** at a command prompt. Figure 12-2 shows the results you receive when you ping a host on the Internet.

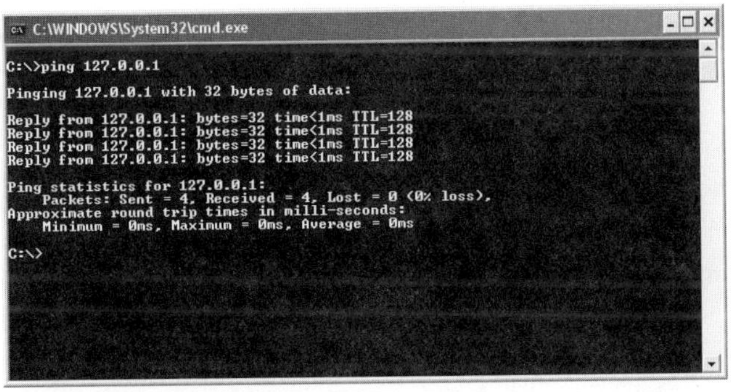

```
C:\WINDOWS\System32\cmd.exe

C:\>ping 127.0.0.1

Pinging 127.0.0.1 with 32 bytes of data:

Reply from 127.0.0.1: bytes=32 time<1ms TTL=128
Reply from 127.0.0.1: bytes=32 time<1ms TTL=128
Reply from 127.0.0.1: bytes=32 time<1ms TTL=128
Reply from 127.0.0.1: bytes=32 time<1ms TTL=128

Ping statistics for 127.0.0.1:
    Packets: Sent = 4, Received = 4, Lost = 0 (0% loss),
Approximate round trip times in milli-seconds:
    Minimum = 0ms, Maximum = 0ms, Average = 0ms

C:\>
```

Figure 12-2 Ping can determine end-to-end connectivity between two IP hosts.

Ping is usually the first test you'd use to troubleshoot a problem with network connectivity. You can use the following troubleshooting sequence with ping to try to isolate a connectivity problem:

1. Type **ping 127.0.0.1.** This will determine whether TCP/IP is configured correctly on the local computer.

2. Ping your own IP address. This doublechecks the first step.
3. Ping a known IP address on the network. It's always a good idea to use the address of a server that you know is up and running. If this is successful, TCP/IP is functioning on the local network.
4. Ping a known IP address on the Internet. If it is successful, you know you have a good Internet connection.
5. Ping a Universal Resource Locator (URL) such as *www.amazon.com*. If this works, Domain Name System (DNS) is functioning.
6. Ping the server that you're attempting to access. If this works, network connectivity to the server is functioning and the problem is now isolated to the application.

Resolving Addresses with ARP

The Address Resolution Protocol (ARP) has an ARP utility that is used at the command line to translate an IP address to its physical address (the MAC address). First we'll go over how the ARP protocol works. A host with an IP address for a destination computer but no physical address will broadcast an ARP request on the network. The host with the matching IP address will then reply and provide its own physical address. The original host will place that physical address, which is resolved to the IP address, into the ARP cache. This will remain in the ARP cache for a period of time to reduce the number of address resolution requests. The ARP cache is periodically flushed of all entries.

The ARP utility can do a couple of different things. It can display the entries currently in the ARP cache, and it can be used to add or delete entries to the ARP cache.

When would you need to use ARP? Well, imagine you're running a computer named host A, and it can't find server B on the network. Also imagine that you've been using Dynamic Host Configuration Protocol (DHCP) and have made server B a DHCP client. The first thing you might do is check the ARP cache by typing **arp –a** at the command prompt. This will display the entries. If there is one for the server's current IP address or for its physical address, you could delete it by typing **arp –d *123.45.67.89***, where *123.45.67.89* is the IP address of the server listed in the ARP cache. This will force the computer to attempt to resolve the address in order to communicate. There shouldn't be a need to add a static entry to ARP, but you can add a static entry to the ARP cache for the server by typing **arp – s *123.45.67.89* 00–*aa–23–f3–c2–bb***, where *123.45.67.89* is the IP address and *00–aa–23–f3–c2–bb* is the physical address of the server's network interface. Another use for ARP is to read the ARP cache to determine which hardware address is being used to send packets. Figure 12-3 depicts the information you might see when you display a computer's ARP cache.

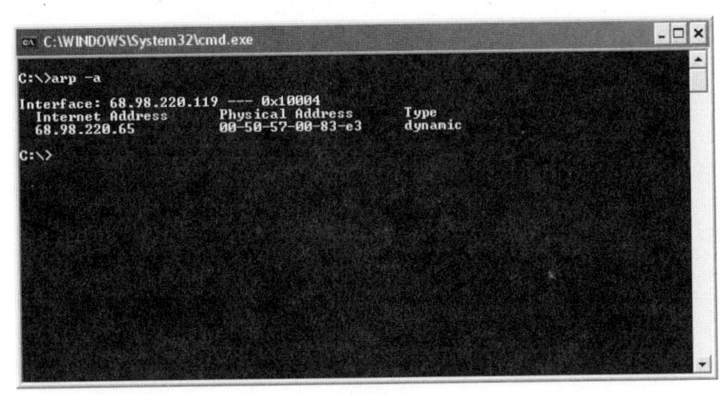

Figure 12-3 The ARP cache is displayed using the arp command.

Netstat vs. Nbtstat

Two commands—Netstat and Nbtstat—are often confused with one another. They're used in similar ways but for different protocols. As you can probably guess, Netstat is used to look at TCP/IP status on the network, while Nbtstat is used to view the NetBIOS over TCP/IP status on the network.

Netstat

Netstat displays the protocol information and connection status, including traffic flow, of TCP/IP across the network. When you use Netstat, you'll need to use the switches that are available for your operating system. Type **netstat /?** at the command prompt to see the switches. On a Windows XP system, you'll see the switches listed in Table 12-2. The command format is netstat [–a] [–e] [–n] [–o] [–s] [–p proto] [–r] [interval].

Table 12-2 Netstat Switches

Switch	Required Information	Function
-a	Nothing	Displays all current connections and listening ports for TCP and UDP.
-e	Nothing	Shows the statistics for Ethernet, such as the number of bytes and packets sent and received.
-n	Nothing	Shows all statistics in numerical format, so that you see the addresses and ports as numbers only, without resolving them to names.
-o	Nothing	Displays which process ID is associated with the connection. You can then use Windows Task Manager, click the Processes tab, and find the application with that same Process ID.

Table 12-2 Netstat Switches

Switch	Required Information	Function
-p	Protocol	Shows the connections associated with the specific protocol, which can be TCP, UDP, TCPv6, and UDPv6. If you also use the –s parameter, you can see the ICMP and ICMPv6 connections too.
-r	Nothing	Displays the routing table on that IP host. This command has basically the same result as using the route print command.
-s	Nothing	Shows all protocol statistics for IP, IPv6, ICMP, ICMPv6, TCP, TCPv6, UDP, and UDPv6.
Interval	Number	The interval is a number you provide that represents seconds of time. When you use this switch, Netstat will display the statistics and refresh them every interval until you stop the process by pressing Ctrl+C.

Netstat will show the IP addresses, the protocol type, and the ports of remote hosts to which your computer is connected. When a port is not established, Netstat shows it with an asterisk (*).

Nbtstat

Nbtstat is a very useful little tool on Windows networks that use NetBIOS names. The Nbtstat command focuses on NetBIOS over TCP/IP connections. It can also update the LMHosts cache.

When users log on to a Windows network, their computers register the name and IP address with Windows Internet Naming Service (WINS). If there is no WINS, the information is probably stored in the LMHosts file. You can use Nbtstat to find out the same type of information that is registered in WINS and LMHosts.

Nbtstat queries NetBIOS over TCP/IP for the NetBIOS names and IP addresses. When a Windows network also uses DHCP, Nbtstat can find the current IP address used by a host simply by knowing its NetBIOS name. The syntax for the Nbtstat command is nbtstat [–a remote_name] [–A IP_address] [–c] [–n] [–r] [–R] [–RR] [–s] [–S] [interval]. The switches are described in Table 12-3.

Table 12-3 Nbtstat Switches

Switch	Required Information	Function
-a	NetBIOS name of a remote host	Displays the remote host's name table.
-A	IP address of a remote host	Displays the remote host's name table. The only difference from the -a switch is that the remote host's IP address is used here, not the NetBIOS name.
-c	Nothing	Shows the remote name cache, including IP addresses.

Table 12-3 Nbtstat Switches

Switch	Required Information	Function
-n	Nothing	Shows the local NetBIOS names—those stored in the LMHosts file.
-r	Nothing	Shows the names that the computer has resolved via broadcast and WINS.
-R	Nothing	Purges and reloads the remote cache name table.
-RR	Nothing	Sends Name Release packets to WINS. Then it refreshes.
-s	Nothing	Displays the sessions table and converts the destination IP addresses to host names by using the hosts file.
-S	Nothing	Displays the sessions table with destination IP addresses.
Interval	A number representing time in seconds	Refreshes and redisplays the Nbtstat statistics using the interval given. This can be stopped using Ctrl+C.

You can use the Nbtstat command to find out who is logged on to a networked Windows PC. The NetBIOS name table will show up on your screen, as shown in Figure 12-4. You can see the type of service is shown by a number. For the logon account, the service type is <03>. The other service type <03> will show the name of the PC. It's usually easy to figure out which is which. From a security standpoint, the problem that this poses is that this Nbtstat command gives an intruder several pieces of information that can be used to log on to the network—the only piece of the puzzle not provided is the password.

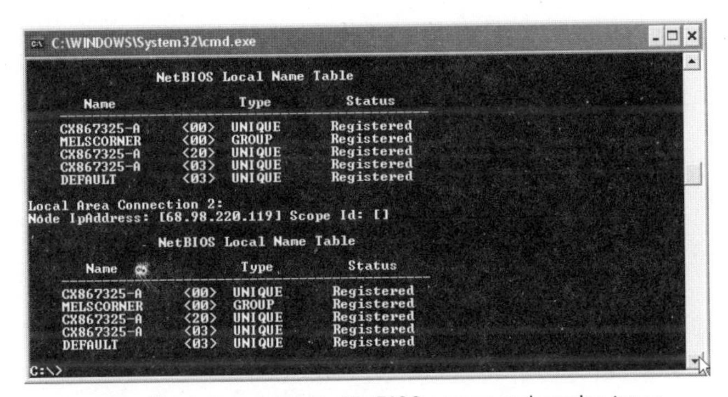

Figure 12-4 Nbtstat provides the NetBIOS names and service types.

Using NSlookup as a Discovery Utility

DNS resolves IP addresses to host names. These are stored in the DNS database on at least one name server. As you might remember, the DNS database is distributed across multiple DNS servers. Each of these name servers can pass queries to other name servers if the information is not contained in the local database. The Nslookup utility takes advantage of this ability. (Nslookup stands for "Name Server Lookup.") Basically, Nslookup queries DNS for information.

One of Nslookup's best uses is its ability to find out a computer's name when all you have is an IP address. Let's say that you've been having a problem on the network, on which a lot of traffic is generated by a host but all you have is the IP address. You can use the Nslookup command to find out the name of that host, and from there you can isolate the problem.

You'll need to use the following information with Nslookup:

- A DNS server's name or IP address
- The IP address that you're querying
- The type of record you're looking for, including (but not limited to):
  - MX: Mail exchanger
  - A: Host name
  - CNAME: Canonical name
  - PTR: Pointer
  - HINFO: Host information

Nslookup can work in two modes: interactive and noninteractive. The tool enters interactive mode if you don't provide any name or IP address. It enters noninteractive mode if you provide the IP address or name of a computer that you want to look up.

When you have more than one query to perform, or if you need some fairly detailed information, it's best to use interactive mode. If you type **nslookup** at the command prompt, you'll be taken to the interactive Nslookup prompt. From there, you can execute specific Nslookup subcommands. If you type a question mark (**?**), you'll see the following text:

```
> Commands (identifiers are shown in uppercase, [] means optional):
NAME            Prints info about the host/domain NAME using default server.
NAME1 NAME2     As above, but use NAME2 as server.
help or ?       Prints info on common commands.
set OPTION      Sets an option.
    all             Prints options, current server, and host.
    [no]debug       Prints debugging information.
    [no]d2          Prints exhaustive debugging information.
    [no]defname     Appends domain name to each query.
    [no]recurse     Asks for recursive answer to query.
    [no]search      Uses domain search list.
```

[no]vc Always use a virtual circuit.
domain=NAME Sets default domain name to NAME.
srchlist=N1[/N2/.../N6] Sets domain to N1 and search list to N1, N2, etc.
root=NAME Sets root server to NAME.
retry=X Sets number of retries to X.
timeout=X Sets initial timeout interval to X seconds.
type=X Sets query type (e.g., A, ANY, CNAME, MX, NS, PTR, SOA, SRV).
querytype=X Same as type.
class=X Sets query class (e.g., IN (Internet), ANY).
[no]msxfr Uses MS fast zone transfer.
ixfrver=X Current version to use in IXFR transfer request.
server NAME Sets default server to NAME, using current default server.
lserver NAME Sets default server to NAME, using initial server.
finger [USER] Fingers the optional NAME at the current default host.
root Sets current default server to the root.
ls [opt] DOMAIN [> FILE] Lists addresses in DOMAIN (optional: output to FILE).
–a Lists canonical names and aliases.
–d Lists all records.
–t TYPE Lists records of the given type (e.g., A, CNAME, MX,
NS, PTR, etc.).
view FILE Sorts an 'ls' output file and view it with pg.
exit Exits the program.

Understanding how to use this list of options is not all that obvious. Let's say you want to find the mail servers on a network. To start doing this, you enter the Nslookup command, and at the Nslookup prompt you'd specify the records you're looking for by typing **set type = mx** at the prompt. Then enter the name of the domain that you're querying for mail servers at the prompt. If you're looking for mail servers on the domainname.net domain, you simply enter domainname.net. The Nslookup command then returns the names and IP addresses of authorized mail servers in the domainname.net domain. If you don't trust the results of this query, you can change the server that you're querying, preferably to one that has authority for that domain. To change the server name to dns.domainname.net, type **server dns.domainname.net** at the prompt. You can type **exit** to quit Nslookup. Figure 12-5 displays the results of Nslookup in a query of address records from Microsoft.com.

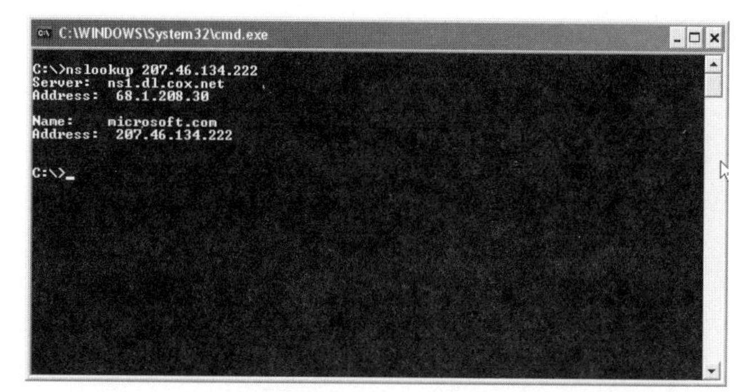

Figure 12-5 Nslookup displays results of queries to the DNS database.

IP Configuration in Microsoft Windows

You can check your computer's own IP configuration on a Windows machine in two ways: One is graphical—Winipcfg; the other is text-based—Ipconfig. Windows 95, Windows 98, and Windows Me offer Winipcfg, while Windows NT, Windows 2000, and Windows XP include Ipconfig.

Winipcfg

The Winipcfg utility lets a user or administrator view the current IP address and other information about the network configuration on the local computer. For a computer that uses DHCP, there might be a need to release the IP address or renew it. Winipcfg makes this easy by providing buttons to Release or Renew a single IP address or Release All or Renew All addresses.

Whenever the Release or Renew button is clicked, the computer communicates with the DHCP server to obtain a new IP address or renew the lease on the existing IP address. If the DHCP server is not reachable, the computer can also assign itself an automatic private IP address (APIPA) if it is using Windows 98 Second Edition, Windows 2000, or Windows XP.

To use Winipcfg, click Start, select Run, and then type **winipcfg** and press Enter. Expand the initial screen if you want by clicking the More Info button. You can see the DNS servers configured for the computer by clicking the button with the ellipsis (...) next to DNS servers. You can also see address information for each individual NIC by selecting the NIC from the list in the Adapter Information box. The Winipcfg utility is shown in Figure 12-6.

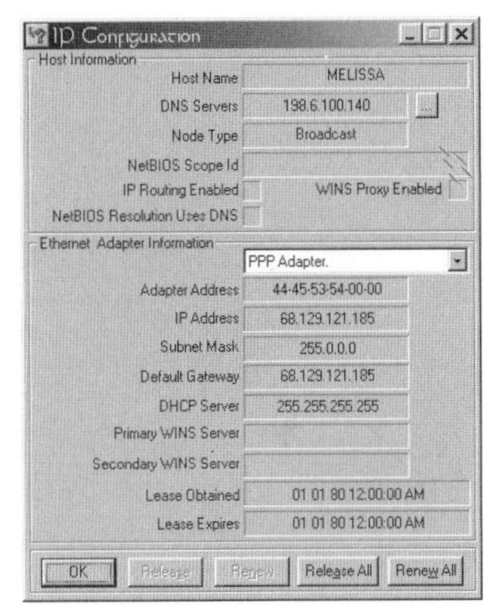

Figure 12-6 Winipcfg displays IP information in a graphical format.

Ipconfig

Like Winipcfg, the Ipconfig command displays TCP/IP configuration settings and allows you to make some small TCP/IP setting changes. The difference is that Ipconfig works only on the command line. You can see the results of Ipconfig in Figure 12-7.

```
C:\WINDOWS\System32\cmd.exe

Ethernet adapter Network Bridge (Network Bridge) 2:

        Connection-specific DNS Suffix  . :
        Description . . . . . . . . . . . : MAC Bridge Miniport
        Physical Address. . . . . . . . . : D2-4C-C4-7B-47-44
        Dhcp Enabled. . . . . . . . . . . : No
        IP Address. . . . . . . . . . . . : 192.168.0.1
        Subnet Mask . . . . . . . . . . . : 255.255.255.0
        Default Gateway . . . . . . . . . :

Ethernet adapter Local Area Connection 2:

        Connection-specific DNS Suffix  . :
        Description . . . . . . . . . . . : Realtek RTL8139/810X Family Fast Eth
ernet NIC
        Physical Address. . . . . . . . . : 00-E0-18-30-85-D1
        Dhcp Enabled. . . . . . . . . . . : No
        IP Address. . . . . . . . . . . . : 68.98.220.119
        Subnet Mask . . . . . . . . . . . : 255.255.255.192
        Default Gateway . . . . . . . . . : 68.98.220.65
        DNS Servers . . . . . . . . . . . : 68.1.208.30
                                            68.2.16.30

C:\>_
```

Figure 12-7 Ipconfig displays IP information in text format from the command line.

The syntax for this command is ipconfig [/all] [/renew [adapter]] [/release[adapter]] [/flushdns] [/displaydns] [/registerdns] [/showclassid adapter] [/setclassid adapter [classid]]. These settings are explained in Table 12-4.

Table 12-4 IPconfig Options

Option	Required Information	Function
/all	Nothing	Shows all the IP configuration information.
/release	Type the name of an adapter, or leave blank	Used when the IP address has been provided through DHCP. Releases the IP address for the specified adapter, or default adapter if left blank.
/renew	Type the name of an adapter, or leave blank	Used when the IP address has been provided through DHCP. Renews the IP address for the specified adapter, or default adapter if left blank.
/flushdns	Nothing	Empties the cache of information collected from resolving IP addresses through DNS (called the DNS resolver cache).
/registerdns	Nothing	Refreshes all DHCP leases and re-registers the DNS names.
/displaydns	Nothing	Shows the contents of the DNS resolver cache.
/showclassid adapter	Nothing	Displays the DHCP class IDs that are allowed for the NIC.
/setclassid	A class ID	Changes the DHCP class ID.

Configuring a UNIX Interface with Ifconfig

Don't confuse Ifconfig with Ipconfig. Ifconfig is a UNIX text-based command used to establish, view, or modify the network interface. When you type **ifconfig** and an interface name at a command prompt, you'll see the current configuration of that network interface. This will provide two lines of output. The first will show the information about the adapter itself, including its status. The second will show the IP configuration for that adapter. Ifconfig is used at the startup of a UNIX computer to configure the network interface with its IP address and other parameters.

For example, a user might call up and ask for help connecting to the network. When you go to the computer, you find out that it is a UNIX machine. You try a number of utilities, such as ping and Tracert, and isolate the problem to the UNIX computer. Then you try the Ifconfig command. If Ifconfig says the interface is Up, it is enabled for use. If it says the interface is Down, you can use Ifconfig to try to bring the interface up by typing **ifconfig *1e1* up**, where *1e1* is the name of the interface. If that doesn't work, you need to check the hardware.

You can check the Ifconfig's second line of output to see if IP is configured correctly. Here you will see the IP address and the subnet mask. A common problem occurs when you have manually assigned IP addresses—either the IP address or the subnet mask has been entered incorrectly. In either case, the computer won't be able to communicate on the network.

Let's take an example of a user who has assigned his own IP address and accidentally entered one that is already in use on the network. In this case, he is using an IP address that should function fine on the network *as long as the other computer is not up and running on the network at the same time.* If you suspect a duplicate IP address problem, you can use Ifconfig to find out the IP address of the nonfunctioning computer and then use Nslookup (from a computer that is functioning on the network) with that IP address to see whether there is another computer with the same name.

Key Points

- TCP/IP utilities are excellent tools for use in troubleshooting network problems.
- Tracert provides information about the path that data takes between two hosts on an IP internetwork.
- Both Tracert and ping use the ICMP protocol.
- Ping will tell you whether another host is reachable.
- ARP uses the Address Resolution Protocol for resolving IP addresses to physical addresses.
- Netstat shows IP network status information.
- Nbtstat shows NetBIOS over TCP/IP status.
- Nslookup is used to query the DNS database.
- On a Windows machine, you can use either Winipcfg or Ipconfig, depending on the Windows version, to show TCP/IP configuration information for the network adapters.
- On a UNIX machine, Ifconfig is used to configure IP information at the startup and can also be used afterward to change adapter configuration.

Chapter Review Questions

1 True or False? You can use Tracert on an AppleTalk network.

a) True

b) False

Answer b, false, is correct. Tracert is a TCP/IP utility, which means that it can be used only on TCP/IP networks. AppleTalk is a different protocol stack.

2 You receive a call from a user who is having trouble connecting to the VPN via her DSL connection, which is connected to her Windows NT computer. What is the first thing you would do to try to isolate the problem?

a) Execute an Nslookup to find out the mail servers on the network.

b) Use Ifconfig interface up to bring up the NIC.

c) Use Nbtstat to find out the user's name and computer name.

d) Try to ping the VPN server's address.

Answer d is correct. One of the first experiments you might try is to have the user ping the VPN server's IP address. Ping will tell you whether the VPN server is reachable from the user's machine. Answer a is incorrect because the mail servers are not part of the problem. Answer b is incorrect because Ifconfig is a UNIX command, and the user is using Windows NT. Answer c is incorrect because this will not help determine connectivity to the VPN server, which is the current problem.

3 Which of the following is a graphical command that displays configuration information for TCP/IP?

a) Winipcfg

b) Ipconfig

c) Ifconfig

d) Nslookup

Answer a is correct. Winipcfg is a graphical command that displays TCP/IP configuration information. Answers b, c, and d are all incorrect because they are text-based commands.

4 Which of the following commands uses ICMP? Select two.

a) Ping

b) Tracert

c) ARP

d) Nbtstat

Answers a and b are correct because both Ping and Tracert use ICMP. Answer c is incorrect because it uses ARP. Answer d is incorrect because it requires NetBIOS over TCP/IP.

5 You have been asked by a user to look at his UNIX workstation, which doesn't appear to be connecting to the network. When you look at the computer, you realize that it's missing a required command in its startup sequence. Which one?

a) Netstat

b) Ipconfig

c) Ifconfig

d) Tracert

Answer c is correct. Ifconfig is required on UNIX machines to configure the interface with TCP/IP parameters and bring it online on the network. Answers a and d are incorrect because they are not required by a UNIX computer at startup. Answer b is incorrect because it is not a UNIX command.

6 You need to find out the name of a host using an IP address on the Internet. Which of the following will provide the name?

a) Nslookup

b) Ifconfig

c) Tracert

d) Ipconfig

Answer a is correct. To find the name of a host on the Internet, you can use Nslookup and query the DNS database. Answer b is incorrect because Ifconfig is used to configure a UNIX interface. Answer c is incorrect because Tracert is used to trace the path between two end points. Answer d is incorrect because Ipconfig is used to display a Windows interface's TCP/IP information.

7 Which of the following can tell you the physical address of a TCP/IP host?

a) Tracert

b) ARP

c) Nbtstat

d) Nslookup

Answer b is correct. ARP can tell you the physical address (also called the MAC address) of a TCP/IP host. Answer a is incorrect because Tracert is used to trace the path of data between two end points. Answer c is incorrect because it provides information about the host's NetBIOS configuration. Answer d is incorrect because it is used to query the DNS database for IP address and host name information.

8 Which of the following tools can tell you which routers a data packet is passing through on its way to its destination?

a) Netstat

b) Ipconfig

c) Tracert

d) Nslookup

Answer c is correct. Tracert is the TCP/IP utility used to trace the path of data, including each router (called a hop) that it passes through. Answer a is incorrect because Netstat is used to provide TCP/IP status. Answer b is incorrect because Ipconfig is used to display TCP/IP configuration information. Answer d is incorrect because Nslookup is used to query the DNS database for information.

9 You're having a problem connecting to a Trivial File Transfer Protocol (TFTP) server. You know that TFTP runs over the User Datagram Protocol (UDP). Which of the following can look at UDP information to help you isolate the problem?

a) Nbtstat

b) Netstat

c) Ipconfig

d) Ifconfig

Answer b is correct. Netstat can be used along with the –p UDP argument to show specific UDP protocol statistics. Answer a is incorrect because Nbtstat looks at NetBIOS over TCP/IP information. Answer c is incorrect because Ipconfig is used to display TCP/IP information for interfaces on a Windows computer. Answer d is incorrect because Ifconfig is used to configure and display TCP/IP information for interfaces on a UNIX computer.

10 You receive a call from a user who can't connect to her ISP through the external DSL adapter that connects to the NIC on her Windows 2000 laptop. She uses the same NIC to connect to the corporate network. You suspect that the problem might be related to DHCP, and you want her to release the DHCP's IP address (123.45.67.89) that is currently assigned to the NIC. Which of the following commands will perform this function?

a) tracert 123.45.67.89

b) ping 123.45.67.89

c) ifconfig /release

d) ipconfig /release

Answer d is correct. To release the DHCP address, the command is ipconfig /release. Answer a is incorrect because tracert 123.45.67.89 will attempt to trace the path between the user's computer and IP address 123.45.67.89. Answer b is incorrect because it will attempt to ping the user's own computer. Answer c is incorrect because it is a UNIX command.

Chapter 13

Linking Up Clients

Users who interact with a network using a PC are aware only of the machine sitting in front of them. Data and applications might be located on a server in a completely different location. But if an application doesn't work, or the computer is unable to access data or just seems slow, the network administrator will probably get a call. The first thing the administrator should do when looking into the problem is determine whether one computer or several are experiencing the problem. When only one person is affected, the first place to look is the user's workstation and the configuration of the client operating system (OS).

Client OSs are not necessarily the same as the server's network operating system (NOS). Usually, you'll run into a network with several different client OS types, and within each type, multiple versions of that OS. For example, Microsoft Windows 98, Windows NT 4.0, and Windows XP might all be running on the same network. In a very large internetwork, you'll find these versions of Windows running alongside different versions of Macintosh and UNIX. It is rare to find a completely homogeneous internetwork.

In this chapter, we'll look at the different types of client OSs and then review how to connect a client to Windows, NetWare, Macintosh, and UNIX networks.

How Clients Work

Which client OS works best with which NOS? That's a question you have to answer when you decide what to install on a network. From a design standpoint, you should consider the following:

- **Connectivity.** Will the server NOS use a different protocol than the client's native protocol or other servers on the network? While clients support multiple protocols, it's best to run as few as possible.
- **Client access application.** Will the client OS require that a special client access application be installed to connect? Additional client access applications require additional installation and configuration work. Client OSs in the same family as the server NOS automatically include the necessary application and protocols to connect to that server NOS. For example, a Macintosh will connect to an AppleShare server using AppleTalk; a Windows 2000 workstation will connect to a Windows 2000 server; and a UNIX client can connect to a UNIX server—all without installing a client access application.
- **Network standards.** Is the client OS different from what the organization's standards call for? The more client OSs on the network, the more administration required.
- **Peer-to-peer or client/server.** Will resources be shared by clients or just servers? Peer-to-peer networks are more difficult to manage. A user can share files and printers or stop sharing at any time. If a user is on vacation, the computer is offline and files are unavailable, which means that productivity can suffer.

Windows 95, Windows 98, and Windows Me

Microsoft Windows 95, Windows 98, and Windows Me are all closely linked from an architectural standpoint. In most cases, an application that works on one will also work on the others; this includes client access applications. All of these OSs include network client access applications and protocols necessary to connect to Windows servers, as well as NetWare servers.

For peer-to-peer networking, or to interact with a Windows server, the Microsoft Client for Microsoft Networks is the correct choice. The second selection that you must make is the protocol. You should select only the protocol (or protocols) that are in use on your network.

Windows NT, Windows 2000, and Windows XP Workstations

Windows NT and Windows 2000 workstations are nearly identical to the server network operating systems of the same name, but they do possess some differences. The server versions can support a more complex set of hardware. For example, a server with four processors can be fully utilized by Windows 2000 Advanced Server, but not Windows 2000 Workstation. In addition, the server versions provide a greater number and type of services, including Dynamic Host Configuration Protocol (DHCP) and Domain Name System (DNS). This is mainly because they were designed to *serve* the network, while workstation versions were designed for use by people who work and use network resources. This explains why the workstation versions include the ability to consume DHCP and DNS services, *but not provide the services to manage DHCP and DNS*. Even in a peer-to-peer network environment, the workstation versions are limited to the number of users that can connect to them at the same time. In fact, both Windows 2000 Professional and Windows NT 4.0 Workstation allow 10 simultaneous client connections.

Windows XP computers (as well as Windows 98 Second Edition and Windows Millennium Edition) have the ability to share an Internet connection. When you configure a computer running Windows XP to share the Internet connection, it loads a DHCP service into memory and provides IP addresses to all clients that want to connect. This is not a DHCP service that you really can configure, however. It is automatic, and it uses a private IP addressing scheme.

One thing that you should be aware of is that while the client applications and protocols are available within the native operating system in any Windows client, they are not usable until configured. For example, a Windows 2000 workstation might be fully capable of serving as a DHCP client, but if the TCP/IP protocol isn't installed and configured to use DHCP, it will not obtain an IP address from a DHCP server.

Windows NT, Windows 2000, and Windows XP Professional all include a client for Microsoft networks and a client for NetWare networks. For access to Macintosh or UNIX servers, you can use either individual applications over the TCP/IP protocol—such as a Web browser—or you must install a proprietary client access application available from the vendor. To gain the ability to fully access file and print services on the server, you must have a true client access application.

Dumb Terminals

People use dumb terminals to access information on mainframes, minicomputers, and UNIX servers. A *dumb terminal* is hardly more than a monitor and a keyboard with some network connectivity. (The lack of a true processor is why the terminal is considered "dumb.") The terminal receives output to display on the monitor and provides input to the server from its keyboard.

Terminal emulation programs can run on a client and access the same information as the dumb terminal. The terminal emulation application mimics a dumb terminal so that a client workstation has the ability to use both the server host's applications and its own local applications.

If you work with Citrix MetaFrame, Windows NT 4.0 Terminal Server Edition, or Windows 2000 Server with Terminal Services, you might also run across Windows terminals. These terminals aren't much more complicated than a standard dumb terminal, but when running a terminal session they look just like a regular Windows workstation. Windows terminals are graphical in nature. They provide the same screen that a regular Windows computer would, and to provide the same graphics as a Windows computer, they have a full-color monitor and some memory for caching the graphics. In addition, Windows terminals have a mouse and a network port for Ethernet connections. Sometimes they also have a floppy disk drive, a printer, or a serial port for dialup purposes.

UNIX and Macintosh Clients I would venture to guess that more than 90 percent of all clients are Windows-based machines. On the Network+ exam, the questions about clients echo this majority–they're all about the various versions of Windows client operating systems. However, you will run into the occasional UNIX or Mac on a network.

If you do encounter UNIX and Macintosh clients, you'll need to consider the interoperability of the servers you use on the network. For example, both Windows and NetWare servers offer services specifically geared to UNIX and Macintosh clients. However, the services must be installed and configured so that these special clients can connect to the servers.

Connecting the Client

In this part of the chapter, we'll review how a client would need to connect to the four different types of servers—Windows, NetWare, Macintosh, and UNIX.

Windows Network

A Windows network with Windows NT and/or Windows 2000 servers is guaranteed to support the following Windows clients (Windows 3.*x* clients require the installation of additional client software to connect to the network):

- Windows for Workgroups 3.11
- Windows 95
- Windows 98
- Windows Me
- Windows NT Workstation 3.*x* and Windows NT Workstation 4.0
- Windows 2000
- Windows XP Professional

Windows networks also support both UNIX and Macintosh clients, and they can interoperate well with UNIX, Macintosh, and NetWare servers.

When a client logs on to a Windows network, two different processes take place:

- **Interactive logon.** This process verifies the user's identification as a domain account or a local account on the local computer.
- **Network authentication.** This process confirms the user's identification to the network. (Windows 2000 with Active Directory will require the use of Kerberos version 5 protocol authentication.)

Windows clients have full access to files and printers on the Windows network, depending on the security that is in place and what has been shared. On the server, you can use the Add Printer Wizard to add network printers for sharing from the server. Windows Explorer is used to share folders. Windows clients can connect to shared folders and printers by browsing the network or through mapping network drives to shared folders.

Mapping a network drive provides a drive letter on the client workstation that appears to be local but is actually a share on the server itself. To map a network drive, an administrator can create a logon script with the Net Use command within it, or a user can right-click the icon representing the network—which is My Network Places in Windows 2000 and Windows XP or Network Neighborhood in Windows 9*x* and Windows NT 4.0—and select Map Network Drive. Then the user can select the drive letter and type in the share name, which is in the Universal Naming Convention (UNC) form of \\server\share.

Users can also browse to the share that they want to map by clicking the browse button, as shown in Figure 13-1.

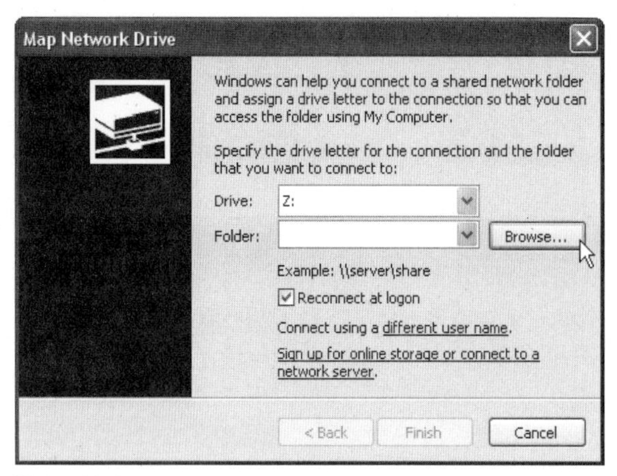

Figure 13-1 Mapping a network share makes the server's share appear as a local drive letter.

The UNC method of identifying shared files enables you to connect to a network share without knowing which drive the directory exists on within the server itself. For example, if a server named MYSERVER has a directory at F:\GORDON\Jk222\ZZFahr\SHARE, the administrator can share that directory as "SHARE" and users don't have to know the path—they simply connect to \\MYSERVER\SHARE. UNC names are helpful because you can use them in place of the standard DOS naming system (which is the drive letter followed by a colon, a slash, and then the path to the file). If you have a Windows computer named JOE and a directory shared as DIR, you can open a file named readme.txt in the DIR share by clicking Start and then Run and typing **Notepad \\JOE\DIR\readme.txt**. The nice thing about UNC is that users can be shielded from the directory structure.

How to Add a Windows XP Professional Computer to an Active Directory Domain

When a computer has not been joined to a domain, it will not provide a logon screen that includes a list of domains. Not only will this list of domains include the one which the computer belongs to, but it will include all the domains that are trusted by that domain. If you have a logon dialog that doesn't allow you to logon to a domain, then the computer is a member of a workgroup, not a domain. To join a computer to an Active Directory domain, complete these steps:

1. Create a computer account to authorize the Windows XP Professional computer within the domain.
2. Join the domain from the Windows XP computer's console so that it knows which domain to log on to.

When installing Windows XP Professional, you have the option to join the computer to the domain at the time of installation. Sometimes this isn't feasible, or you might need to move a computer from one domain to another, which requires roughly the same administrative process.

You can create a computer account (which appears as an object) in the Active Directory Users and Computers console on a Windows 2000 domain controller. This console is shown in Figure 13-2. You can also use this utility on a computer on which the console version of Active Directory Users and Computers has been installed, but you are guaranteed to find the utility on a domain controller. To create the computer object in the Active Directory, follow these steps:

1. Open the Active Directory Users and Computers console.
2. Navigate the tree structure until you reach your destination organizational unit (OU).
3. Right-click the OU and select New, followed by Computer, from the context menu.
4. Provide the computer object with the same name as the computer that will be joining the domain.

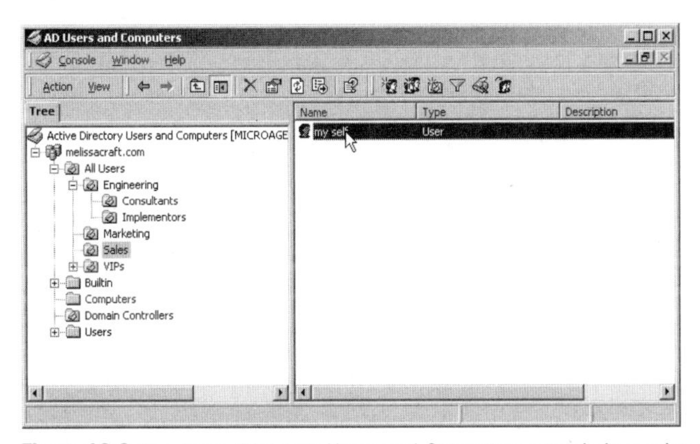

Figure 13-2 The Active Directory Users and Computers console is used to create computer objects in the domain.

Next you must ensure that the person who joins the computer to the domain has the authority to do so. By default, any member of the Authenticated Users special group has permission to join a computer to the domain. If the default hasn't been changed, you don't need to do anything special. However, if this permission has been removed, you must apply the permission to a user account or group of which that user is a member. (Remember, it's always best to apply rights to groups rather than directly to users because a group is so much easier to keep track of.) Right after you provide a name for the computer object, you simply need to click Change for the item This Computer Can Be Joined To A Domain By. Then select the user or group, and click OK.

To join the Windows XP Professional computer to the domain, start by logging on to the computer using an account that has local administrative privileges. Then do the following:

1 Right-click My Computer, and select Properties from the context menu.

2 Click the Computer Name tab. You'll see the screen shown in Figure 13-3.

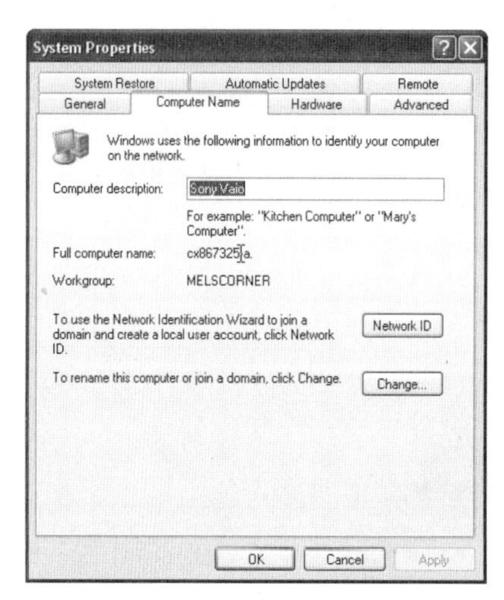

Figure 13-3 The Computer Name tab in System Properties allows you to join the computer to a domain.

3 You can select either the Network ID button or the Change button to join the computer to the domain. Selecting the Network ID button will begin the Network Identification Wizard, which will walk you through the process of creating a computer account (as opposed to having an administrator create the account first) and joining the domain simultaneously. If you select the Change button, you'll see the dialog box shown in Figure 13-4.

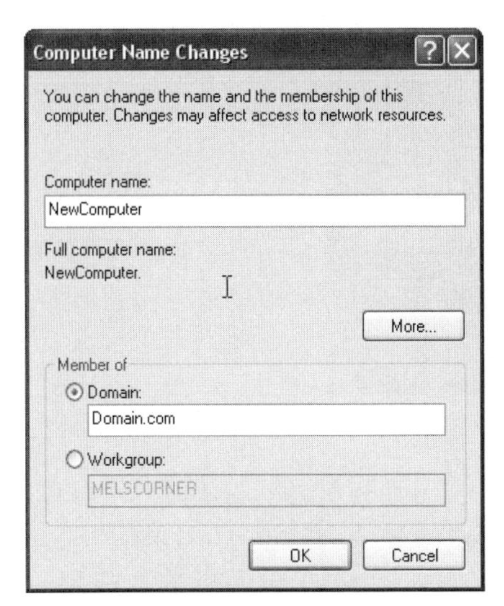

Figure 13-4 The Change button leads you to a dialog box in which you can join a domain or a workgroup or change the computer name.

4 Type the name of the computer account that you created in the Active Directory in the Computer Name box.

5 Click the radio button next to Domain and type in the full DNS name of the Active Directory domain. Then click OK until all dialogs are closed. Reboot, when prompted to do so.

The Net Use Tool

The Net command is a command-line tool that is handy in all sorts of situations. I've used it in scripts for installing large numbers of computers on a network as well as in logon scripts and while troubleshooting network applications. The Net command has many applications, one of the most useful of which is Net Use.

Net Use is the command that enables network resource access, including both printers and files. Net Use allows you to connect a drive letter on a local computer to a share on a network server. If you type **Net Use /?** on the command line, you will see a screen similar to the one shown in Figure 13-5. The minimum information that you need to connect a drive letter to a share is this, where X: is the drive letter and \\server\share is the share name:

Net Use X: \\server\share

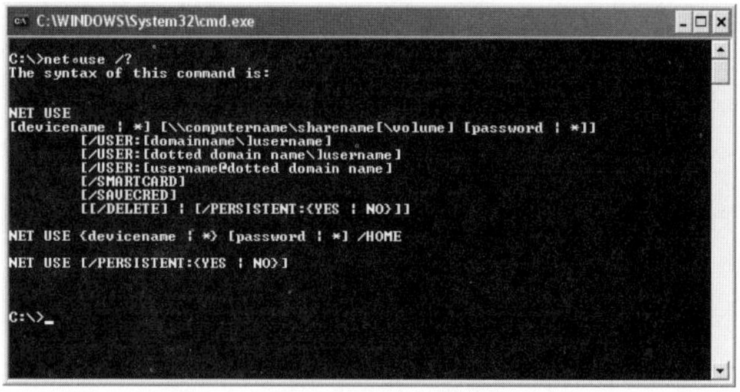

```
C:\WINDOWS\System32\cmd.exe

C:\>net use /?
The syntax of this command is:

NET USE
[devicename | *] [\\computername\sharename[\volume] [password | *]]
        [/USER:[domainname\]username]
        [/USER:[dotted domain name\]username]
        [/USER:[username@dotted domain name]
        [/SMARTCARD]
        [/SAVECRED]
        [[/DELETE] | [/PERSISTENT:{YES | NO}]]

NET USE {devicename | *} [password | *] /HOME

NET USE [/PERSISTENT:{YES | NO}]

C:\>_
```

Figure 13-5 The Net Use command offers several options.

When you use Net Use, the server will authenticate you before connecting the drive. If your user account is the same as the server's, with matching passwords, you'll be able to connect to the share. If not, you' will receive an error message stating that access has been denied. This is one of the reasons why the command allows you to specify a user name and password. In order to connect to a server share from a computer that is not participating in the domain or which is logged on with a user account that does not have access to that share, you'd type the following command, where the user name is *username*, located in the domain named *domain.com*:

Net use X: \\server\share /USER:*domain.com\username*

NetWare Server

NetWare servers are built to provide services to all types of clients. While Windows 95 and later clients each include a Microsoft Client for NetWare Networks, Novell also offers a NetWare client that can be installed on Windows workstations. The NetWare client software provides full access to all NetWare resources.

On NetWare version 5 and later, you can install the client to use IP only, IP along with IPX, or IPX only. While previous versions supported IP, they did not do so natively, and IPX had to be present in some form.

Except for servers running NetWare 3.*x* (or earlier) or NetWare 4.*x* in Bindery mode, all clients must log on to Novell Directory Services (NDS) to access the resources provided by NetWare servers. (The earlier NetWare Bindery allowed users to log in to a specific server. Users had to authenticate to each server individually because the Bindery was neither shared nor distributed across the network.) Authentication with a NetWare server appears to be the same as the Windows server when you use the same ID and password for the local workstation as you do in NDS. However, if you use different passwords or IDs, you will be presented with multiple logon screens. The local logon will have no effect on the client's access to resources on the NetWare network.

NetWare's file services—whether the Novell Storage Services (NSS) or the NetWare File System—allow users to store, manage, and retrieve data on the NetWare network. Both of these systems create volumes that are available to users as though they were shares. Printers are provided either through Novell Distributed Print Services or the older versions of NetWare print queues, and can be accessed via IP or IPX. Users can browse the network and connect to resources or map network drives and network printers. Mapping is performed through the login script, or the user can right-click the network icon and select the mapping option from the context menu.

The use of IPX requires that services be accessible to the Service Advertising Protocol (SAP). SAP is used to make certain that every server knows of the services (file, print, or other) that are available on the network. While SAP is incredibly bandwidth-intensive, it allows clients to find services by listing them in every server's SAP database and then updates those listings fairly frequently. For example, a client workstation trying to connect to a printer would contact the nearest server to query the database for printers. Then the client workstation would contact the server sharing that particular printer to connect to the printer. This is all transparent to the user. IP-based services do not need SAP.

You can configure a Windows 2000 workstation to use IPX by following these instructions:

1. Right-click My Network Places, and select Properties from the context menu.
2. Right-click the connection that will soon be using IPX, and select Properties from the context menu.
3. Click Install, select Protocol, and click Add.
4. Select NWLink IPX/SPX/NetBIOS Compatible Transport Protocol, and click OK.

NWLink is the Microsoft version of IPX/SPX that is fully compatible with IPX on a NetWare server. NetWare servers that are version 4 or earlier can be accessed only using IPX.

Setting the Context for the NDS Tree (or Using Bindery)

Logging on to a NetWare network is not as easy as it might first appear. Users who are not familiar with NDS logins can have difficulty logging in if the default context on their computers is not set to the same context that their user account belongs in. In NDS, a context is a location in the NDS tree. Remember that the NDS tree is an organization filled with organizational units (OUs) that contain either more OUs or leaf objects such as user accounts. The context walks the tree from the location of the user account to the root of the tree. For example, a user named JGrande that is located in the Accounting OU, which is in the Western OU, which is in the GlobalCorp Organization, has the context Accounting.Western.GlobalCorp.

A computer that logs in to NetWare is configured with a default context. Most users simply type in their own user ID, which, in the above example, would be JGrande. If the user's account is not within that computer's context or the context of the server that the login executable contacted, the user can't log in at all. To overcome this, the user can type in their entire context: **JGrande.Accounting.Western.GlobalCorp**, or change the context in the Login box.

Bindery logins are also an option for logging in. Bindery logins are used for any NetWare server prior to version 4.*x*, and they can be used for NetWare servers that have been set with a Bindery context. The Bindery is a flat file database of users and computers that exist on a single server. When using a Bindery login, the user connects to a single server only. This function can be enabled in the NetWare Client 32 using the Advanced Login option in the NetWare Client 32 Properties sheet. After enabling this option, the next time the login screen is shown, a Connection tab is available that allows you to select the option to log in to a server and use a Bindery connection.

Using the Map Command

On a NetWare network, users can use the Map command from the command line in much the same way that the Net Use command works on a Windows network, given that it allows you to connect a local drive letter to a NetWare server volume. The command line offers the ability to place the Map command in a login script to use in batch files and even during scripted installation procedures.

The Map command allows you to view a workstation's current drive mappings simply by typing **Map** at the command prompt. In addition, you can use the Map command to create or change drive mappings in a variety of ways.

You can type **Map Del X:** to delete a mapping of a drive that follows the X: drive letter. You can map the next available drive letter on a computer without specifying the exact drive letter by typing **Map N \\server\volume**.

Making a Windows Server Pretend It's NetWare Windows NT (and Windows 2000) servers can pretend that they are NetWare servers. This functionality was built into the NOS so that a company could use existing NetWare client access applications to use Windows servers without making major changes to client software. This was partially a stop-gap measure, meaning that during the transition from NetWare to Windows servers, the Windows servers would temporarily pretend they were NetWare. But now the functionality can be used for ongoing interoperability between Windows and NetWare networks.

To emulate NetWare, a Windows server must use the File and Print Services for NetWare (FPNW). FPNW is not built into the operating system by default. Instead, you must obtain it as an add-on from Microsoft. The file and print resources that the server provides can be managed using NetWare tools and are accessed from the NetWare client.

The other services that a Windows server can use to interact with NetWare servers is the Gateway Service for NetWare (GSNW). GSNW is available within the Windows server's operating system. This service does the opposite of FPNW, in that it provides access to NetWare servers without requiring the NetWare clients to be installed on workstations. In addition, the Directory Service Manager for NetWare (DSMN) enables interoperability among NetWare servers and Windows NT domains, with the goal of providing a single logon to clients for authentication.

Macintosh Network

On a Macintosh network, the AppleShare servers provide services to Macintosh clients via TCP/IP or AppleTalk. AppleShare servers can interoperate with Windows clients through Windows Server Message Block (SMB) file sharing over TCP/IP. However, the Macintosh clients access files via Apple Filing Protocol (AFP) over TCP/IP or over AppleTalk.

Macintosh servers can provide Web services using Hypertext Transfer Protocol (HTTP) and mail services using Simple Mail Transfer Protocol (SMTP), Post Office Protocol (POP), or IMAP. These services are the same as those provided by other NOSs, plus Macintosh servers have the ability to organize computers into workgroup-type structures called zones. When computers are assigned to zones, they automatically see the resources available in their own zone when browsing the network. This method is very different from Windows NT domains, Active Directory, or Novell Directory services because it merely groups resources in logical groups and does not incorporate authentication as part of the zone structure.

UNIX and Linux Servers

When you connect clients to UNIX or Linux servers, you'll need to configure the server to use Server Message Block (SMB) file sharing. You can do this using open source software such as Samba. When SMB file sharing is enabled on a UNIX or Linux server, the server appears to be a Windows server to Windows clients.

When clients connect to the server, they'll likely be using the Network File System (NFS), which is a distributed file system that enables users to access remote files as though they were local. However, when clients use

UNIX-based printers, they typically use the LPR/LPD utility. The LPR component of this utility is the client portion that sends commands to the printer such as "print the job." The LPD component runs on the print server and responds to the client commands.

Key Points

- Client operating systems consume services. Servers provide them.
- Clients can network in a peer-to-peer mode and in a client/server mode.
- Most Windows clients include client services for Microsoft networks as well as for NetWare networks.
- For a Windows client to access a UNIX or Macintosh server, the server must be set up for Server Message Block (SMB) shares, or the Windows client must be installed with a special client access application that will work with the server.
- Both the client and the server must use the same protocol for communication to take place.
- Terminal emulation programs can access mainframes, minicomputers, and UNIX hosts across a network.

Chapter Review Questions

1 An administrator has asked you to help select an OS for new computers being installed. Currently all computers are running Windows 98, and all servers are Windows 2000. The administrator is considering Windows XP, UNIX, and Macintosh client OSs. Which client do you tell the administrator to buy, given that it is required by the servers?

a) Windows XP

b) UNIX

c) Macintosh

d) All of the above

e) None of the above

The correct answer is e, none of the above. Windows servers, whether Windows NT or Windows 2000, will work with all of these clients—Windows XP, UNIX, and Macintosh. The administrator would probably have less trouble using Windows XP because it is similar to Windows 98 in its desktop, usage, and file structure, but it is not required by the server.

2 Which of the following is the rough equivalent of a dumb terminal?

a) Windows 95

b) UNIX

c) Terminal emulation application

d) TCP/IP

Answer c is correct. A terminal emulation application is equivalent to a dumb terminal because it provides the same access to a host—whether it is a UNIX server, minicomputer, or mainframe—as a dumb terminal. Answers a and b are incorrect because they are operating systems. Answer d is incorrect because it is a protocol stack.

3 Windows servers include network client access applications for Macintosh computers.

a) True

b) False

False. Windows servers do not provide special client access applications for Macintosh computers.

4 You've been called in to troubleshoot a network problem. There is only one segment, and the network does not connect to the Internet. The servers are all Windows NT 4.0, and the clients are a mix of Windows 95, Windows NT 4.0, and Macintosh. The administrator is new and is installing his first new computer, a Windows XP computer, on the network. He has selected the TCP/IP protocol and Client for Microsoft Networks. He sees that the computer is receiving an Automatic Private IP Address (APIPA) of 169.254.1.88 with a network mask of 255.255.0.0, without any default gateway address or DNS server address. The computer does not connect to anything on the network. What is likely the problem?

a) The network is probably using a different protocol stack, most likely NetBEUI.

b) Windows XP computers are too advanced to connect to Windows NT servers.

c) There is no default gateway address for the data to route across.

d) The DNS server does not include a name for the client.

Answer a is correct. While there can be many reasons why a computer is not connecting to the network, of these four scenarios, the most likely is that the new administrator selected the wrong protocol for connecting to the network. Answer b is simply not true. Answer c is also incorrect; there is no need for a default gateway address. Answer d is incorrect; the client's name doesn't have to be in DNS for it to connect to network resources.

5 You have been asked by a company to install a network in its small office. The manager wants to use Windows XP computers, dedicating one of them to provide DHCP addresses in the range of her Class C address and to run DNS, but not to share an Internet connection. What problem will you run into if you attempt to do this?

a) The dedicated Windows XP computer will need Samba.

b) All of the Windows XP computers will need client access applications installed to communicate with each other.

c) The dedicated Windows XP computer will not have the DHCP or DNS services available to configure this way.

d) An Internet connection is required to run DNS.

Answer c is correct. The dedicated Windows XP computer does not have DHCP or DNS services that can be configured as the client requires. Answer a is incorrect because Samba is not required. Answer b is incorrect because the client access application (Client for Microsoft Networks) is already installed within the computer. Answer d is incorrect because DNS does not require Internet connectivity.

6 Which of the following commands can be used to make a Windows 2000 server share named \\server\share appear as the drive letter Y on a Windows XP computer?

a) Net \\server\share Y:

b) Map Y: \\server\share

c) Net use Y: \\server\share

d) Map use Y: \\server\share

Answer c is correct. The Net Use command is used to connect a share to a drive letter on a Windows-based client. The proper syntax is net use Y: \\server\share. Answer a is incorrect because the syntax is wrong. Answers b and d are incorrect because the map command is not used for Windows servers.

7 You have received a call from a user who is unable to log onto the network. The user has a Windows XP Professional laptop that he purchased outside the company, and this is the first time that he has tried to connect to the network. The user can access the Internet without problems, but he doesn't see his domain in the logon dialog box even though that domain appears on his workstation when it is plugged into the same network outlet. Given this information, what is the most likely problem?

a) The user has tried the wrong username and password for the domain.

b) The laptop has not joined the domain.

c) The user needs to try the Net Use command.

d) The cable is bad.

Answer b is correct. If the computer doesn't have an account within a domain, it will not provide you with a dialog box that can log on to a domain. Answer a is incorrect because the user hasn't reached the point where a logon can take place. Answer c is incorrect because Net Use is used to connect to network resources. Answer d is incorrect because the cable works when it is used with the workstation.

8 Which of the following clients doesn't require a hard drive?

a) Windows terminal

b) Macintosh

c) Linux

d) Windows 2000 Professional

Answer a, Windows terminal, is correct. This type of client does not require a hard drive because it has a locally installed operating system. Answers b, c, and d are incorrect because each of these clients must have a hard drive on which to store the operating system.

9 Which of the following operating systems can share an Internet connection?

a) Windows 95

b) Windows XP Professional

c) Windows NT 3.1

d) Windows 98 First Edition

Answer b is correct. Windows XP Professional has the ability to share an Internet connection. Answers a, c, and d are all incorrect because these OSs cannot share an Internet connection. Note that Windows 98 Second Edition can share an Internet connection.

10 A Windows 98 computer must be a member of an Active Directory domain before the user can logon.

a) True

b) False

False. Windows 98 machines are not treated as true computer accounts by the domain and cannot join a domain.

Chapter 14

Tools and Troubleshooting

When part of a network goes down, whether it's a workstation, a server, or an entire group of network hosts, troubleshooting involves asking where the problem is and how to fix it. True troubleshooting can appear to be an art, but it's really a system that methodically tests and then isolates the problem to a specific cause. Troubleshooting often requires tools and utilities—the right tool for the right job, and the problem is easily solved.

In this chapter, we'll look at the tools used for troubleshooting network wiring problems. Then we'll go over a methodical system for troubleshooting that can help in any situation. And finally we'll review how to troubleshoot common network problems.

Test Smart Memorize the troubleshooting steps; they might be on the test, and they'll definitely help you in your networking career.

Tools of the Trade

A few years ago, as director at an IT company I was tasked with putting together a Microsoft Business Practice in collaboration with 140 Microsoft Certified Systems Engineers (MCSEs). The group was spread over the entire U.S., and one of my goals was to find a way to welcome them all and make them feel like part of the team. I decided to put together a welcome kit, complete with a laptop bag, notebook, pens, software, and a small toolkit. When I proposed this to the folks in the marketing department and they saw the toolkit, I swear they started to spit nails. They protested, "These are engineers! They will never be opening up a PC—that's a technician's job." (You have to have patience with people who don't understand a networking engineer's job.) I convinced them that the toolkit was a good move, and we all compromised on a small one that had the essentials for networking, including a flashlight, screwdrivers, a media tester, and a variety of other handy instruments. As it turned out, the toolkit was a smash hit. Let's face it, the best engineers like to tinker around and can't help but investigate a problem on the network.

You'll need several tools to fix cabling. The logic behind network cabling is fairly simple. In UTP, for example, the wires are color coded; you send signals using one set of tools across the cable to find the distance to the problem, and eventually you uncover the problem and use a different set of tools to fix it. Here are some tips for working with cables.

- **Tip 1.** Always use the correct cable, and make certain you have a few extra lengths of it on hand. Many users think that because the telephone jack (RJ-11) looks so similar to an Ethernet jack (RJ-45), the two cables can be interchanged. Not so. The network will not communicate across a cable just because it looks similar to the correct cable; it has to have the right number of wires in the right places and the ability to carry the electrical signal across the right distance. For example, an Ethernet network very likely has Category 5 (Cat5) cabling installed. If you try to replace a drop cable (between the wall and the PC) or a patch cable (between the patch panel and the hub) with a telephone wire, which might be Category 3 (Cat3) or flat telephone satin, you'll lose connectivity. A quick comparison between the two will reveal that the telephone cable has fewer wires than the Cat5 cable.
- **Tip 2.** Always make sure your cables are the correct length and are in the right place. A pinched cable—especially a fiber optic cable—isn't likely to carry a signal because the strands of copper or optical media can be broken. A cable too close to power lines or electrical fixtures might suffer from interference; and one that is too long will lose the signal before it has a chance to reach its destination. Make certain that you add the length of the cable's entire distance, including the patch cable, drop cable, and the cable behind the walls.
- **Tip 3.** Write down your network's configuration. If you know where your cables are running behind the walls and above the ceilings, you'll find it easier to troubleshoot, not to mention that you'll probably appreciate less time spent above the ceiling and fixing holes in the walls.

- **Tip 4.** If you perform your own network termination, make certain that your wires are in the correct wiring order. For example, two wiring standards created by AT&T—T658A and T658B—apply for Cat5. You can employ either one because they use the same pairing scheme (just different colors), but you have to make certain that both ends of the wire use the same standard. Another problem to watch out for is accidentally using a crossover cable in place of a standard cable. A crossover cable is wired in such a way that the data is transmitted directly to the receiving pins on the other side. It literally crosses the wires. You would use a crossover cable only to connect two computers directly without a hub in between, or possibly to connect two hubs together. If you do use a crossover cable with a hub on a standard port, you won't receive any connectivity.

Now let's look at the tools that you'll need when working with your cables.

Wire Crimper

A wire crimper connects a cable to its terminating connector. While there are different types of crimping tools for different types of connections, the most common one that you'll need is for an RJ-45 connector on a Cat5 cable.

The crimper helps attach the terminating connector, which is sometimes called an ice cube because it is made of clear plastic and is roughly cube-shaped. A Cat5 wire has four pairs of wires inside it, each corresponding to the eight leads in the RJ-45 connector. The crimper locks the RJ-45 connector onto the wire by pushing the copper feet into the connector, so that they pinch down onto the wires as the wires are inserted and the connector is clamped securely on the cable.

You can use these steps to terminate an Ethernet cable with an RJ-45 connector.

1. Strip the cable using the crimping tool's cable-stripping blades. (Most crimping tools have this feature.) Simply insert the cable so that the end of it hits the "stop" at one side. (This ensures that you don't strip too far down the wire. When the wire is stripped too far, it untwists and there can be interference across the line.) Close the clamp so that it slices into the outer insulating plastic of the wire, and then pull the cable out of the crimper, leaving the plastic coating behind.
2. Select your pin-out color sequence based on your cabling scheme. If you're unsure of your network standard, simply select a cable and copy the colors. The term *pin-out* refers to the sequence of colors for

the eight pairs of wires. For a standard cable, this sequence must be identical at each end.

3 Move the individual wires so that they're in the correct order, and then insert the cable into the connector. The wires should slide into the grooves inside the connector. RJ-45 connectors are transparent, so you should be able to identify the pin-out easily.

4 Insert the RJ-45 connector into the wire crimper, with the cable wires still inserted. The copper (or gold) teeth of the connector should face the crimper's teeth, and the connector should be inserted fully into the crimper.

5 Close the crimper to secure the connector to the cable, but don't crimp the wire too much. Pinched wires will not carry a good signal.

6 Test the cable to see whether it will work. You can do this by replacing an existing "good" wire with the one you just created. If the station connects to the network, you were successful.

Punch Down Tool

The punch down tool (sometimes simply referred to as "punch tool") is used to connect a cable to a wall jack. While a crimper adds a male terminator to the wire, the punch tool adds a female terminator, which is simply the wall socket. A punch down tool literally punches a wire inside a metal junction that slices through the cable's insulation to let the wall jack's pins make contact with the wire.

Media Tester

You can use several types of tools to test your cables. All are forms of a media tester.

The digital voltmeter (DVM) is a fairly simple and inexpensive tool that helps determine whether you have a faulty cable. It can also be used to test the power supply. As the name suggests, this tool measures voltage as it passes through resistance. In the network, a DVM can measure continuity to ensure that the cable can carry current.

A time domain reflectometer (TDR) tests whether the media can carry a specific signal, not just current. The TDR sends a signal across the wire and then waits for it to echo back from a discontinuity in the cable. Changes in impedance will be displayed to help locate the distance to the fault. The cumulative amount of echo, or reflection, caused by the variations in impedance is called *return loss*. The impedance and return-loss measurements will vary depending on the frequency of the signal, therefore, the use of a TDR requires that you test at the same frequency as your network. It's best to use a TDR that has been designed for your specific network type. TDRs can help you find shorts and shield defects, kinks, and other interruptions in the cable.

Advanced cable testers usually combine the capabilities of a DVM and a TDR and have additional abilities for analysis of network traffic. One type of cable tester is an oscilloscope. The oscilloscope is somewhat different from a TDR because it provides output in a graphical format, but it is used for the same purpose—to find shorts and faults in the wire. Oscilloscopes are used for a wider range of testing than just network media.

Protocol analyzers are extremely advanced, and pricey to boot. They can look at each passing packet on the cable. A protocol analyzer is effective at finding faulty network interface cards (NICs), bridges, switches, and routers.

An external loopback adapter tests NICs. The loopback adapter can be connected directly to the card (while it's inside the PC) and used to test the card to see whether it functions. This works without the NIC having to be hooked up to the network. Don't confuse a physical loopback adapter with the software version mentioned in Windows NT computers. Windows NT included a software loopback adapter for installing network protocols when there wasn't a physical NIC in the machine.

Before you begin to test cabling using any tool, you should be able to answer these questions:

- **Is the wire connected to the correct port?** On a few occasions, I've found cables hanging out of the patch panel, uncoupled from the hub or switch that they should have been connected to. I've also come across servers that have been mysteriously disconnected from the network. (Most organizations lock their cabling closets and server rooms to avoid the above problems.)
- **Are the visible cables free from defects?** Make certain that the cables don't have any nicks, cuts, kinks, or other strange flaws. Users shouldn't hang holiday lights or ornaments on their drop cables, no matter how convenient it might seem.
- **Do you have a map of the wires behind the walls?** If you can roughly follow the pattern of the wiring, make certain that there are no fluorescent lights, new fixtures, new power cables, or other sources of interference that pass near your network cables.
- **Are your cables tested before they're used in the network?** If you have just put a new cable in the network and the machine or

machines connected to that cable are not able to communicate with the rest of the network, it's likely that your new cable is the culprit.

- **Are your cables the correct length?** Any cables that are longer than the maximum for the network might appear to connect but then lose connectivity intermittently. Always add up the length of the patch cable, the drop cable, and the one between to find the maximum length. If you can't measure some of the cables, you can use the TDR to find their total lengths.

Optical Tester

An optical tester is simply a media tester made specifically for fiber optic networks. Instead of concerning itself with voltage, the optical tester verifies and validates light signals.

Two properties of fiber optics make it difficult to work with:

- **Fiber optic cable can be installed over long distances.** Long lengths of cabling are usually buried in trenches. If there is a problem with the cable, not only must you locate the problem without seeing the cable but you must also dig it up to fix it.
- **Fiber optic cables are extremely brittle.** They can break if bent just a bit too far. This means that breakage can result from any movement, such as the shifting of the soil that the cables are buried in. When fiber optic cables are buried, they're usually placed within a heavy plastic watertight conduit.

Overall, fiber optic cabling can be a headache, but it's worthwhile when you consider the much higher speed it offers, the long distance it provides for high speed traffic, and its resistance to interference. To test fiber optic cables, look at each fiber segment individually. An optical light source and power meter can be used to check for end-to-end attenuation of the signal. An optical time domain reflectometer (OTDR) locates and measures signal losses (similar to a TDR used on copper cables).

Tone Generator

A tone generator, which is sometimes referred to as the *fox and hound*, tests the cabling by producing different tones or electronic signals from a wall socket, or it can be clamped directly into the wire. You use the tone generator with a signal amplifier to determine which wire is connected to the jack the tone generator is connected to. With the tone generator sending signals from the wall socket, you use the signal amplifier as a probe at the other end of the wires. As you use the probe on the wire, which is actually attached to the tone generator, the sound that the signal amplifier generates will be louder. Usually, the sheath that protects the other cables from the one that you're testing will not let the signal through, and you won't hear a tone on those other cables. However, if the cable sheathing is somehow stripped or broken, you might hear some crosstalk that generates a tone at the probe's end.

You'll find a tone generator exceedingly useful in two specific situations. The first is in dealing with a tangle of wires. Over time and with lack of care, a wiring closet can become extremely disorganized. A tone generator makes it a simple task to find which wire belongs to which wall jack.

A tone generator also becomes invaluable when you suspect that two wires have gotten crossed and you're receiving crosstalk. Crosstalk can happen when wires come into close contact, such as when the insulation is cut on two neighboring cables.

Visual Indicators (Link and Collision Lights)

While the link and collision lights on network equipment are not tools per se, you can definitely use them for troubleshooting connectivity problems. Link lights are found on networking equipment such as NICs, hubs, routers, and switches.

Typically, if a port is not connected to a cable, the light emitting diodes (LEDs) don't light up, whether they are for the purpose of detecting a link or collisions. When the port is connected to a cable, which is connected to a workstation that is turned off, the LEDs are either off, or the LED for the link is red while the others are off. When the computer is connected to the network, the link LEDs shine amber or green. When a computer is connected to the cable, which is connected to the port, but the computer can't communicate, a quick check of the link lights will tell you whether the hub detects a signal coming from the computer. The color of the link light can also provide additional information. There are no standards applied to link and collision lights. Each manufacturer tends to develop its own interpretation for the lights' colors and whether they blink or are solidly lit. For this reason you should always check with the manufacturer's documentation to see what a particular light may mean. When there is no link light and everything should be working, try checking for a loose cable.

Collision lights tell you how many collisions are taking place on the network. The more collisions on a network, the slower the network. In some cases, collisions can get so heavy that network traffic slows to a crawl. If the network has become sluggish, this is one of the first places to check. Keep in mind that collisions are normal on an Ethernet network, but a collision light that is continuously on indicates a problem with a workstation on the segment.

The Troubleshooting Method

Knowing the proper method to troubleshoot can save you a tremendous amount of time (plus make you look like a hero to those who have suffered an outage). The troubleshooting method is fairly logical, and it is diagrammed in Figure 14-1.

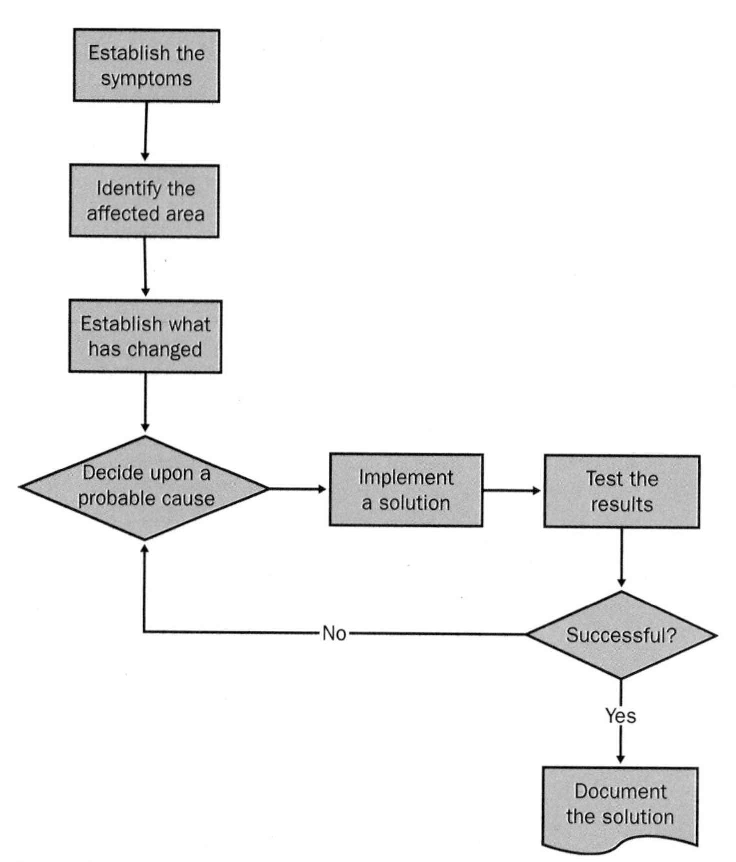

Figure 14-1 The troubleshooting method follows a progression that repeats until a solution is found.

The first step in troubleshooting is to be absolutely certain about the symptoms of the problem. Many times a user will provide the exact same description for two entirely different problems. I worked on a migration from Microsoft Windows 95 with Office 95 to Windows 2000 Professional with Office 2000. One office complained bitterly about the slow network performance. The office was fairly small, and after speaking with each user on the network it turned out that only one person was affected by the "slow network" and had complained so much that all the people in the office decided to report the problem.

I soon found out from the original user that only a handful of documents were actually slow. It turned out that the user hadn't converted her Word document templates by opening each template and resaving it as instructed by my project team in the migration documentation, and as a result each of the old templates was exceedingly slow. This is a classic case of a problem being blown out of proportion. Unfortunately, it's very common. To ensure that you don't get caught up in any user's distress, you must ask a lot of questions to find out the exact symptoms. If at all possible, you should try to replicate the problem yourself.

The second thing that you must do is narrow down the problem to a specific area of the network. One network problem can affect the entire wide area network (WAN), a building with several local area networks (LAN), a single hub and all its attached devices, a server, a workstation, or a single application.

After establishing the symptoms and finding out which part of the network has experienced them, you must then find out what has changed on the network since the first symptom of the problem appeared. Was a logon script changed? Did a user upgrade an application on his own workstation? Did an administrator change permissions for a group? A million seemingly innocuous changes can take place on a network and cause disruption.

Knowing the symptoms, the part of the network that has been affected, and what has changed, you can probably guess at a cause. You might come up with several theories about the cause, and it's always a good idea to list them on paper in order of probability.

Your next step is to select a solution and implement it. For example, if users can't connect to the network, you isolate that problem to a single network segment with all the users on the same hub, but there have been no changes since the outage occurred, you can probably guess that the hub is not functioning. The solution to this might be to cycle the power on the hub. Another solution might be to replace the cables connecting it to the rest of the network. Yet another would be to replace the hub with a new one. You should simply select one of these solutions and go ahead with it.

Let's assume that you've decided to cycle the power on the hub. Once you complete that process, you should check to see whether it has been successful. Do the link lights come on? Can workstations connect? What are your results if you try to ping a server from one of the affected workstations?

The final step is to decide whether your solution was successful. If so, document the results. Otherwise, select a different solution and implement it.

Test Smart You must know the troubleshooting method steps in order on the exam. 1. Establish the symptoms of the problem. 2. Discover what part of the network has been affected. 3. Confirm what has changed on the network. 4. Decide on a probable cause. 5. Try a solution. 6. Test the results. 7. Determine whether the solution was successful. If not, try the next probable solution.

Common Problems

A wired network (as opposed to a wireless network) suffers many of the same problems that other wired networks suffer. The cabling itself fails, or there is interference, or the termination is wrong. In this section, we'll review how to isolate and correct problems that you might find in the media.

Wiring Problems

Cables are prone to failure. One day they work fine; the next they don't. Because of this, checking the cabling when a workstation can't connect to the network is usually one of the first, if not the very first, solution that is tested. The most common cable to fail is the drop cable, which is the one that connects a workstation to the wall jack. Drop cables receive the most abuse, whether from being kicked, yanked, bent, or twisted.

When you determine that a cable might be the culprit of a network-related problem, your first test of this theory is to replace the cable with a known good cable. If the replacement cable works, you've solved the problem.

If you replace a drop cable and a patch cable but still haven't gained connectivity for that workstation, rather than try to re-cable behind the walls, you still need to determine whether the problem is the cable. In this case, move the affected workstation to a different location where you know the cable is in good working condition. If the workstation functions properly at the new location, you know that the cable behind the walls is at fault and you can work on fixing it. If the workstation fails to communicate with a known good cable, the workstation itself has a different problem to troubleshoot. Table 14-1 shows some cable problems and their probable causes.

Table 14-1 Cabling Problems

Cable Type	Problem	Probable Cause
ThinNet or ThickNet coax	None of the workstations can connect.	The cable has been severed and should be reconnected.
UTP	A new cable is used to connect a workstation, which suddenly stops working.	Either the cable is not terminated correctly or it is a crossover cable.

Table 14-1 Cabling Problems

Cable Type	Problem	Probable Cause
Any type	A known good workstation was moved to a new location that functioned previously, and now the workstation cannot communicate.	The drop cable might have been damaged in the move.
UTP	A workstation intermittently loses communication.	Check for loose cable terminators or kinked cable.
UTP	No workstations can connect after you have installed a new piece of equipment.	The new equipment is using a foreign protocol that interrupts communication and should be disconnected.

Interference with the Network

Interference is what I call an "icky" problem. Icky is my own acronym for "it can kill your" network (or project, depending on your situation). I once worked on a token ring network that had recently been rewired from shielded twisted pair (STP) to unshielded twisted pair (UTP). There were several ports on the network that, after the rewiring, were wrapping at the Multistation Access Unit (MAU). After several weeks of intermittent problems, a buzzing fluorescent light clued me in, and a quick test with a media tester confirmed the problem. The wiring contractors had laid the new wiring in the ceiling by following the old wiring paths. While this led the new wires to their correct destinations, it also placed a lot of unshielded wire near fluorescent lights, so the wires suffered interference. Fixing the problem required new cabling to be installed at the expense of the poor contractor.

Fiber optic cables do not suffer interference. This is one of the benefits of using fiber optics. You can place them in locations that have high magnetic fields without fear of network failure.

When wiring a new office or planning a wiring layout with copper cabling, you should avoid placement of the following near your cables:

- Power outlets
- Uninterruptible power supplies (UPS)
- Fluorescent lighting
- Strong magnets

Bad Wiring and Connectors

Network performance is affected by several factors that involve cabling: the connecting hardware, the patch and drop cables, the cable's characteristics, the number of connecting hosts, and the care with which the cable has been installed. You can end up with bad cabling from simply not taking notice of what's on the network.

A properly installed cable is flexible and can be moved or curved with no effect on network performance. However, if you replace a Cat5 patch cable with a Cat3 cable, the network performance will degrade. If you replace a patch cable with a Cat3 cable that is longer than the cable it replaced, network performance will further degrade.

Sheathing on the cable is important to retain the twists in the wire, which ensure that there is no crosstalk. If you remove the cable sheathing, the network's performance degrades. This happens if even only one inch of cabling is in question. If the wires untwist and the cable sheathing is removed, the network performance degrades much faster. (Some tests have shown 5 to 10 times the rate of degradation, depending on the length of the cable.)

Kinked cables can cause network performance degradation as well. A UTP cable can be wound up in a roll up to six inches in diameter with no effect on the network. However, if that cable is wound more tightly around a three-inch diameter, the performance breaks down. Any time a cable is wound more tightly or kinked, connectivity loss will occur.

Connectors are the source of termination problems. When a jack is loose or sheathing is exposed for too long a length, the network will show degraded performance or no performance at all. Incorrect termination—in which the wires are not in the correct sequence—will result in no connectivity.

To avoid bad wiring, you can perform a few tasks at the time that you cable the network:

- Mark each cable end before cabling.
- Pull the cables throughout the building, making certain that there are no kinks, twists, or exposed wires, and that the cable is away from any electrical or magnetic fixtures.
- Punch down the cables at each wall jack and at the corresponding patch panel.
- Test the cable between the wall jack and patch panel to ensure it can carry a signal. Check for the following:
    - Wire map: validate whether the cable termination has been done with the correct pairings in the correct sequence.
    - Signal attenuation: ensure that the signal loss is within acceptable limits for the cabling specification.

    - Length: verify whether the physical length is below the maximum for the cabling specification, such as 100 meters for Ethernet over Cat5 cable.
    - Propagation delays: check for the time that it takes for signals to travel from one end of the cable to the other.
- Install the networking equipment and workstations.

Key Points

- Tools are used to troubleshoot networks, especially when cabling problems occur.
- A wire crimper is used to add a male jack at the end of the cable.
- A punch down tool is used to add a female jack (a wall jack or patch panel jack).
- There are several types of media testers. A digital voltmeter tests signal power. A time domain reflectometer tests for cable problems.
- Optical testers are used strictly for fiber optic cabling.
- A tone generator is useful in finding crosstalk and identifying unmarked cables.
- Visual indicators will display whether a link has been successfully made and whether there are too many collisions.
- The troubleshooting method consists of these steps:
    - Establish the symptoms.
    - Discover what has been affected.
    - Determine what has changed.
    - Decide on a probable cause.
    - Implement a solution.
    - Test the results.
    - Determine whether you've been successful.

Chapter Review Questions

1 You've been assigned the task of determining which cable is connected to a workstation that is not functioning. The cables aren't marked. Which tool is the best choice for the job?

a) Wire crimper

b) Punch down tool

c) Time domain reflectometer

d) Tone generator

Answer d is correct. The tone generator is used to send a signal down a wire with a signal amplifier at the other end. The correct wire will generate the loudest sound at the signal amplifier. Answer a, wire crimper, is incorrect because it's used to terminate cables with connectors. Answer b, punch down tool, is incorrect because it is used to terminate cables at wall jacks and patch panels. Answer c, time domain reflectometer, is incorrect because it is used to detect problems in the cable.

2 A periodically flashing collision light on a NIC indicates that the card has failed.

a) True

b) False

False. Collisions are normal on an Ethernet network, and you should see a collision light flash periodically. The collision light will indicate a problem if it is always lit, showing that there is nothing but collisions on the wire.

3 You're instructed to find a problem with some network cabling. All the workstations, servers, and printers have failed to connect. The network consists of three Ethernet hubs that are all linked to a fiber optic backbone. Which tool do you select?

a) Wire crimper

b) Tone generator

c) Digital voltmeter

d) Optical tester

Answer d is correct. Because all of the hosts can't connect, the problem is most likely the backbone, which is a fiber optic network. The only tool listed for fiber optics is the optical tester. Answer a is incorrect because a wire crimper is used to terminate cables. Answer b, a tone generator, is incorrect because it is used to discover cables that are not marked. Answer c, a digital voltmeter, is incorrect because it is used to discover the electrical signal power, and a fiber optic cable uses light signals.

4 Your network has three hubs, each supporting 16 users. Each hub is dedicated to a department in the organization. Hub A is used by Accounting; hub B is used by Sales; and hub C is used by Production. Three users in Sales can't connect to the network. Which of the following should you use to help diagnose the problem?

- **a)** Link lights on hub A
- **b)** Link lights on hub B
- **c)** Collision lights on hub A
- **d)** Collision lights on hub C

Answer b is correct. Given that all of Sales connects to hub B, the link lights on hub B should give you an indication of whether the signal is getting through to the hub. Answers a, c, and d are all incorrect because none of these is the hub that Sales uses.

5 You've finished interviewing three users who have had problems on the network and have listed the symptoms of the problem. You've also narrowed down the problem so that you know it's affecting only the users who are trying to use resources on a certain server. What is your next step in the troubleshooting method?

- **a)** Implement a solution.
- **b)** Determine what has changed.
- **c)** Test your results.
- **d)** Use a media tester on the cable connected to the server.

Answer b is correct. You should determine what has changed that might have caused this problem to occur. Answer a is incorrect because you have not decided a probable cause in order to determine a solution. Answer c is incorrect because you have not yet implemented a solution to test. Answer d is incorrect because it is not a troubleshooting method step.

6 A company has hired you to replace a contractor in the middle of a wiring job. The contractor used Cat5 cabling, but not all of the cables are connected and none of the wires are labeled. You have to wire up the server room first, and after testing you find that one of the wall jacks is not hot, while all the others are. What is the best tool for finding the cable that attaches to that wall jack?

a) Time domain reflectometer

b) Tone generator

c) Digital voltmeter

d) Optical tester

Answer b is correct; a tone generator sends a signal across a wire, and a signal amplifier (a probe) at the other end generates a tone when the signal is received. Answer a is incorrect because the time domain reflectometer tests the cable for its viability. Answer c is incorrect because the digital voltmeter tests the power of the signals on the cable. Answer d is incorrect because optical testers will not test a Cat5 cable.

7 A cable that is 94 meters long connected to a seven-meter drop cable and a three-meter patch cable is too long to carry a good signal.

a) True

b) False

True. The signal will attenuate when it reaches beyond 100 meters; the total length of this cable is four meters too long.

8 You've heard that three people are having problems connecting to the network. You immediately replace the hub that one of the people is connected to. What steps have you missed in the troubleshooting method? Select all that apply.

a) Test your results.

b) Determine what has changed.

c) Determine what area of the network is affected.

d) Establish the symptoms of the problem.

Answers b, c, and d are correct. You've skipped establishing the symptoms of the problem (step 1), determining what area of the network has been affected (step 2), and establishing what has changed (step 3). Answer a is incorrect because it's the step that follows implementing a solution, which is what was done first by replacing the hub.

9 Which of the following indicates a possible bad cable?

a) The cable is terminated at both ends by an RJ-45 male connector.

b) The cable is wound in a roll that is eight inches in diameter.

c) The cable is missing six inches of its plastic sheathing.

d) The cable is 80 meters in length from patch panel to wall jack, using Cat5.

Answer c is correct. A cable that is missing even a small portion of its sheathing can indicate that it might fail. Answers a, b, and d are all incorrect because they are features of a cable that might be in good condition.

10 You're trying to figure out what is happening with a workstation that constantly disconnects from the network. None of the other workstations are affected. You can't find a problem with the network, so you finally move the user to a known good location with existing functioning cables. The disconnecting problem follows the user to the new location, and the problem disappears from the old one. Which of the following could be a cause of this problem?

a) The cable at the old location is bad.

b) The user has been using the drop cable to string lights and magnets in his cubicle.

c) The cable at the new location is bad.

d) The hub has a bad port.

Answer b is correct. These are symptoms that indicate interference, which can be caused by electromagnetic fields. Answer a is incorrect because the cable worked after the user moved. Answer c is incorrect because the cable worked before the user moved to that location. Answer d is incorrect because the two cables would be attached to two different ports in a hub.

Chapter 15

Troubleshooting a Network

Problem resolution, troubleshooting, and just getting the job done demand that you be creative, inquiring, and logical in your process. Many people who enter the IT industry, whether they are engineers or administrators, have a knack for this, and I think it's mainly because they've embraced the troubleshooting method.

Even seemingly impossible problems can be resolved if you work them out logically. The worst problems are actually opportunities for growth. Once you have fixed something difficult, the next time the same or a similar problem appears it has become fairly easy to fix. You should keep one thing in mind when you're troubleshooting. Spend 20 percent or less of your time describing the problem and 80 percent or more of your time actively solving it.

Applying the Troubleshooting Method to Logical Problems

As you'll recall from Chapter 14, the troubleshooting method consists of the following steps:

1. Establishing the symptoms of the problem
2. Identifying the area affected by the problem
3. Determining what has changed

4. Selecting a probable cause for the problem
5. Implementing a solution
6. Testing the result
7. Recognizing whether the solution is successful, and if it is unsuccessful, selecting the next most probable cause and solution
8. Documenting the solution

So many situations might require your troubleshooting skills—users who have created their own errors, applications that won't work without the correct permissions, connections that don't connect, and configuration errors, to mention a few. These are not always physical problems, in which a cable is bad or a power supply is unplugged, for example. The majority of troubleshooting involves logical problems. In following the troubleshooting method, your first job is to find out exactly what is happening. (Sometimes it's not a bug, it's actually how the software or operating system is supposed to behave. It just doesn't meet the expectations of the user.)

When troubleshooting errors involving an individual user and workstation, in step 1, you'll need to find out what steps the user is following when the error takes place. This will not only show you whether the user is causing the problem inadvertently, it will also show you how to re-create the problem and what components are involved in the user's process.

In step 2, you need to verify what area is affected. Are other users experiencing the same problem? Is this problem limited to a certain application? A workstation? Does the problem follow the user around the network if the process is tried at a different station?

The next course of action is to find out what has changed since the process last worked correctly. If it is the first time the user has tried it, you should always consider the possibility that the user lacks the correct rights and privileges, which we will discuss later in this chapter.

Step 4 instructs you to decide on a probable cause. This will depend entirely on the circumstances of the problem, and there might be more than one possible cause.

In step 5, you implement a solution, in step 6 you test it, and in step 7 you recognize whether the solution was successful. If not, you try another solution.

Once you've fixed the problem, you should document it. A well-documented network means that others who follow in your footsteps are able to keep the network running smoothly when you're away.

Troubleshooting the Connectivity Protocols

You should be aware of three connectivity services: Dynamic Host Configuration Protocol (DHCP), Domain Naming System (DNS), and Windows Internet Naming Service (WINS). Each can cause workstations to stop working with the network. While the symptoms that indicate a problem with a particular service differ, the results are similar—the user (or users) can't connect to the network.

The PC Can't Obtain an IP Address

Transmission Control Protocol/Internet Protocol (TCP/IP) is an intimidating protocol suite because it appears to be a very complex and comprehensive collection of protocols and utilities. Some of the protocols are necessary, others simply make administration of the network easier, and yet others are cool utilities. DHCP is the protocol that enables a server to provide an IP address to a computer on a leased basis—that is, the host that receives the IP address is usually granted its use on a temporary basis. The fact that a computer can lose its lease means that it can also lose connectivity if no DHCP server is available to renew the IP address or deliver a new one.

DHCP problems are usually driven by insufficiently distributed DHCP servers. I once did some consulting for a large financial institution whose head office was located in a different state. The managers had decided that they wanted to centralize everything regarding IT at the head office, and as a result most servers were relocated long before I was called in to help. Centralizing the DHCP servers in one location caused 90 percent of the entire company's computers to grab their IP addresses from across the wide area network (WAN), as shown in Figure 15-1. One day, the WAN connection to the offices in Phoenix went down, and because the head office had designated a short lease duration, thousands of computers in the city could not communicate. It was fairly easy to determine the cause of that problem; the only fixes involved waiting for the WAN connection to return and installing a temporary DHCP server with short leases. (As you can see, network design plays an important part in network uptime.)

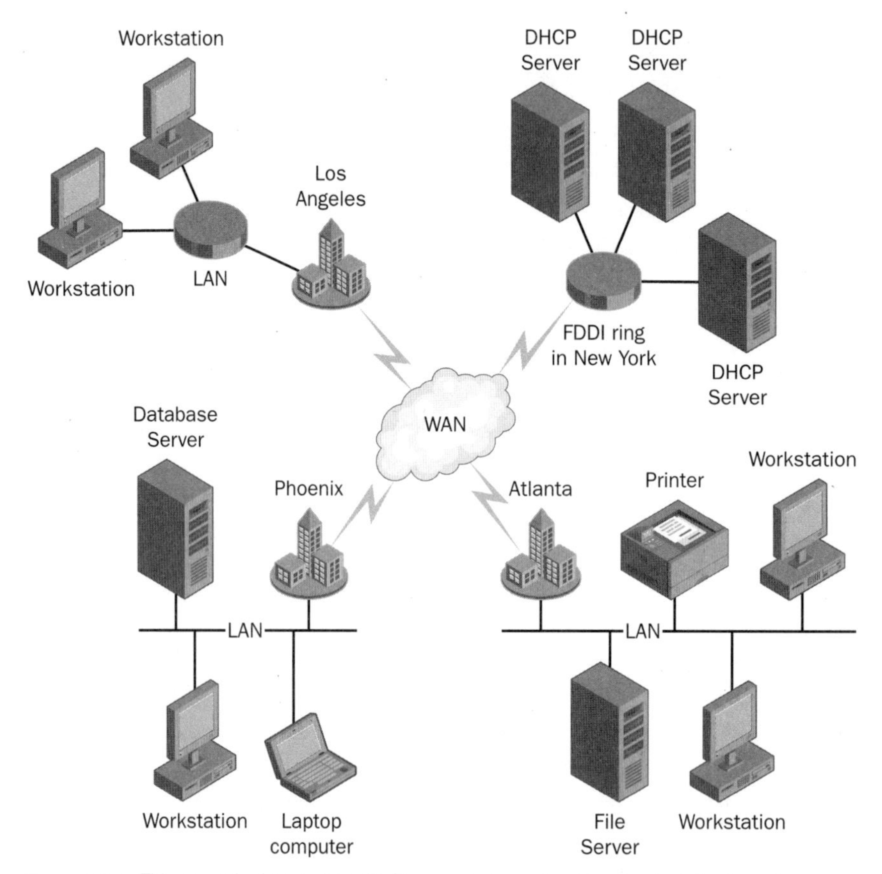

Figure 15-1 This network placed all its DHCP servers in one location, forcing all other sites to communicate across the WAN to get IP addresses.

While an inability to communicate with the DHCP server is the cause of DHCP problems, another trouble is simply misconfiguration. If the DHCP server delivers an incorrect subnet mask or default gateway address, communication suffers.

Ipconfig is an extremely useful tool for diagnosing a DHCP problem. This tool will tell you exactly what you have received from the DHCP server, such as an IP address, default gateway address, subnet mask, or DNS server addresses, to mention a few. To run Ipconfig (substitute Winipcfg for Ipconfig if you are using a Microsoft Windows 95, Windows 98, or Windows Me machine), open a command prompt, type **Ipconfig**, and then press Enter. This will display a summary of the IP addresses that have been assigned to the adapters on the machine, as shown in Figure 15-2.

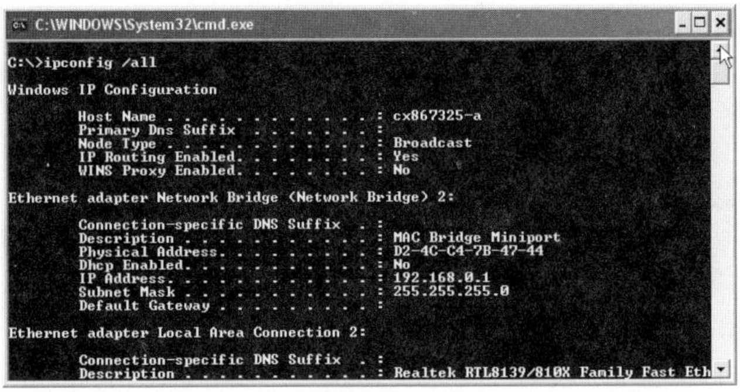

```
C:\WINDOWS\System32\cmd.exe

C:\>ipconfig /all

Windows IP Configuration

        Host Name . . . . . . . . . . . . : cx867325-a
        Primary Dns Suffix  . . . . . . . :
        Node Type . . . . . . . . . . . . : Broadcast
        IP Routing Enabled. . . . . . . . : Yes
        WINS Proxy Enabled. . . . . . . . : No

Ethernet adapter Network Bridge (Network Bridge) 2:

        Connection-specific DNS Suffix  . :
        Description . . . . . . . . . . . : MAC Bridge Miniport
        Physical Address. . . . . . . . . : D2-4C-C4-7B-47-44
        Dhcp Enabled. . . . . . . . . . . : No
        IP Address. . . . . . . . . . . . : 192.168.0.1
        Subnet Mask . . . . . . . . . . . : 255.255.255.0
        Default Gateway . . . . . . . . . :

Ethernet adapter Local Area Connection 2:

        Connection-specific DNS Suffix  . :
        Description . . . . . . . . . . . : Realtek RTL8139/810X Family Fast Eth
```

Figure 15-2 Ipconfig is one of the tools useful for diagnosing DHCP problems.

When your workstation can't contact a DHCP server and the IP address lease has expired, Ipconfig will display an address of 0.0.0.0 *unless* that machine is using Automatic Private IP Addressing (APIPA), in which case the machine will have an IP address beginning with 169.254.x.x. In either case, you'll know that the communication between the DHCP server and the client is not working.

To see whether an address was statically assigned or provided by a DHCP server, type the command **Ipconfig /all**. When you add the /all switch, Ipconfig will display detailed information, including whether an address was delivered via DHCP.

Ipconfig can be used to obtain a new lease on a different IP address, or to renew a lease on an existing IP address. You might want to use this command and switch if you discover that the machine is using an IP address meant for a different subnet. To obtain a new lease, type **Ipconfig /release** at the command prompt. This will release the DHCP lease, and upon reboot the computer will attempt to obtain a new lease from the first available DHCP server. (In Winipcfg, clicking a button will release the IP address lease.)

If you want to renew an existing lease, you can do so by typing **Ipconfig /renew** at the command prompt. Keep in mind that a client will automatically attempt to renew the lease on a periodic basis, but you can try to renew the lease manually to see whether connectivity to the DHCP server still exists. (Again, in Winipcfg, clicking a button will renew an IP address lease.)

When a DHCP client is not communicating with the server, it's likely a problem with the network connection for that computer. You should always check the cable and the network interface card (NIC). You should also ping the loopback address to see whether TCP/IP is installed correctly.

You could have a problem involving a conflicting IP address, which occurs when a person has assigned a static IP address that is identical to one in the DHCP server's pool of IP addresses. A conflicting address is usually indicated by a message on the client computer, but a user who doesn't recognize the message might forget to tell you. You can check to see whether the IP address is in use elsewhere by pinging that address from another computer.

DHCP problems that affect more than one computer usually indicate that the problem is the DHCP server(s) or one of the links between the DHCP servers and the DHCP clients. Before looking further, you should make certain that any router, hub, switch, or bridge is up and running, as well as the cables connecting the back-end networking equipment.

Note Ipconfig and Winipcfg are both useful tools for troubleshooting DHCP problems on Windows machines.

DHCP messages do not route without a little help. Any router sitting between a DHCP server and a DHCP client will need a DHCP Relay Agent (on a Windows NT or Windows 2000 server running the DHCP server service) or a helper address on another type of router in order to forward BOOTP type messages, or even a router that is compliant with Request for Comment (RFC) 1542. If DHCP was functioning before and has since stopped functioning, you can check to see whether any routers have been reconfigured.

You should check the status of the DHCP server itself. On Windows 2000 Server, for example, you can check the status of the DHCP Server Service, and also check to see if the DHCP server has been authorized within the Active Directory. All DHCP servers installed on Windows 2000 must be authorized within the Active Directory before they will provide DHCP services. You should also look at the event viewer to see if there are any messages regarding the DHCP server service. Verify that the cable connecting the server to the network and the NIC are both good.

One possible problem could be a rogue DHCP server. Someone on the network could have installed a new DHCP server and configured it incorrectly, and it could have then begun servicing DHCP clients with the incorrect TCP/IP information.

Internet Names Aren't Recognized

DNS is the naming system used by TCP/IP hosts to resolve names to IP addresses. This means that a human can type a Uniform Resource Locator (URL) such as *http://www.microsoft.com* in a browser address window and trust it to be translated to the correct IP address so that the Web page will download to the browser. So, what do you do when this system breaks down?

Users will probably report that the Internet, or the intranet, is not working at all because they don't know that they can use an IP address in place of the URL. Before you determine that DNS is the problem, you might find that you have checked the TCP/IP configuration of the machine, the cables, the NIC, and the hub. After that, you can simply use ping to see if DNS is functioning. Try pinging a URL, and when that doesn't work, ping a known good IP address. If pinging a known good IP address works, the problem is with DNS.

While it's unlikely that the client resolver cache is the cause of the problem, you might try emptying the cache, which will force DNS to resolve each name and IP address again. The client resolver cache holds DNS names resolved to IP addresses on a temporary basis. To do this on Windows 2000 and later machines, type **Ipconfig /flushdns** at the command prompt.

You might be troubleshooting a problem related to a computer whose IP address has recently been changed. When you use Dynamic DNS, the computer should register the IP address with the DNS server upon reboot. But if you don't reboot the computer, you must manually register the DNS information. You can invoke Dynamic DNS to work on Windows 2000 and later machines by typing **Ipconfig /registerdns** at a command prompt.

If the DNS server is at the root of the problem, you can use the Nslookup command to verify its responsiveness. For example, if the DNS server's IP address is 192.168.1.1, at a command prompt type **Nslookup 192.168.1.1 127.0.0.1**. This command then responds with the name of *localhost* if the DNS server is working, or there will be no response at all and you'll know that the client can't reach the DNS server or that the server itself is not functioning.

You should make certain that, even if the DNS server is running, the DNS service is functioning correctly. On a Windows NT or Windows 2000 Server, look at the event viewer to determine whether there are any critical errors regarding the DNS service. If the DNS service is down, you can type **net start dns** at a command prompt on the server to start the service.

It's possible that the DNS problem lies with your Internet Service Provider (ISP). If your DNS server is functioning correctly but can't forward recursive queries to the Internet, you won't be able to resolve names outside your own DNS domain(s). For example, if your network is mydomain.local, you'll be able to resolve any names in mydomain.local, but not any Internet names such as microsoft.com. There are three possible causes for this problem:

- Your DNS server is incorrectly configured.
- The ISP's DNS server(s) are down.
- The network link to the ISP is down.

The most likely cause is a failed connection to the ISP. The next most likely cause is the failure of the ISP's DNS servers. In both instances, you must work with your ISP to fix the failure. Even though the least likely cause is that the configuration within your own DNS servers is wrong, this could have happened if you've recently made changes to your DNS configuration. Part of the troubleshooting method includes a step to verify what has changed, which will ensure that you will have that information to help lead you to the correct conclusion.

Computers Can't Find Local Names

WINS can cause a computer to seem as though it's not connected to the network at all. Servers stop connecting to each other, workstations don't find network servers, and there is an extreme lag time between connections that do work. Oddly enough, the network still retains Internet connectivity. The key to determining that this is a WINS problem is the fact that TCP/IP and DNS are still functioning, while connecting to a local computer is difficult at best. WINS problems usually fall into the following categories:

- The WINS server or the WINS service is down.
- An incorrect static entry exists in the WINS database.
- Clients have incorrect IP addresses for the WINS servers.
- The WINS database is corrupted.

When you troubleshoot WINS problems, your first step is to see if the WINS server is up and running the WINS service. You can use the event viewer to find out whether there were any WINS errors. If the service is down, you can start the WINS service on the Windows NT or Windows 2000 Server (WINS runs only on Windows servers) by typing **net start wins** at a command prompt on the server.

The next likely possibility is that static WINS entries are preventing the replication of dynamically created WINS entries for hosts with more than one NIC. When a machine is connected to two different subnets, it might not appear as a member of a domain if the static entry points to an IP address that is on an unreachable subnet. Replication of the dynamic WINS entries is prevented because static entries take precedence over dynamic entries. WINS doesn't require static entries to work. So, when troubleshooting WINS, you should remove all static entries and see if that solves the problem. It is always a good idea to write down anything that you will be deleting so that you can add it back later, if it is a required static entry.

If clients have the wrong IP addresses for WINS servers in their TCP/IP configuration, they won't be able to resolve NetBIOS names to IP addresses. It's fairly easy to fix a problem with a client that has a statically assigned IP address. As you can see in Figure 15-3, the WINS server's IP address can be changed in the properties for a network connection. This figure is from a Windows XP Professional computer, and most Windows machines have similar dialog boxes. However, a client that receives its TCP/IP configuration from a DHCP server will need to have the DHCP server's scope parameters changed to reflect the correct IP address of the WINS server.

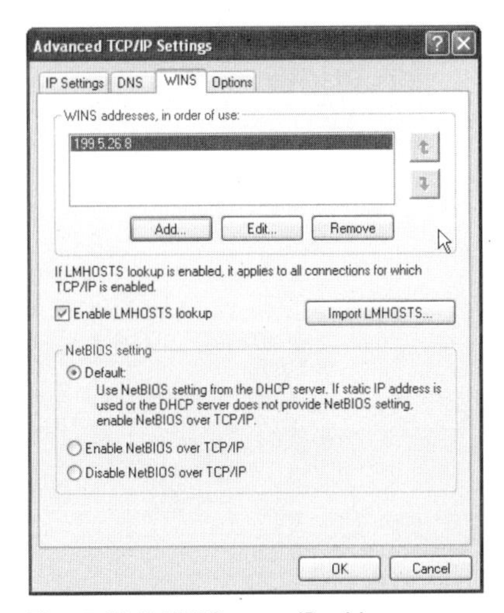

Figure 15-3 WINS server IP addresses are assigned in Network properties dialogs.

You can use Nbtstat to troubleshoot WINS because it will show you how the computer is resolving NetBIOS names. The command can also delete or rectify existing entries.

At a command prompt, you can type **Nbtstat –c** to display the contents of the NetBIOS name cache. If you discover that this cache has a problem, you can purge and reload the cache from a WINS server by typing **Nbtstat –R**.

Physical Topology

Troubleshooting the physical topology of the network is a process best performed using network diagrams. Sometimes a problem is most easily identified if you use a network diagram and simply circle all the affected locations (or use a highlighter) on the network map. This allows you to isolate the problem on the map to a certain area of the network.

Figure 15-4 shows a network map for a network for which all users have called and complained that they can't connect to the Internet. By looking at this map, you can see that router A, proxy server B, and WAN connection C are all affected areas.

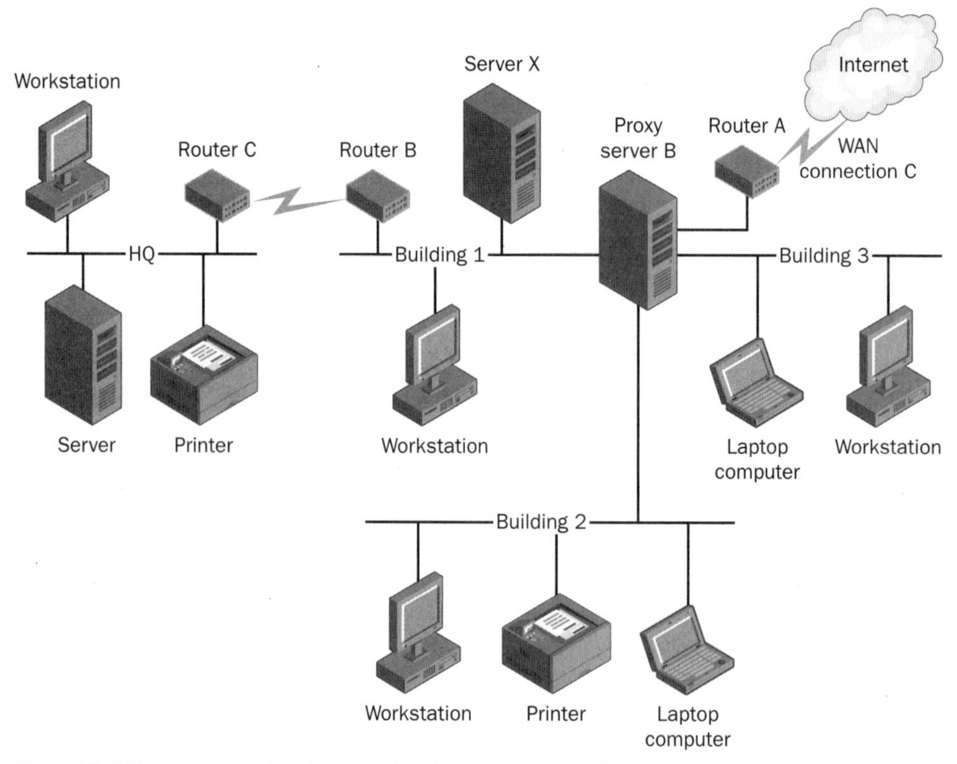

Figure 15-4 The area connecting the network to the Internet is the affected portion of the network.

Figure 15-5 displays a network map in which the people using computer A, computer B, and computer C are unable to connect to the network. When you look at this network map, you can see that these computers share a single network segment, and the likelihood is that the hub that is common to each of these computers is at fault.

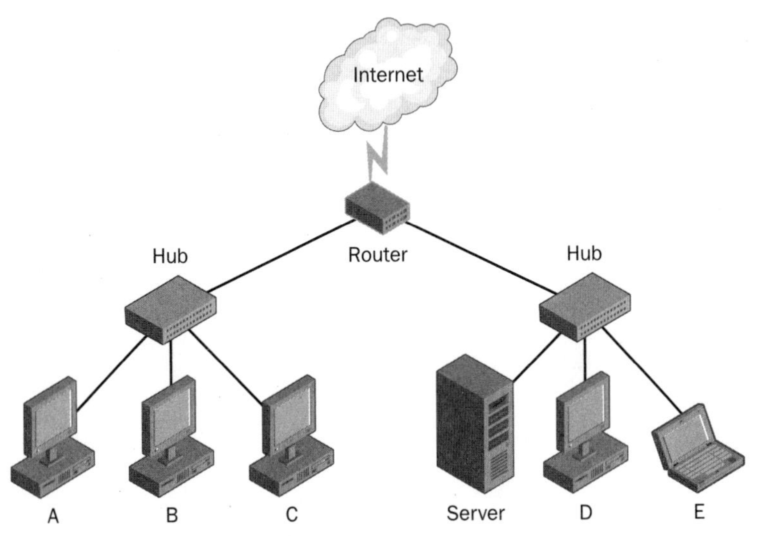

Figure 15-5 The network map will help isolate the affected network area.

Enterprise WAN Topology

The WAN is usually the link that connects multiple offices. Many WAN protocols (especially the older ones) are unreliable and slow, which means that you'll probably encounter WAN outages on a periodic basis.

The first step is to understand how a WAN outage will affect your network. When WAN connections go down, each separate LAN will be able to contact other computers within the network, but not any computers outside the local area. Often an organization will centralize its servers in one (or a few) offices, and place only workstations and peripheral equipment at the smaller offices.

A bank, for instance, with many branch offices and one location serving as headquarters would likely place the majority of its database servers at the headquarters. If the WAN connection were to fail at headquarters, all the branches would fail to have access to the database servers, but the headquarters would continue to have access to the data.

If a failure occurs in a WAN connection, do the following:

- Contact the WAN provider to see if the failure is at their end.
- Check the cabling between the WAN equipment and the local area equipment.
- Ensure that the router hardware is functional.
- Check the router configuration.

In addition to an outright failure, you might find that your WAN connection's speed degrades. In this situation, all the physical functions of the WAN appear to be in working order, however any browsing or downloading activity will be extremely slow. The link lights on the WAN equipment will tell you whether the WAN link is up. If the network link is extremely slow, the network interface might be losing packets on the router.

The LAN Topology

When you administer a LAN, you'll notice intermittent failures of network connectivity, which are annoying but normal. You can use a variety of troubleshooting tools to help diagnose and repair network problems.

One of the problems you'll likely encounter is an incorrect protocol. By the time you've looked at the cabling, the NIC, and the hub for a workstation that does not connect, you'll know that the problem is likely related to configuration.

The first thing you should check is the network protocol configuration for the workstation. If you're using TCP/IP, you should also use the Ipconfig or Winipcfg utilities to display current information. If you find that the computer is not configured with the correct protocol, install and configure the correct protocol.

If you're using Internetwork Packet Exchange/Sequenced Packet Exchange (IPX/SPX), ensure that the computer is configured for the correct frame type. Frame type refers to the header that encapsulates the data-link layer packet. Three frame types can be used:

- Ethernet II
- Ethernet 802.2
- Ethernet 802.3

When the frame type in use is not the same across the rest of the network, the computer will not communicate. The Microsoft implementation of IPX/SPX is called NWLink. NWLink will allow you to set the frame type, and it will autodetect the frame type, or it will default to Ethernet 802.2. The Ethernet 802.2 frame type is the most commonly used, but not in every network.

Key Points

- Use the same troubleshooting method whether the problem you've encountered has to do with physical or logical problems.

- Troubleshooting includes the following steps:
    1) Establish the symptoms.
    2) Identify the area affected by the problem.
    3) Determine what has changed.
    4) Select a probable cause for the problem.
    5) Implement a solution.
    6) Test the result.
    7) Recognize whether the solution is successful, and if it was unsuccessful, select the next most probable cause and solution.
    8) Document the solution.
- When a computer does not obtain an IP address, the problem is related to DHCP.
- When a computer does not recognize the names of URLs on the Internet, it is usually a DNS problem.
- WINS problems are sometimes caused by static entries for hosts with more than one NIC.

Chapter Review Questions

1 You've been called in to troubleshoot a network in which all the PCs on a single network segment are not receiving IP addresses. The DHCP server is up and computers on other network segments are receiving IP addresses. Which of the following is most likely the problem?

a) The DHCP server service is not running.

b) The router between the DHCP server and the segment with the computers not receiving IP addresses isn't forwarding DHCP packets.

c) The DNS server is down.

d) The WINS server is down.

Answer b is correct. The router is not forwarding DHCP packets between the DHCP server and the computers that are not functioning. Answer a is incorrect because other computers were receiving IP addresses via DHCP. Answers c and d are incorrect because neither of these services would affect the DHCP clients.

2 A WINS problem will affect people who are using Web browsers to access Web sites using URLs on the Internet.

a) True

b) False

False. A WINS problem will not affect people who are trying to browse the Internet. WINS is used only for NetBIOS name resolution, which doesn't apply to URLs.

3 Which of the following is a tool you could use to diagnose and repair WINS problems?

a) Netstat

b) Ifconfig

c) Nbtstat

d) Trblsht

Answer c is correct. Nbtstat is a tool used to diagnose and repair WINS problems. It can also be used to look at NetBIOS over TCP/IP. Answer a, Netstat, is incorrect because it is used for pure TCP/IP problems. Answer b, Ifconfig, is incorrect because it is used to configure a Unix interface. Answer d, Trblsht, is incorrect–Trblsht is a fictitious tool.

4 Which of the following utilities can tell you whether the DNS service has undergone an error on the DNS server?

a) Event viewer

b) Ping

c) Netstat

d) Ipconfig

Answer a is correct. Event viewer displays errors that take place on a Windows server. Answers b, c, and d are all incorrect because these command-line utilities test and diagnose TCP/IP problems.

5 You've been called in to troubleshoot a network. The database server at site A is working, and all users at site A can access it. Users at site B, which is connected to site A using a WAN link, can't access the database server but can use the resources at site B. Users at site C, which is connected to site A via a WAN connection, can access all network resources at site C and site A. Which of the following should you check? Select all that apply.

a) The link lights on the WAN equipment at site A

b) The link lights on the WAN equipment at site B

c) The link lights on the WAN equipment at site C

Answers a and b are correct. Because the affected areas are site A and site B, the WAN equipment at each of these sites should be checked to see whether the appropriate link lights are up.

6 You are troubleshooting a network problem. A workstation can ping 127.0.0.1 and can ping 192.168.1.1, which is the address of a local server, but cannot ping name.contoso.com, which is the FQDN of the server with the address 192.168.1.1. What is the problem?

a) WINS is configured incorrectly.

b) DHCP is configured incorrectly.

c) DNS is configured incorrectly.

d) TCP/IP is configured incorrectly.

The correct answer is c. DNS is configured incorrectly. The name that is not being resolved is one that should be supplied by a DNS server or a hosts file. Answer a is incorrect because WINS is not concerned with FQDNs. Answer b is incorrect because DHCP does not handle name resolution. Answer d is incorrect because you are able to ping both a loopback address as well as a TCP/IP address on the network.

7 Which of the following is a tool that you should select when troubleshooting DNS problems on a Windows XP client?

a) Nslookup

b) Ifconfig

c) Tracert

d) Winipcfg

Answer a is correct. Nslookup is used to troubleshoot DNS problems. Answer b is incorrect because Ifconfig is used to configure a network interface on a Unix machine. Answer c, Tracert, is incorrect because it's used to test the path between two IP addresses. Answer d, Winipcfg, is incorrect because it isn't available on a Windows XP machine.

8 If a Windows 95 computer can connect to *http://www.microsoft.com*, but can't connect to the local server named "SERVER01", it is having a problem with DNS.

a) True

b) False

False. When a Windows 95 computer can connect to *http://www.microsoft.com*, then it is able to resolve a FQDN to an IP address, which is the service provided by DNS.

9 You're working at the help desk which provides support for a network of 2,000 users. All 20 of the users located in a sales office have called to complain that they are unable to connect to the Internet, they cannot connect to their e-mail server, and they cannot connect to the database server. The users state that they can connect to a local file server and can print to local printers. When you look at a network diagram, you see that the users connect to the Internet through the HQ office and that the database server and e-mail servers are also located at the HQ office. What is the most likely problem?

a) The DNS server is not working at the sales office.

b) The WAN link is down between the sales office and HQ.

c) The users have been infected by a virus.

d) There are 20 bad cables connected to the machines.

Answer b is correct. The WAN link is down between the sales office and HQ. Even though the users report three different problems, the common denominator for all of them is the inability to connect to or through HQ. Answer a is incorrect because there is not enough information to tell you whether DNS is involved. Answer c is incorrect because it isn't likely that a virus would affect only applications that go through a certain office. Answer d is incorrect because the users are able to connect to local servers.

10 A user reports that he is unable to use any network resources. All clients on your network use DHCP addressing. All servers are Windows NT 4.0. You check the link light on the hub and the link light on the user's NIC—they are both green. You sit down with the user and find that you can ping 127.0.0.1 on his workstation. You are able to ping *http://www.contoso.com*. You are able to ping the local server's IP address, but you cannot execute a Net Use \\server\share command. What is the most likely problem?

a) TCP/IP is configured incorrectly.

b) DHCP is not delivering an IP address to the client.

c) DNS is not working.

d) WINS is not functioning.

Answer d is correct. WINS is not functioning because you cannot use a NetBIOS name to access a share on the network. Answer a is incorrect because you are able to ping the loopback address and a local address. Answer b is incorrect because you are able to ping another IP address. Answer c is incorrect because you are able to ping *http://www.contoso.com*.

Check Yourself

(Before You Test Yourself)

Domain 1.0: Media and Topologies

1.1 Recognize the following logical or physical network topologies given a schematic diagram or description.

- The term *network topology* describes how a network is connected. A physical topology defines the physical connections. A logical topology defines the shape of how data travels throughout the network.
- Logical and physical topologies are not necessarily the same. For example, a token ring network uses a physical star topology, although data travels around the star in a logical topology shaped like a ring.
- A topology can be applied to a local area network (LAN), which is a typical network inside a building or office; a metropolitan area network (MAN), which is a network consisting of several LANs within a metropolitan area such as a city; or a wide area network (WAN), which is a network spanning a long distance and consisting of two or more LANs.

Table A-1 Network Topologies

Topology	Description	Uses	Advantages	Disadvantages
Star	Each device is connected to a central hub.	10BaseT, 100BaseT.	Easy to troubleshoot, install, and add to. Cabling is cheap.	Centralized point of failure (hub).
Bus	Each device is connected to a common backbone.	10Base2, 10Base5, cable TV.	Easy to install, very cheap, uses the least amount of cable.	Difficult to troubleshoot, no fault tolerance.
Mesh	Each device has a point-to-point connection to every other device.	Used in some WAN implementations.	Very fault tolerant.	Expensive and difficult to set up.
Ring	Each device is connected to two others to create a closed loop.	Token-ring, FDDI.	High speed, easy to troubleshoot.	Expensive hardware.
Wireless	Each device accesses other devices via radio transmission.	802.11b.	Mobile solution.	Difficult to troubleshoot, slow speed.

Star/Hierarchical

- Star topologies are physical only. They consist of many cables connected to a central hub, as shown in Figure A-1.
- A failure of a single cable, workstation, or hub port will not affect other devices.
- A failure of the hub will affect all connected devices.

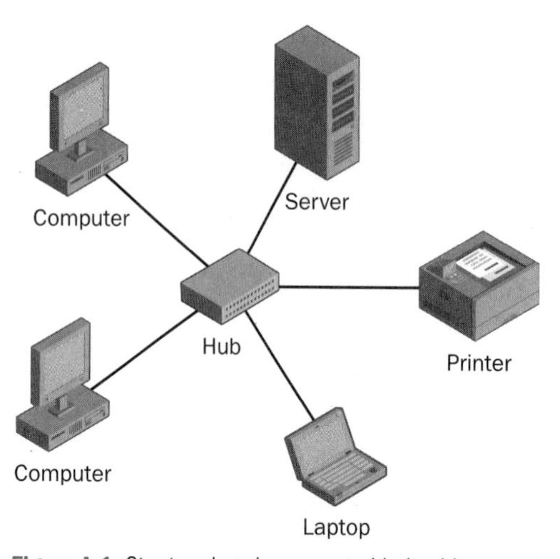

Figure A-1 Star topology has a central hub with separate cables to each device.

Bus

- The bus topology consists of a single cable trunk to which all nodes are connected, as shown in Figure A-2.
- The length of the trunk is used to measure the total length of the cable; the short drop cables connecting the nodes to the trunk are not added into the total length of the cable.
- Each end of the cable trunk must be terminated.
- 10Base5 and 10Base2 networks use both a physical and logical bus topology. A 10BaseT network uses a physical star topology with a logical bus. In a logical bus topology, a node sends data to all other nodes simultaneously.
- The advantage of a physical bus topology is that it is cheap and simple to install.

- The disadvantage of a physical bus is that a failure affects all other nodes, and excess traffic can occur because data is passed to all nodes.

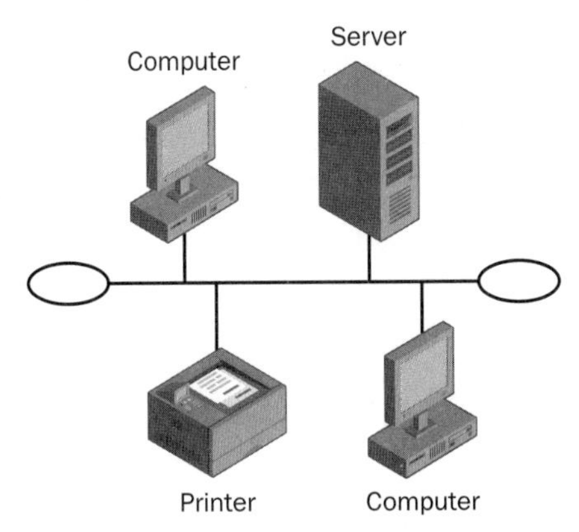

Figure A-2 A physical bus topology has a main cable with each node attached to it by short drop cables.

Mesh

- The mesh topology, which is depicted in Figure A-3, consists of each node connected to all other nodes.
- The main advantage of a mesh topology is the redundancy it provides, which preserves connectivity even in the event of a failure between links.
- The disadvantage of a mesh topology is that it is very expensive to set up.

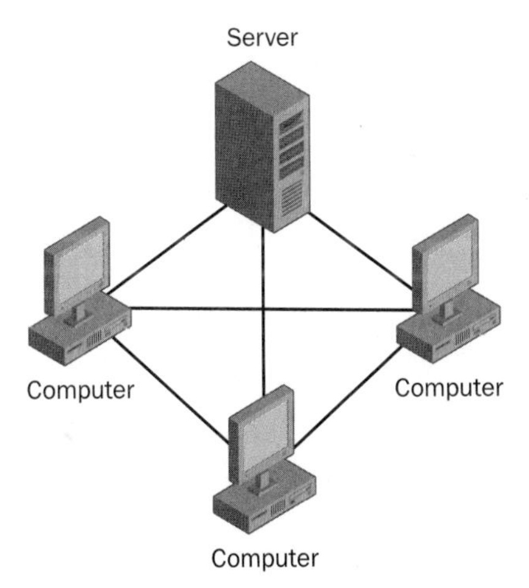

Figure A-3 Each node is connected to all other nodes in a mesh topology.

Ring

- The physical ring topology is a group of nodes that are joined in daisy-chain fashion until the last is connected to the first, creating a ring of communication, as shown in Figure A-4.
- A logical ring topology transmits data in a ring from node to node, even if the physical topology is a star topology, as in a token ring network.
- The advantage of a ring topology is that it provides equal access to the network.
- One disadvantage of a ring topology is that it is difficult to troubleshoot; another is that a failure in the ring will affect all other nodes in the ring.

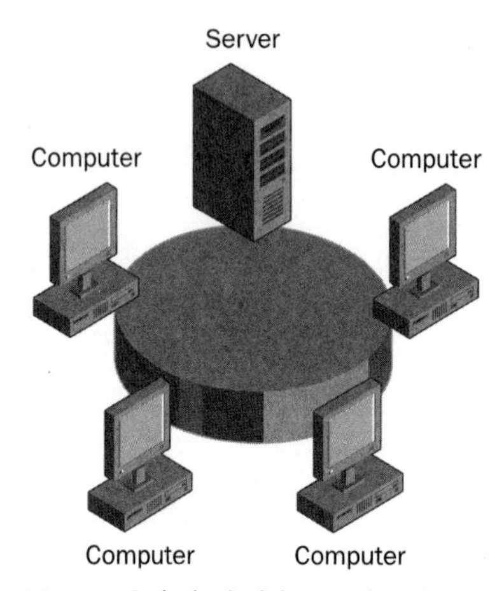

Figure A-4 A physical ring topology is created by connecting each node to the next until the last is connected to the first.

Wireless

- Wireless topology (also called *cellular*) is one in which a wireless access point (WAP) creates a *cell* within which any wireless device can connect and WAPs can be interconnected, as shown in Figure A-5.
- An ad-hoc wireless network can be created by connecting two wireless devices directly to each other without the use of a WAP.
- The advantage of wireless networks is that devices can easily move about and retain connectivity.

- The disadvantage of wireless networks is that they're subject to interference. In addition, wireless networks will degrade in speed as a device moves farther away from the WAP (or other device, in the case of an ad-hoc network).

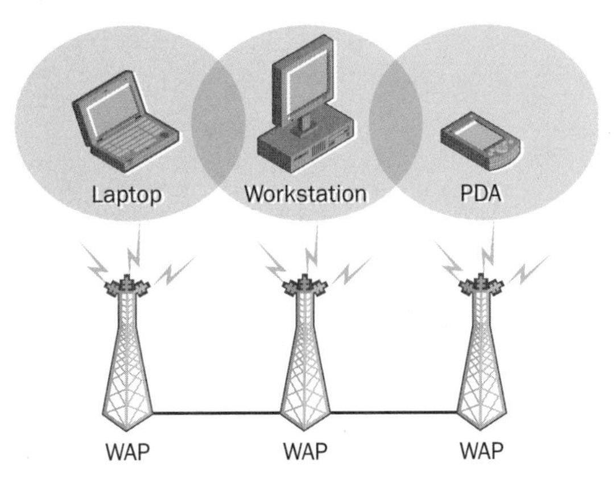

Figure A-5 Wireless topology is typified by cells of connectivity centralized at the WAP.

1.2 Specify the main features of 802.2 (LLC), 802.3 (Ethernet), 802.5 (token ring), 802.11b (wireless), and FDDI networking technologies.

- The Institute of Electrical and Electronics Engineers (IEEE) 802 specifications are for frame types. A frame type is the format of the packet that is used to communicate across the network.
- The 802.2 specification provides for the Logical Link Control (LLC) header information to identify upper layer protocols.

Table A-2 802 Specifications

Specification	Speed	Access Method	Topology	Media
802.3 or Ethernet	5 Mbps, 10 Mbps, 100 Mbps	Carrier Sense Multiple Access with Collision Detection (CSMA/CD)	Physical bus, physical star, logical bus	Copper unshielded twisted pair (UTP), copper coaxial
802.5 or token ring	4 Mbps, 16 Mbps	Token passing	Physical star, physical ring, logical ring	Copper UTP, copper shielded twisted pair (STP)
802.11(b) or Wireless	11 Mbps	Carrier Sense Multiple Access with Collision Avoidance (CSMA/CA)	Wireless	Air
802.12 or 100VGAnyLan	100 Mbps	Demand priority	Physical star, logical bus	Copper UTP
FDDI, or Fiber Distributed Data Interface	100 Mbps	Token passing	Physical ring, logical ring	Fiber optic

1.3 Specify the characteristics (i.e., speed, length, topology, cable type) of the following 802.3 (Ethernet) standards.

- The Ethernet IEEE 802.3 specification can run across fiber optic, coaxial, and UTP copper cabling.
- To use coaxial cabling in a physical bus topology (which can be either 10Base2 or 10Base5), you must follow the 5-4-3 rule, which states that there can be five (5) cable segments connected with four (4) repeaters, but only three (3) of the segments can be populated with nodes.

10BaseT

- 10BaseT uses copper UTP cabling of Cat3 or better grade, connected to a central hub, to create a physical star topology.
- The UTP cables in a 10BaseT network use a registered jack 45 (RJ-45) type of connector.
- 10BaseT runs at 10 Mbps, and cables can be a maximum of 100 meters in length. This length includes the spans of the patch cable and drop cable.

100BaseTX

- On the exterior, 100BaseTX looks the same as 10BaseT because 100BaseTX uses copper UTP cabling of Cat5 or better grade connected to a central hub. Cat5 has two pairs of copper wires within the cable.
- All cables use an RJ-45 connector and have a maximum cable length of 100 meters.
- 100BaseTX provides speeds of up to 100 Mbps.

10Base2

- 10Base2 uses a physical bus topology in which all computers connect to a main cable, which is RG-58 coaxial cable, also called ThinNet.
- The maximum length of the ThinNet coax cable is 185 meters.
- A 10Base2 network allows only 30 computers per segment and provides a speed of 10 Mbps.

10Base5

- 10Base5 uses a physical bus topology in which all the computers connect to a main cable, and it requires RG-8 or RG-11 coaxial cable, which is known as ThickNet.

- The maximum length of a 10Base5 segment is 500 meters, and it offers 10 Mbps.

100BaseFX

- 100BaseFX provides Ethernet over fiber optic cable at a 100-Mbps speed.
- The maximum length of the fiber optic segment is 6,562 feet, or 2000 meters.

Gigabit Ethernet

- Gigabit Ethernet uses Cat5 or better grade cabling in a physical star topology and requires RJ-45 connectors.
- Gigabit Ethernet is also called 1000BaseT, and it offers 1 Gbps of speed.

1.4 Recognize the following media connectors and describe their uses: RJ-11, RJ-45, AUI, BNC, ST, and SC.

Table A-3 Connectors

Connector	Cabling	Usage
RJ-11	Telephone unshielded twisted pair (UTP)	Used for telephones
RJ-45	UTP, Cat3, or better grade	Used in Ethernet 10BaseT, 100BaseT, 1000BaseT, and token ring over UTP networks
AUI	ThickNet RG-8 or RG-11 coaxial cabling	Used in 10Base5 networks
BNC	ThinNet RG-58 coax	Used in 10Base2 networks
ST and SC	Fiber optics	Used in 100BaseFX networks

1.5 Choose the appropriate media type and connectors to add a client to an existing network.

- Each type of media, whether it is UTP, fiber, or coax, varies from the others from the standpoint of cost and installation.
- You can't mix cables. In a 10Base5 network, you can't use ThinNet coax for all or even part of the backbone cabling (although you might use ThinNet for the drop cables with the appropriate transceivers). In a network that requires a certain grade of UTP (for example, 100BaseT requires Cat5 or better grade), you must use only that grade of UTP throughout the network.

- The cabling must be correct for the network interface card (NIC).
- The cabling must be correct for the hub and network type. For example, you can't use telephone wire with a 100BaseT network.
- The connectors must be appropriate for the cable, NIC, and network type.

1.6 Identify the purpose, features, and functions of the following network components.

Hubs

- A repeater copies a signal received from one port out through its other port. This effectively lengthens a cable.
- Hubs are multiport repeaters that replicate a signal received from one port out through all other ports. There is no filtering.
- A hub is the central piece of equipment in a physical star topology. It used in 10BaseT, 100BaseTX, and 1000BaseT networks.
- A passive hub will repeat a signal but won't regenerate it. An active hub will regenerate and amplify the signal.
- Each computer connects to the hub with its own cable.

Bridges

- A bridge connects two LAN segments but only passes through network packets that are destined for the other segment.
- A bridge can reduce traffic problems by filtering out the network packets between the two connected segments.
- A transparent bridge is used in Ethernet networks. This type of bridge lists the Media Access Control (MAC) address of all network devices by way of a tag that states which segment the device is located on.
- A source route bridge is used in token ring networks. This type of bridge expects all network devices to store a copy of the table of devices in their memory.

Switches

- A switch is essentially a multiport bridge. Each port on a switch can be connected to a single device or an entire network segment.

- Data received from one port on a switch is forwarded out only on the port that has the device for which that data is destined. (Compare this to a hub that forwards data to all ports.)
- Both bridges and switches forward data based on the destination MAC address of the data.

Routers

- A router connects two or more networks and forwards only the data that is destined for a particular network.
- A router will forward data based on the logical network segment and node address of the data packet.
- Routers can connect dissimilar network architectures.
- A *Brouter* provides the same function as a router but also bridges non-routable protocols. A Brouter is a combination of a router and a bridge.

Gateways

- A gateway connects networks that use entirely different protocols.
- A gateway acts as a translator between the two protocol stacks. For example, a gateway can translate between IPX/SPX on one network and TCP/IP on the other.

CSU/DSU

- A CSU/DSU is used in WANs and converts digital data frames for the WAN to the protocols used on the LAN.
- CSU/DSUs are often used with leased lines.

Network Interface Cards/ISDN Adapters/System Area Network Cards

- NICs, ISDN (Integrated Services Digital Network) adapters, and SAN adapters are expansion boards that are placed within a computer to enable it to communicate with external equipment.
- NICs provide the ability for the computer to transmit data in packet form onto a network.
- ISDN adapters allow a computer to communicate across the ISDN, which is a digital network used by the telephone company.

- SAN adapters allow a computer to communicate with a storage area network (or system area network) to access or exchange stored data.

Wireless Access Points

- A WAP is similar to a hub for an Ethernet network except that it transmits in radio frequencies across air, rather than as digital signals across copper cables.
- WAPs are used in wireless networks in which more than two devices connect to the network via wireless transmissions.

Modems

- A modem is a modulator/demodulator that modulates digital signals into analog signals, transmits them across the plain old telephone system (POTS), and then demodulates analog signals received from the telephone system.
- External modems are connected to the computer via an RS-232 serial connector.

Domain 2.0: Protocols and Standards

2.1 Identify and give an example of a MAC address.

- MAC addresses are assigned to each network interface by the manufacturer. They must be unique.
- The IEEE developed a MAC address assignment process to ensure that all network devices would have unique MAC addresses. The IEEE supplies manufacturers with the first 3 bytes of the address and allows them to assign the remaining 3 bytes of the address.
- An example of a MAC address is 00-A1-F3-27-6A-EB. In this example, the manufacturer has been given the 00-A1-F3 prefix and has assigned the remaining portion.

2.2 Identify the seven layers of the OSI model and their functions.

- The Open Systems Interconnection (OSI) model was developed by the International Organization for Standardization (ISO).
- The OSI model is provided as a reference for the development of protocols to promote interconnectivity and interoperability. Each layer works with the others to provide full connectivity.

- Seven layers are listed in order from layer 7 to layer 1: application, presentation, session, transport, network, data-link, and physical. To remember these layers in order from layer 7 to layer 1, you can memorize this sentence: All People Seem To Need Data Processing.
- Physical, or layer 1, specifies the hardware connection, including the topology, cabling, wiring, and signaling of the data.
- Data-link, or layer 2, has two sublayers: Media Access Control (MAC) and Logical Link Control (LLC). The data-link layer creates and interprets frames based on the network type used. The MAC address is assigned at the MAC sublayer. The LLC sublayer maintains connections between devices. At the data-link layer, datagrams received from the network layer are broken down into frames with headers appropriate for the network type.
- Network, or layer 3, provides the logical network and node addresses that are assigned by the network administrator. This layer enables data to be routed among different segments. Segments received from the transport layer are broken down into datagrams with appropriate network-layer headers.
- Transport, or layer 4, ensures flow control and error handling and is involved in correcting transmission problems. Because data is broken down into smaller packets, the transport layer adds sequence numbers to ensure that data can be reassembled in the correct order at its destination. Sockets are provided by this layer to identify which applications that data is coming from and which it is destined for.
- Session, or layer 5, manages the dialogs between computers. It provides for simplex, half-duplex, or full-duplex communications. A session dialog is established, data is transferred, and then the session layer terminates the dialog.
- Presentation, or layer 6, is mainly concerned with the format of the data. This means that it will look at encryption and decryption, compression and expansion, character-set conversion and graphics.
- Application, or layer 7, provides the application interface for file transfers, Web browsing, database access, e-mail messages, and more. It handles flow control, error recovery, and a consistent interface through which software can access the network.

2.3 Differentiate among the following network protocols in terms of routing, addressing schemes, interoperability, and naming conventions.

- Protocols are sets of rules that govern how network devices communicate.
- Protocol suites, or protocol stacks, are sets of protocols that are designed to work together to ensure full communication between devices, even if they are dissimilar.

TCP/IP

- Transmission Control Protocol/Internet Protocol (TCP/IP) is the protocol suite used by the Internet as well as many private networks.
- The TCP/IP suite consists of numerous different protocols with their own functions. The main protocols are TCP and IP.
- IP provides the logical node address for a device and the logical network address for a network segment. These addresses are used in routing data among different networks.
- An IP address takes the form of a 32-bit number, which in its written form comprises four decimals separated by decimal points. An example of an IP address is 172.11.88.254.

IPX/SPX

- Internetwork Packet Exchange/Sequenced Packet Exchange (IPX/SPX) is the protocol suite developed by Novell for use in Novell NetWare networks.
- An IPX address consists of three components: the network address, the node address, and the socket.
- IPX network addresses are made up of eight hexadecimal characters. The IPX node address is copied from the MAC address of the NIC. The socket number provides a virtual circuit for communications between the client and server.
- IPX is a routable protocol.

NetBEUI

- NetBIOS Enhanced User Interface (NetBEUI) is the standard protocol used by Microsoft Windows NT servers. (Today Windows 2000 servers use TCP/IP as the standard.)

- *NetBEUI is not routable.* To transmit NetBEUI between two different segments, it must be bridged or tunneled within a different routable protocol, such as TCP/IP.
- NetBEUI uses NetBIOS naming. It provides for node addresses but not network addresses.

AppleTalk

- AppleTalk was developed by Apple Computer for communication among Macintosh computers.
- AppleTalk is routable and provides for automatic assignment of addresses.

2.4 Identify the OSI layers at which the following network components operate.

Hubs

- Hubs (in addition to repeaters) work at the physical layer of the OSI model.
- Both hubs and repeaters merely retransmit data signals without looking at packet addresses.

Bridges

- Bridges operate at the data-link layer of the OSI reference model.
- A bridge looks at the MAC addresses of packets, which are in the data-link frame header, in order to make forwarding decisions.

Switches

- Because switches are basically multiport bridges, they too act at the data-link layer of the OSI model.

Routers

- Routers operate at the network layer of the OSI reference model.
- Routers connect two or more networks and route data between them based on the logical network and node addresses, which are found in the network layer header of a datagram.
- A routing protocol maintains routing tables and makes routing decisions about how to forward packets.

- Examples of routing protocols include Routing Information Protocol (RIP) and Open Shortest Path First (OSPF), which are both in the TCP/IP suite.
- TCP/IP, IPX/SPX, and AppleTalk are routable protocols. NetBEUI is nonroutable.

Network Interface Cards

- A NIC operates at the physical layer of the OSI reference model and is assigned a MAC address at the data link layer.

2.5 Define the purpose, function, and use of the following protocols within TCP/IP.

IP

- Internet Protocol (IP) is essentially a network-layer protocol that handles addressing packets and routing data.
- IP is connectionless, meaning that it does not guarantee delivery nor acknowledge lost packets or packets that have been sent improperly. It depends on higher layer protocols to provide these functions.

TCP

- Transport Control Protocol (TCP) is responsible for sequencing packets and tracking their source and destination.
- TCP is connection-oriented, which means that it ensures guaranteed delivery of the data.
- TCP guarantees the delivery of data through acknowledgment (ACK) packets.

UDP

- User Datagram Protocol (UDP) is a connectionless protocol. It does not guarantee delivery.
- Because UDP does not guarantee delivery, its header contains less information, which makes it a faster protocol than TCP.
- TCP/IP packets use either UDP or TCP, but not both.

FTP

- File Transfer Protocol (FTP) is used for transferring files between nodes. It requires that one node be an FTP server, with files available for download, and that the other node be a client.
- FTP is an application-layer protocol that uses TCP port 21 for control and port 20 for data transfer.
- FTP is connection-oriented and guarantees delivery of the data.

TFTP

- Trivial File Transfer Protocol (TFTP) also provides for the transfer of files between a client and TFTP server.
- TFTP is connectionless, and uses UDP as the transport protocol.

SMTP

- Simple Mail Transfer Protocol (SMTP) is used for reliable transport of electronic mail over the network.
- SMTP is an application-layer protocol.

HTTP

- Hypertext Transfer Protocol (HTTP) is used for the transfer of Hypertext Markup Language (HTML) files that can include text, graphics, sound, and other multimedia elements.
- HTTP is an application-layer protocol used by Web browsers.
- In a Uniform Resource Locator (URL), an address beginning with http:// uses the HTTP protocol.

HTTPS

- Hypertext Transfer Protocol over Secure Socket Layer (HTTPS) is a secure version of HTTP.
- When data is sent using HTTPS, the URL of the address begins with https://.

POP3

- Post Office Protocol version 3 (POP3) is a mail server that is used for holding an e-mail message until a user downloads it to read it.

- POP3 and SMTP are used together; SMTP in the sending of mail and POP3 in its retrieval.

IMAP4

- Internet Message Access Protocol version 4 (IMAP4) is an alternate protocol for retrieving e-mail messages. Just like POP3, a client can download e-mail messages from an IMAP4 server, however IMAP4 is intended for a centralized storage of e-mail messages, whereas a POP3 server is considered a temporary storage container of e-mail messages.

TELNET

- Telnet provides a virtual terminal emulation across the network.
- Telnet is an application-layer protocol. For Telnet to function, the client must be running a Telnet service, which allows clients to connect and log in remotely.

ICMP

- Internet Control Message Protocol (ICMP) is a network-layer protocol used on a TCP/IP network for status and error information. Both ping and Tracert commands use ICMP as the underlying protocol to retrieve status information.

ARP

- Address Resolution Protocol (ARP) will resolve an IP address that exists at the network layer to a MAC address, which exists at the data-link layer.
- ARP is available as a command to view the ARP cache of IP address to MAC address resolutions that are temporarily stored on a machine, as well as to add, edit, and delete ARP table entries.

NTP

- Network Time Protocol (NTP) is an application-layer protocol that is used to synchronize computer clock times within the network.

2.6 Define the function of TCP/UDP ports and identify well-known ports.

- An important feature of TCP/IP protocols is their use of sockets. The socket is the combination of an IP address and a port number.

- Upper-layer protocols use different port numbers, which enable multiple applications to use the same network connection. For example, a user can browse the Web using the HTTP protocol, which uses port 80 and simultaneously sends e-mail messages using Simple Mail Transport Protocol (SMTP), which uses port 25.

Table A-4 Well-Known TCP/UDP Ports

Protocol	Name	Port
NTP	Network Time Protocol	UDP port 123
POP3	Post Office Protocol version 3	TCP port 110
IMAP4	Internet Message Access Protocol version 4	TCP port 143
HTTP	Hypertext Transfer Protocol	TCP port 80
HTTPS	HTTP over Secure Sockets Layer (SSL)	TCP port 443
FTP	File Transfer Protocol control	TCP port 21
FTP	File Transfer Protocol data transfer	TCP port 20
TELNET	Telnet	TCP port 23
SMTP	Simple Mail Transfer Protocol	TCP port 25
TFTP	Trivial File Transfer Protocol	UDP port 69
SNMP	Simple Network Management Protocol	TCP port 161

2.7 Identify the purpose of each of the following network services (i.e., DHCP/BOOTP, DNS, NAT/ICS, WINS, and SNMP).

- Dynamic Host Configuration Protocol (DHCP) is closely related to BOOTP. BOOTP is used to boot a machine remotely. DHCP provides an IP address to a machine from a remote server.
- DHCP requires a server with a pool of IP addresses, called a *scope*, to deliver the IP addresses to DHCP clients.
- Domain Name System (DNS) enables a client to automatically resolve an IP name to an IP address.
- Before DNS, people used *hosts files* on local machines. A hosts file is a text file that listed the names and IP addresses of known hosts. Administrators had to update hosts files on any computer on which a user wanted to use human-friendly names.
- DNS provides for a primary name server, which contains DNS information for a zone and is used by the administrator to update DNS information, as well as a secondary name server, which contains a read-only copy of the zone.

- Names are organized into a hierarchical structure of domains.
  - Root-level domain: the top of the tree.
  - Top-level domains: well-known domains such as .com for commercial, .mil for military, .edu for education, .gov for United States government, and .org for non-profit organizations.
  - Second-level domains: the individually managed domain names assigned to networks. For example, microsoft.com and comptia.org are both second-level domains. These can be further subdivided into subdomains such as sub.microsoft.com.
  - Hosts: hosts are the final level of the hierarchy. Each host has a name that when added to its location in the hierarchy creates a fully qualified domain name (FQDN). For instance, host.microsoft.com is the FQDN of a computer in the Microsoft.com domain.
- DNS records are the entries in a DNS zone. Each record provides a different type of information.
  - An A record associates a host name with an IP address.
  - CNAME records are provided for aliases so that a computer can have several names simultaneously. Web servers often use CNAME records so that the name www.microsoft.com can be as easily resolved as webserver.microsoft.com to the same IP address.
  - MX records identify mail servers.
  - NS records identify DNS servers.
  - PTR records are the reverse of A records in that they associate an IP address to a FQDN. The PTR record is used for a reverse lookup, such as trying to find out what the name of the computer is for IP address 123.45.67.89.
- Windows Internet Naming Service (WINS) is used on Windows networks to resolve a NetBIOS name to an IP address.
- LMHosts files are text files that contain entries of the known computer NetBIOS names and IP addresses.
- Network Address Translation (NAT) is the translation of internal private IP addresses to one or more public external IP addresses used on the Internet. Using NAT enables multiple machines to share the same IP address on the Internet but have unique IP addresses on the local network.

- Internet Connection Sharing (ICS) is Microsoft's implementation of NAT for a small network sharing the Internet connection of a single computer.
- Simple Network Management Protocol (SNMP) is used to monitor and maintain status information of network devices and services.
- SNMP uses a Management Information Base (MIB) to organize the types of information that a service or device can provide via SNMP.

2.8 Identify IP addresses (IPv4, IPv6) and their default subnet masks.

- IPv4 addresses can be broken into two parts—the network portion and the host portion. Each host on the same network must have a unique host portion and the same network portion of the IP address.
- IPv4 addresses are 32 bits in length, with four octets separated by decimal points. For example, 123.45.67.89 is a typical IP address.
- IPv4 addresses are divided into the classes shown in the following table.

Table A-5 IP Address Classes

Class	First Octet Range	Default Subnet	Number of Subnets	Number of Hosts per Subnet
A	1-126	255.0.0.0	126	16,777,214
B	128-191	255.255.0.0	16,384	65,534
C	192-223	255.255.255.0	2,097,152	254
D	224-239	N/A	N/A	N/A
E	240-254	N/A	N/A	N/A

- The IPv4 address 127.x.x.x is reserved for loopback testing.
- The IPv4 addressing scheme is running out of addresses. IPv6 is a new addressing scheme that provides for a 128-bit address using a hexadecimal numbering method.
- An example of an IPv6 address is 3FFB:A00:8AB1:2:1A:C.

2.9 Identify the purpose of subnetting and default gateways.

- Subnetting allows you to divide a class A, B, or C network into multiple logical smaller networks.
- An example of a subnet is a class B address 172.10.8.234 with its default mask of 255.255.0.0. This shows you that the network portion is 172.10 and the host portion is 8.234. If you subnet this class B address, you could make the subnet mask 255.255.255.0, which would create 254 subnets of 254 hosts each.

- A default gateway is configured for an IP host to provide a location to send all data that is not destined for the local network.
- Most often a default gateway address is configured on a computer to point to the nearest router that will lead to the Internet.

2.10 Identify the differences between public and private networks.

- A public network is one that interacts directly with the Internet. Its hosts can be contacted by other Internet computers and have addresses that are on the Internet.
- A private network is one that uses NAT so that internal IP addresses are not the same as the IP addresses used to interact with Internet services, or uses another mechanism that secures the private network nodes from direct contact by Internet entities. A private network uses private IP addresses within its boundaries.
- Private IP address ranges are reserved as follows:
  - Class A, 10.0.0.0 through 10.255.255.254
  - Class B, 172.16.0.0 through 172.31.255.254
  - Class C, 192.168.0.0 through 192.168.255.254
  - Automatic Private IP Addressing (APIPA) provides a class B address range of 169.254.0.1 through 169.254.255.254. APIPA is a special type of private addressing scheme used when a DHCP server is unavailable to provide an IP address to a DHCP client.

2.11 Identify the basic characteristics (i.e., speed, capacity, media) of the following WAN technologies.

Packet Switching versus Circuit Switching

- Packet switching refers to a routing method in which messages are divided into packets and then sent along the first available route to the destination. Each packet can travel along a different path and is then reassembled at the destination.
- Circuit switching refers to the routing method in which a dedicated path, or circuit, is created and maintained until all data has been transmitted. Circuit switching is used in telephone calls.

ISDN

- ISDN is the digital telephone service provided by telephone companies. ISDN allows voice, data, text, graphics, and multimedia data to be transmitted over existing telephone wires.
- ISDN provides a Basic Rate Interface (BRI) that consists of 2 B (bearer) channels of 64 Kbps and 1 D (data) channel of 16 Kbps.
- ISDN also provides for Primary Rate Interface (PRI) that consists of 23 B channels and 1 D channel of 64 Kbps. PRI runs over a leased T1 line of 1.544 Mbps.

FDDI

- Fiber Distributed Data Interface (FDDI) offers high-speed data networks over fiber optic media.
- FDDI is a dual ring using token-passing methods. It offers a data-transfer rate of 100 Mbps and can reach a length of up to two kilometers.

ATM

- Asynchronous Transfer Mode (ATM) is a high-speed packet-switched network that uses short packets of a fixed length called *cells*.
- ATM cells are 53 bytes in length.
- ATM can transmit voice, data, and multimedia traffic over variable-speed LAN and WAN connections. It can reach as high as 622 Mbps.

Frame Relay

- Frame relay is a WAN technology that provides packet switching through virtual circuits established across a "cloud" of multiple interconnected routers within the frame relay provider's network.
- Frame relay offers speeds from 56 Kbps to a full T1 of 1.544 Mbps.
- Frame relay is a common WAN protocol used to connect remote offices.

SONET/SDH

- Synchronous Optical Network (SONET) and Synchronous Digital Hierarchy (SDH) represent a set of related standards for data transmission across fiber optic networks.

- SONET is the United States version, and SDH is the international version.
- The base rate of SONET is 51.84 Mbps, and it can reach up to 40 Gbps.

T1/E1

- T1 lines are the United States version of a digital leased line offering 1.544 Mbps.
- E1 lines are the international version offering 2.048 Mbps.
- T1 lines consist of 24 individual channels, each called DS0, which transmit at the rate of 64 Kbps.
- E1 lines have 32 channels of DS0.

T3/E3

- T3 lines are faster circuits than T1 lines and have speeds of 44.736 Mbps.
- E3 lines are the international version offering speeds of 34.368 Mbps.

OCx

- Optical Carrier (OC) levels are the data rates provided over fiber optic networks offered by a WAN provider.
- OC-1 = 51.85 Mbps
- OC-3 = 155.52 Mbps
- OC-12 = 622.08 Mbps (notice that this is the same as the maximum rate for ATM)
- To find the correct data transmission rate, memorize the OC-1 rate and then multiply it by the number following OC.

2.12 Define the function of the following remote access protocols and services.

RAS

- Remote access service (RAS) is a service providing dialup and virtual private network connections from remote users to a private network.
- RAS runs on Windows NT and Windows 2000 servers.

PPP

- Point-to-Point Protocol (PPP) allows a computer to connect to the Internet or to a private network via a standard telephone line and modem.
- When connected via PPP, a remote machine can use network services just as though it were connected locally to the network.
- PPP can support multiple protocols, including TCP/IP, IPX, and AppleTalk.
- PPP supports dynamic IP address assignment to take place after the remote machine connects.

PPTP

- Point-to-Point Tunneling Protocol (PPTP) offers a virtual private network (VPN) connection across the Internet.
- PPTP encapsulates PPP packets into IP datagrams, encrypting the data that is transferred.

ICA

- Independent Computing Architecture (ICA) is a protocol used with Citrix MetaFrame servers to provide Windows terminal emulation sessions to remote clients.

2.13 Identify the following security protocols and describe their purpose and function.

IPSec

- Internet Protocol Security (IPSec) is a set of protocols used for secure, encrypted communications between two computers across a public network.
- IPSec creates an end-to-end secure data transmission by encrypting the packets before transmitting them and then allowing them to be decrypted only by the recipient computer.

L2TP

- Layer 2 Tunneling Protocol (L2TP) is a tunneling protocol for VPN connections that is similar to PPTP.

- L2TP provides authentication and header compression. It is used with IPSec for data encryption.

SSL

- Secure Sockets Layer (SSL) is a public-key cryptography method that provides encryption for secure communication.

Kerberos

- Kerberos is an Internet standard security protocol that enables mutual authentication between two network nodes.
- In Kerberos, the two nodes share a cryptographic key in order to verify each other's identities.

Domain 3.0: Network Implementation

3.1 Identify the basic capabilities (i.e., client support, interoperability, authentication, file and print services, application support, and security) of the following server operating systems.

- A peer-to-peer network lacks a dedicated server. Each computer can act as both a client and a server. Peer-to-peer networks are beneficial for small networks.
- A client/server network uses dedicated servers. Administration is centralized, as are resources. This method is beneficial for large networks in which ease of administration is a necessity.

UNIX/Linux

- UNIX and Linux are nearly identical operating systems with similar functionality and are able to act as peer-to-peer networks and as client/server networks.
- They support all clients using TCP/IP but work best with UNIX and Linux clients.
- UNIX and Linux use TCP/IP to interoperate with dissimilar networks.
- Applications are sometimes installed in the form of daemons. Daemons, much like Windows services, run on top of the UNIX operating system and provide services such as Web browsing or e-mail service, both locally and across the network.

- One of the more common file systems used by UNIX is the Network File System (NFS).
- Security method is based on an object model with entries for each user and access to resources.

NetWare

- Novell NetWare is a client/server network operating system. NetWare provides the server portion and supports clients of all types—Windows, UNIX, and Macintosh.
- NetWare can use multiple protocols to communicate across different networks.
- Applications are installed in the form of NetWare Loadable Modules (NLMs), which provide services to clients across the network.
- NetWare provides namespaces to support different types of file systems. For example, to support 8.3 names, the DOS.NAM namespace must be loaded; and to support long filenames on OS/2 and Windows 95 or later computers, the LONG.NAM namespace must be loaded.
- NetWare 3.*x* and older versions use a security system called the Bindery. The Bindery was dedicated to each server.
- NetWare 4.*x* and newer versions use Novell Directory Services (NDS), a distributed directory service that is accessible by all servers. NDS uses a tree structure to organize the objects (such as printers, servers, and users).

Windows

- Windows servers are usually dedicated as servers in a client/server fashion; however, the Windows operating systems are able to function as peer-to-peer networks.
- Windows servers can support all types of clients, including UNIX, Windows, and Macintosh.
- The default protocol on Windows NT 3.*x* servers is NetBEUI. The default protocol on later versions is TCP/IP.
- Applications run as executables to provide services both locally and across the network.
- Windows servers typically use the NT File System (NTFS).

- Windows NT servers used a domain system to contain security information limited to a single domain. This information was stored on a primary domain controller (PDC), with read-only copies on backup domain controllers (BDCs).
- Windows 2000 servers use the Active Directory service, which is a distributed, hierarchical directory service. The tree structure within each Active Directory domain is created using Organizational Units (OUs).

Macintosh

- Macintosh servers are called AppleShare servers when they are dedicated. Macintosh networks are typically installed as peer-to-peer.
- Macintosh networks usually support Macintosh clients but can provide services via TCP/IP.
- The two main protocol suites used are AppleTalk and TCP/IP.

3.2 Identify the basic capabilities of client workstations (i.e., client connectivity, local security mechanisms, and authentication).

- Windows 3.1 does not have any networking capabilities and must have a client application installed to access any type of server.
- All Windows client peers are limited to 10 simultaneously connected users.
- Windows 3.11 has the basic capabilities for peer-to-peer networking as well as accessing a Windows server. Local security is provided via share-level security in which any user with the correct password can connect to the share.
- Windows 95, Windows 98, and Windows Me all have basic peer-to-peer networking and client access software for both Windows and NetWare servers. Sharing files is accomplished through share-level security. Windows 98 Second Edition can share its Internet connection.
- Windows NT, Windows 2000 Professional, and Windows XP Professional have the fullest networking capabilities with share-level and user-level file sharing and Internet connection sharing.

3.3 Identify the main characteristics of VLANs.

- Virtual local area networks (VLANs) are used with switches.

- The VLAN maps a workstation and dedicates it to a particular group of computers so that it exchanges information amongst that group regardless of its physical location.

3.4 Identify the main characteristics of network attached storage.

- Network attached storage (NAS) consists of data storage systems that are connected directly to the network.
- When a client accesses the data storage, the client uses a server that then requests the data from the NAS.
- An example of a NAS is a CD-ROM tower.
- NAS is not the same as a SAN, which is a separate network of data storage connected outside the normal network.

3.5 Identify the purpose and characteristics of fault tolerance.

- Fault tolerance is a computer with components that are designed so that if a component fails, a backup component can take its place.
- Redundant Array of Inexpensive Disks (RAID) is the most commonly used form of fault tolerance.
- RAID 1 is the process of mirroring two disks so that if one disk fails the other takes its place.
- RAID 5 is the system of striping disks with parity so that if a single disk fails, the data is available on the remaining disks.

3.6 Identify the purpose and characteristics of disaster recovery.

- Disaster recovery is the ability to recover from a failure on the network (or of the entire network) with little or no downtime.
- Backups of data stored on the network are indispensable to all disaster recovery plans.
- Backup tapes should be stored offsite so that a building disaster will not ruin the tapes. The disaster recovery plan should be tested periodically, such as every six months.
- A full backup will copy all the data from the server to the tape. When executing a full backup, the archive bit on each file is reset to indicate that the file has been backed up.
- A copy backup will store the data but will not reset the archive bit.

- An incremental backup will back up only the files that have been changed since the previous backup. When you restore incremental backups, you first restore the last full backup and then each subsequent incremental backup.
- A differential backup will store all the files that have been changed since the previous full backup.

3.7 Given a remote connectivity scenario (i.e., IP, IPX, dial-up, PPPoE, authentication, physical connectivity, and so on), configure the connection.

- When you configure a VPN connection using PPTP or L2TP, you will need to have a connection configured for the Internet in addition to the VPN connection.
- Setting up a computer for Internet access via dial-up is similar to configuring a computer for network access. The modem simply takes the place of the NIC.
- In Windows 95, open the Dial-Up Networking folder, which is in My Computer. Use the Make New Connection wizard to create a connection.
- In Windows XP, open My Network Places and click View Network Connections in the left pane. Then click Create a New Connection in the right pane and use the wizard to select and configure the appropriate connection.

3.8 Identify the purpose, benefits, and characteristics of using a firewall.

- A firewall can be configured as hardware, software, or a combination of both.
- Firewalls are generally routers that have two or more network interfaces. When receiving data packets on one interface, the firewall will apply access and filtering rules to the data packets before forwarding them to the next interface.
- A firewall can deny access by protocol and port number, and some can deny specific IP addresses and Web sites.
- Firewalls are essential to securing the network.

3.9 Identify the purpose, benefits, and characteristics of using a proxy.

- A proxy server makes Web requests on behalf of client Web browsers.
- In general, the proxy server has two interfaces, one that is local to the clients and the other that leads to the Internet.

- Proxy servers provide enhanced security and can increase speed to Internet access for frequently used Web sites if the proxy server caches Web site data. For proxy servers that do not cache data, and for infrequently accessed Web sites, the proxy server will slow down access.

3.10 Given a scenario, predict the impact of a particular security implementation on network functionality (i.e., blocking port numbers, encryption, and so on).

- When network security is implemented, it affects all users on the network.
- If a port is blocked on the firewall, the applications that require data across that port won't function when it's used across the firewall.
- When encryption is implemented, data might not be readable by those who haven't implemented it on the other side. For example, if an encrypted e-mail message is sent outside the network and the recipient doesn't have a way to decrypt the message, the recipient is out of luck.

3.11 Given a network configuration, select the appropriate NIC and network configuration settings (DHCP, DNS, WINS, protocols, and NETBIOS/host name).

- When you connect a Windows client to a NetWare server on a token ring network and the NetWare server uses IPX/SPX, you'd select a token ring NIC and then configure the client with IPX/SPX and the NetWare client software.
- When you connect a Windows client to a Windows server on a Fast Ethernet network and the Windows server uses TCP/IP, you'd select a 100BaseT NIC and follow it up by configuring the client with an IP address, subnet mask, default gateway, NetBIOS name, host name, and DNS server address.

Domain 4.0: Network Support

4.1 Given a troubleshooting scenario, select the appropriate TCP/IP utility from among the following.

Tracert

- Tracert is used to trace a path from the source workstation to a destination IP address.
- Tracert uses ICMP.
- Tracert is good for troubleshooting network response time problems.

Ping

- Packet Internet groper (ping) uses ICMP packets to determine the responsiveness of a destination IP address.
- Pinging the loopback address (127.0.0.1) can help test local TCP/IP configuration.
- Ping can determine whether the local computer can "talk" to another TCP/IP computer based on its address.

ARP

- ARP resolves IP addresses to MAC addresses.
- The ARP utility can check for invalid or duplicate entries in the ARP cache and can be used to delete those entries.

Netstat

- Netstat displays TCP/IP statistics and information about the TCP and UDP port connections.
- Netstat can be used to see whether an unauthorized person is connected to your computer without your knowledge.

Nbtstat

- Nbtstat displays the NetBIOS protocol statistics and any connections using NetBIOS over TCP/IP.
- Nbtstat can troubleshoot NetBIOS name resolution problems.

Ipconfig

- Ipconfig is used on Windows NT, Windows 2000, and Windows XP to display the TCP/IP configuration of network interfaces.
- Ipconfig can be used to reset and renew DHCP leases.

Ifconfig

- Ifconfig is the Unix utility for initializing and configuring network interfaces.

Winipcfg

- Winipcfg is the graphical version of Ipconfig and will display TCP/IP configuration information in addition to allowing you to reset and renew DHCP leases.
- Winipcfg is used on Windows 95, Windows 98, and Windows Me.

Nslookup

- Nslookup is used to query DNS servers in either interactive mode or non-interactive mode.

4.2 Given a troubleshooting scenario involving a small office/home office network failure (i.e., xDSL, cable, home satellite, wireless, POTS), identify the cause of the failure.

- When a small network has a failure, you must first determine whether the failure is with the Internet or the local network.
- If the Internet connection has failed, you should make sure the cables are connected, external modems are turned on, the TCP/IP configuration is correct, and the ISP is not having problems.

4.3 Given a troubleshooting scenario involving a remote connectivity problem (i.e., authentication failure, protocol configuration, physical connectivity), identify the cause of the problem.

- When you troubleshoot a remote connectivity problem, you should see at which point the process breaks down. For example, if the client can dial the connection but is then prevented from connecting, the problem is likely an authentication failure.
- Always check the username and password, the configuration for network protocols, and whether the equipment is plugged in and running.
- On Windows 95, you can look at remote connections by clicking Start, followed by Programs and then Accessories, and finally Dial-Up Networking.
- The Dial-Up Networking function for Windows 98 is located in the Communications folder within the Accessories folder, which is in the Programs folder on the Start menu.
- In Windows 2000, you can view remote connections by clicking Start, Settings, and finally Network and Dial-Up Connections.

4.4 Given specific parameters, configure a client to connect to the following servers: Unix/Linux, NetWare, Windows, and Macintosh.

- When configuring a NIC, the cabling protocol (such as Ethernet) and the network protocol (such as TCP/IP) must match the network that it will join.
- You can use an administrative account to test whether a computer can log on. To test whether a regular account can use the computer, you should have a test account with standard user rights available.
- Unix/Linux uses the TCP/IP protocol and can exist on any type of network, but you most typically will configure the workstation to connect to an Ethernet network.
- NetWare servers will require a special client access application to be installed, along with the correct protocol (usually TCP/IP or IPX/SPX), on top of the correct cabling protocol.
- Windows servers will either use NetBEUI or TCP/IP but can also use NWLink (Microsoft's own form of IPX/SPX).
- When you connect a Windows client to an AppleShare server (Macintosh), you'll use TCP/IP.

4.5 Given a wiring task, select the appropriate tool (i.e., wire crimper, media tester/certifier, punch down tool, tone generator, optical tester, and so on).

- Wire crimpers connect cables to the appropriate connectors.
- Media testers, such as Time Domain Reflectometers (TDR), send a signal down a cable to determine whether the signal bounces back because of a break or problem in the cable. They also show how far down the cable the break is located.
- Punch down tools are used to connect a cable to a wall jack.
- A tone generator tests cabling to identify which cable is the one being sought. A tone is sent down the wire, and a sound amplifier is used at the other end of the cable. When the correct cable is used with the amplifier, it creates a sound.
- An optical tester tests the signaling of fiber optic cables.

4.6 Given a network scenario, interpret visual indicators (i.e., link lights and collision lights) to determine the nature of the problem.

- Power lights glow when the unit is on.
- Link lights glow when data can transmit across the network on a particular port. If there is a problem, the link light either changes color or remains unlit.
- Collision lights blink for normal collisions but are solid if there are too many collisions. Too many collisions indicate a NIC problem.

4.7 Given output from a diagnostic utility (i.e., Tracert, ping, Ipconfig, and so on), identify the utility and interpret the output.

- The test will require you to recognize the output of ping, Ipconfig, Tracert, and Winipcfg. This output can vary somewhat. In order to recognize the output, you should practice with the tools and familiarize yourself with the results of the commands.
- When you use ping, you will see several lines stating that a reply was received in a certain time period, or that the request timed out, as shown in Figure A-6.

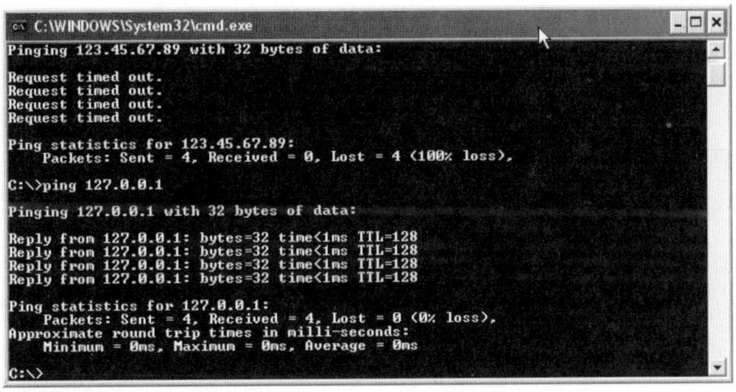

```
C:\WINDOWS\System32\cmd.exe
Pinging 123.45.67.89 with 32 bytes of data:

Request timed out.
Request timed out.
Request timed out.
Request timed out.

Ping statistics for 123.45.67.89:
    Packets: Sent = 4, Received = 0, Lost = 4 (100% loss),

C:\>ping 127.0.0.1

Pinging 127.0.0.1 with 32 bytes of data:

Reply from 127.0.0.1: bytes=32 time<1ms TTL=128
Reply from 127.0.0.1: bytes=32 time<1ms TTL=128
Reply from 127.0.0.1: bytes=32 time<1ms TTL=128
Reply from 127.0.0.1: bytes=32 time<1ms TTL=128

Ping statistics for 127.0.0.1:
    Packets: Sent = 4, Received = 4, Lost = 0 (0% loss),
Approximate round trip times in milli-seconds:
    Minimum = 0ms, Maximum = 0ms, Average = 0ms

C:\>
```

Figure A-6 Ping displays reply times and timeouts.

- When you use Tracert, you will see the words "tracing route to" followed by the name of the destination or the IP address that you specified, and then several lines stating the time that it takes for packets to reach each hop on the path to the destination.
- When using Ipconfig, you will see the IP information for each network interface in the computer, and it will be similar to the output shown in Figure A-7.

```
C:\WINDOWS\System32\cmd.exe

C:\>ipconfig /all

Windows IP Configuration

        Host Name . . . . . . . . . . . . : cx867325-a
        Primary Dns Suffix  . . . . . . . :
        Node Type . . . . . . . . . . . . : Unknown
        IP Routing Enabled. . . . . . . . : Yes
        WINS Proxy Enabled. . . . . . . . : No

Ethernet adapter Network Bridge (Network Bridge) 2:

        Connection-specific DNS Suffix  . :
        Description . . . . . . . . . . . : MAC Bridge Miniport
        Physical Address. . . . . . . . . : D2-4C-C4-7B-47-44
        Dhcp Enabled. . . . . . . . . . . : No
        IP Address. . . . . . . . . . . . : 192.168.0.1
        Subnet Mask . . . . . . . . . . . : 255.255.255.0
        Default Gateway . . . . . . . . . :

Ethernet adapter Local Area Connection 2:

        Connection-specific DNS Suffix  . :
        Description . . . . . . . . . . . : Realtek RTL8139/810X Family Fast Eth
```

Figure A-7 Ipconfig with the /all switch will display extended IP information.

4.8 Given a scenario, predict the impact of modifying, adding, or removing network services (i.e., DHCP, DNS, WINSs) on network resources and users.

- A network is somewhat symbiotic in nature. All nodes are interconnected in some way. If a network repair interferes with the network, all users will be affected. If a network repair is involved with a single node, only one computer is affected.

4.9 Given a network problem scenario, select an appropriate course of action based on a general troubleshooting strategy. This strategy includes the following steps.

1. Establish the symptoms.
2. Identify the affected area.
3. Establish what has changed.
4. Select the most probable cause.
5. Implement a solution.
6. Test the result.
7. Recognize the potential effects of the solution.
8. Document the solution.

    - When the solution is not successful, you must return to the earlier steps of selecting a probable cause and implementing a solution for that problem.
    - Do not forget the final step—documenting the solution is an important final step.
    - The entire process is shown in Figure A-8.

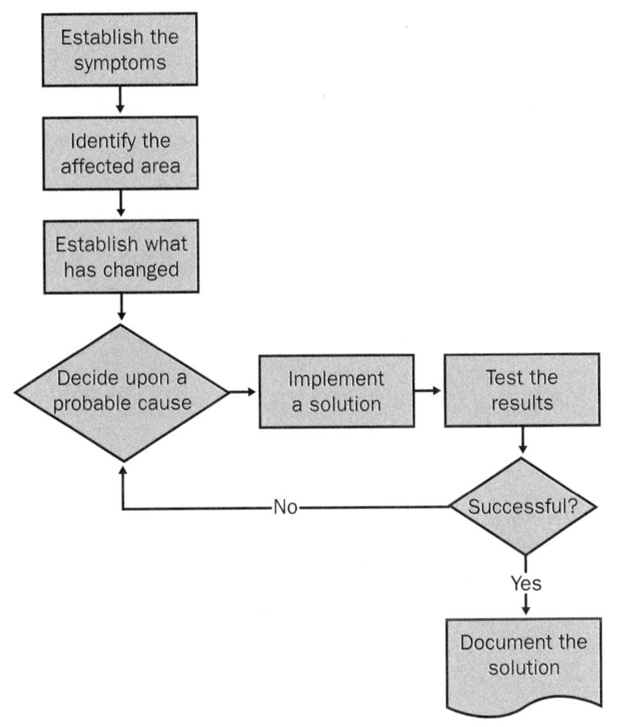

Figure A-8 The troubleshooting method is important to both physical and logical problems on a network.

4.10 Given a troubleshooting scenario involving a network with a particular physical topology (i.e., bus, star/hierarchical, mesh, ring, and wireless) and including a network diagram, identify the network area affected and the cause of the problem.

- One of the first steps to take when you discover a network problem is to isolate the affected area.
- If you can isolate the problems to a certain section of the network, then you can determine whether the hub, switch, bridge, router, or server in that area is the cause of the problem.
- Practice diagramming networks and familiarize yourself with the equipment that may govern a network area.

4.11 Given a network troubleshooting scenario involving a client connectivity problem (i.e., incorrect protocol/client software/authentication configuration or insufficient rights/permission), identify the cause of the problem.

- When a client is using the incorrect protocol, the client won't be able to log on, nor will it "see" any other computers on the network.

- If the client is using the incorrect client software but the correct protocol, the client should be able to "see" other computers; it should be able to ping other addresses on the network, for example.
- If the client has incorrect authentication information (name and password), the client will receive an error that states that the password is wrong or the username was not found.
- If the client can log on to the network but can't access a certain share, file, or application, the problem is most likely insufficient rights. To test this, try accessing the same information using an account that has administrative access, and then use a test account to which you can try adding or removing various rights until you discover which rights are required.

4.12 Given a network troubleshooting scenario involving a wiring/infrastructure problem, identify the cause of the problem (i.e., bad media, interference, network hardware).

- You should understand how to configure a NIC, whether it is an older card that uses jumpers and EPROMs or one that is plug and play.
- You should be able to perform network diagnostics on the NIC, such as a loopback test or vendor-supplied diagnostics.
- You should be able to look at the event viewer to determine whether there are IRQ, DMA, or I/O Address conflicts.
- You should be able to determine whether cabling has been damaged by checking the link lights on the NIC and the hub and performing tests using a media tester.
- When using new cables, make sure you didn't inadvertently use a crossover cable rather than a straight-through for a drop or patch cable; the crossover cable would prevent connectivity.
- Keep in mind that environmental factors, such as room conditions, building contents, and abnormal physical conditions, can cause unusual equipment breakdown.

Glossary

Numbers

10Base2 The implementation of Ethernet, or IEEE 802.3 specification, at 10 Mbps over RG-58 coaxial cable, which is also known as ThinNet.

10Base5 The implementation of Ethernet, IEEE 802.3 specification, at 10 Mbps over RG 8 coaxial cable, which is also known as ThickNet.

100BaseFX The implementation of Fast Ethernet at 100 Mbps over fiber optic cabling.

10BaseT The implementation of Ethernet, or IEEE 802.3 specification, at 10 Mbps over Category 3 or better unshielded twisted pair (UTP) cabling using a hub.

100BaseTX The implementation of Fast Ethernet at 100 Mbps over Category 5 or better UTP using a hub.

A

Active Directory The directory service used on a Microsoft Windows 2000 Server network that provides user, group, computer, resource, and security information. This directory service is a database that is distributed across the Windows 2000 servers and organized into a hierarchical tree.

adapter A board that interfaces with a computer and extends the computer's capabilities.

Address Resolution Protocol *See* ARP.

APIPA (Automatic Private IP Addressing) A failover mechanism. In the event that a DHCP server cannot be found, APIPA self-assigns an IP address from the range of 169.254.0.1 to 169.254.255.254 to a computer so that it can function on the local network.

AppleTalk The protocol stack that was developed by Apple Computer for connecting Macintosh computers in a peer-to-peer network.

application layer Layer 7 of the OSI model. The application layer provides user interface information for applications and processes.

ARP (Address Resolution Protocol) The protocol in the TCP/IP stack that resolves the IP address to a physical address. ARP is available as a text-based command to work with the cache of addresses that have been resolved.

Asynchronous Transfer Mode *See* ATM.

ATM (Asynchronous Transfer Mode) A high-speed protocol used for WANs and backbone networks. ATM uses a small packet called a cell that is 53 bytes in length, which enables it to reach up to 622 Mbps in speed.

Automatic Private IP Addressing *See* APIPA.

B

B channel (bearer channel) As used in ISDN, a B channel carries 64 Kbps of data or digital voice traffic.

backbone The main network link that interconnects smaller networks.

Backup Domain Controller *See* BDC.

bandwidth The maximum rate of data transmission supported by a network media type.

Basic Rate Interface *See* BRI.

BDC (Backup Domain Controller) In Windows NT networks, a BDC provides redundancy because it holds a read-only copy of the security information in the SAM, or security account manager.

bearer channel *See* B channel.

binary The numeric system of base 2, consisting of zeros and ones.

Bindery The flat file database of user, group, and security information that is used on NetWare 3.*x* and earlier servers. The Bindery is a server-centric database, which means that users must have a username and password for each different server.

BRI (Basic Rate Interface) The ISDN specification, with two B channels of 64 Kbps each and one D channel of 16 Kbps, that runs over standard telephone wiring.

bridge A device used to connect segments of a LAN and forward data packets based on the destination MAC address in the frame header.

Brouter A device that connects two segments and either forwards the data based on the MAC address of the frame header or routes the packets based on the logical network address in the network-layer header. Brouters combine the functions of a bridge and a router.

bus topology A physical network topology that uses a main cable trunk to connect all workstations. Logically, the bus topology forwards data to all connected stations.

C

cable modem A piece of equipment that connects a network to the Internet via cable TV service.

Carrier Sense Multiple Access/Collision Avoidance *See* CSMA/CA.

Carrier Sense Multiple Access/Collision Detection *See* CSMA/CD.

cell A fixed-length packet that is used to transport data, voice, or video. In ATM, the cell is 53 bytes in length.

cellular topology A wireless network that provides for wireless access points that are connected together, providing "cells" of physical area in which data transmission is available.

Challenge Handshake Authentication Protocol *See* CHAP.

Channel Service Unit/Data Service Unit *See* CSU/DSU.

CHAP (Challenge Handshake Authentication Protocol) The authentication protocol that challenges a user to provide a username and password.

CIDR (classless interdomain routing) Developed in response to a shortage of IPv4 addresses, CIDR uses addressing space by removing the boundaries of class A, B, and C networks. CIDR notation is in the form of an IP address followed by a slash and then the number of digits (between 1 and 31) that are assigned to the subnet mask, such as 123.45.67.89/24.

class A address The TCP/IP address classification of 126 networks with addresses in the range of 1.x.x.x to 126.x.x.x.

class B address The TCP/IP address classification of 16,384 networks with addresses in the range of 128.x.x.x to 191.x.x.x.

class C address The TCP/IP address classification of 2,097,152 networks with addresses in the range of 192.x.x.x to 223.x.x.x.

classless interdomain routing *See* CIDR.

client/server The type of network system that involves dedicated servers providing resources to be accessed by client workstations.

coax (coaxial cable) A type of cabling made of concentric conductors and insulation. The central conducting wire is surrounded by insulation, which is then surrounded by a copper mesh conductor and finally covered by the outer insulation.

coaxial cable *See* coax.

connection-oriented The method of data transmission that guarantees delivery using extra information in data headers and provides a virtual connection between sending and receiving devices.

connectionless The method of data transmission that does not guarantee data delivery but through the reduced overhead (because of streamlined header information) provides for a fast transmission.

CSMA/CA (Carrier Sense Multiple Access/Collision Avoidance) A contention media access method that looks for potential collisions before transmitting data in order to avoid them.

CSMA/CD (Carrier Sense Multiple Access/Collision Detection) A contention media access method that recognizes when a collision has taken place and provides for retransmission of the data based on a random waiting period.

CSU/DSU (Channel Service Unit/ Data Service Unit) The device used to transmit to a frame relay WAN link.

D

D channel (data channel) As used in ISDN, a D channel carries either 16 Kbps or 64 Kbps of control information. The D channel carries the information that builds, maintains, and terminates B channel connections.

data channel *See* D channel.

data-link layer Layer 2 of the OSI model, with two sublayers: the Media Access Control (MAC) and Logical Link Control (LLC). This layer provides the framing of data and physical addressing.

default gateway The address of the router where packets are sent to devices that are not on the local segment.

DHCP (Dynamic Host Configuration Protocol) The protocol that enables a computer to obtain an IP address on a leased basis from a server that holds a pool of addresses, along with extended IP information such as the subnet mask, default gateway address, and so on.

Domain Name System *See* DNS.

DNS (Domain Name System) The hierarchical naming system that provides resolution of a computer's name to its IP address.

DSL (digital subscriber line) A high-speed digital network available over standard telephone wiring that integrates both voice and data transmission.

Dynamic Host Configuration Protocol *See* DHCP.

E

electromagnetic interference *See* EMI.

EMI (electromagnetic interference) The noise or static created by a current-producing apparatus such as fluorescent lighting, which affects the quality of electric signals on a network.

encapsulation The process of enclosing data from an upper-layer protocol within new header information.

encryption The process of securing data packets prior to transmission using an encoding method. At the receiving end, the data must be decrypted using a decoding method.

Ethernet The transmission specification at the physical and data-link layer that can extend up to 1 Gbps (for Gigabit Ethernet) as a data rate, and transmit across copper coax, UTP, and fiber optic media using CSMA/CD.

F

Fast Ethernet *See* 100BaseTX.

fault tolerance The use of redundant components to make a network, or server, resistant to failures.

FDDI (Fiber Distributed Data Interface) A network technology using token-passing media access across a dual ring topology running on fiber optic media, with a speed of 100 Mbps.

Fiber Distributed Data Interface *See* FDDI.

fiber optic cabling A network media cable made of a glass core enclosed within a glass tube and covered by an insulating plastic cover.

File Transfer Protocol *See* FTP.

firewall A router that filters data packets as they enter a private network in order to protect the network from unauthorized access.

frame relay A WAN specification that works at the physical and data-link layers of the OSI model. Frame relay provides connection-oriented data transmission at rates up to 1.544 Mbps.

FTP (File Transfer Protocol) This protocol in the TCP/IP stack enables a client to transfer files to and from an FTP server.

G

gateway A network device that translates between dissimilar network protocols and types. A gateway operates at the upper layers (application, presentation, session, and transport layers) of the OSI model.

Gigabit Ethernet The Ethernet specification that provides 1000 Mbps (or 1 Gbps) transmission speed over fiber optic or Cat5 or better grade copper cabling.

H

hexadecimal The numeric system using base 16. Hexadecimal uses the numerals 0 through 9 and the characters A, B, C, D, E, and F to provide all 16 values.

hop A router that data packets must pass through on their way to their destination.

HTTP (Hypertext Transfer Protocol) The application-layer protocol in the TCP/IP stack that provides the communications to download Hypertext Markup Language (HTML) documents from a Web server into a Web browser.

HTTPS (Hypertext Transfer Protocol over Secure Sockets Layer) This protocol is an extension of HTTP that ensures secure communications across the World Wide Web.

hub A repeater that connects multiple cables in a physical star topology. Most hubs are Ethernet using UTP with RJ-45 connectors.

Hypertext Transfer Protocol *See* HTTP.

Hypertext Transfer Protocol over Secure Sockets Layer *See* HTTPS.

I

ICMP (Internet Control Message Protocol) An error and status information protocol in the TCP/IP suite.

IEEE (Institute of Electrical and Electronics Engineers) The professional organization that develops various computer networking standards.

Institute of Electrical and Electronics Engineers *See* IEEE.

Integrated Services Digital Network *See* ISDN.

interface The part of a network device that connects the device to the network. Interfaces can be integrated or installed as separate adapters.

Internet Control Message Protocol *See* ICMP.

Internet Protocol *See* IP.

Internet Protocol Security *See* IPSec.

Internet Service Provider *See* ISP.

Internetwork Packet Exchange *See* IPX.

IP (Internet Protocol) The connectionless network-layer protocol in the TCP/IP stack. IP provides the logical network and node addressing that allow data to be routed.

IPSec (Internet Protocol Security) A protocol within the TCP/IP stack that provides strong security through encryption and authentication. Designed mainly for use in virtual private networks.

IPX (Internetwork Packet Exchange) The connectionless network-layer protocol used primarily by Novell NetWare networks.

ISDN (Integrated Services Digital Network) A network technology that utilizes existing telephone wiring and provides digital transmission of voice and data at two speeds: the BRI offers 128 Kbps, and PRI offers 1.544 Mbps over a T1 leased line.

ISP (Internet Service Provider) A company that provides Internet connectivity.

L

L2TP (Layer 2 Tunneling Protocol) A protocol in the TCP/IP stack that provides virtual private network tunneling. L2TP is used with IPSec to ensure secure communications.

LAN (local area network) A network of computers that are in close proximity to each other.

LLC sublayer (Logical Link Control sublayer) The subset of the data-link layer of the OSI model that prepares data frames for translation to the network layer.

local area network *See* LAN.

Logical Link Control sublayer *See* LLC sublayer.

loopback address An IP address of 127.0.0.1 representing the same PC. A packet sent to 127.0.0.1 circles back to the same interface from which it was sent. This process can be used to test a network device to validate that TCP/IP has been installed correctly.

M

MAC address (Media Access Control address) The physical address of the network interface assigned at the data-link layer of the OSI model. MAC addresses are unique and consist of six octets written in hexadecimal. An example is 00-FE-A1-02-89-3C.

MAC sublayer (Media Access Control sublayer) A subset of the data-link layer that provides for the physical address of a network interface.

MAU (Multistation Access Unit) The central unit in a token ring physical star topology. The MAU resembles a hub but provides a logical ring of data transmission, including a ring-in and ring-out port for extending the logical ring to another MAU.

media The transport system, whether wiring, fiber optic, or air, that enables signals to be sent from one location to another in a network.

Media Access Control address *See* MAC address.

Media Access Control sublayer *See* MAC sublayer.

media access method The way in which a computer prepares to transmit data across the network media.

mesh topology A physical network topology that provides for redundant network connections from each network device to all other network devices.

modem (modulator/demodulator) The piece of equipment that converts data from digital form to analog to transmit it across the telephone network. Modems transmit at speeds up to 56 Kbps.

modulator/demodulator *See* modem.

MSAU *See* MAU.

Multistation Access Unit *See* MAU.

N

NAT (network address translation) The system whereby IP addresses on the private network are translated to the IP addresses used on the public network.

NDS (Novell Directory Services) The directory service of users, groups, computers, and security information that is distributed across NetWare 4.*x* and later servers. The directory service is organized into a hierarchical tree.

NetBEUI (NetBIOS Enhanced User Interface) The nonroutable protocol used traditionally in Windows NT networks and LAN Manager networks. NetBEUI is not scalable and is recommended only for small networks.

NetBIOS Extended User Interface *See* NetBEUI.

NetWare Link *See* NWLink.

network address translation *See* NAT.

Network File System *See* NFS.

network interface card *See* NIC.

network layer Layer 3 of the OSI model, which provides for packet addressing. Routers operate at the network layer.

NFS (Network File System) The file system developed for Unix that provides transparent access to remote files.

NIC (network interface card) The adapter card that extends the functionality of a computer so that it can access a network.

node A computer, host, or other device on the network that provides or consumes network services.

Novell Directory Services *See* NDS.

NT file system *See* NTFS.

NTFS (NT file system) The file system used on Windows NT and Windows 2000 computers.

NWLink (NetWare Link) A protocol compatible with IPX/SPX that is provided on Windows computers and can be used to connect to networks that use IPX/SPX, especially NetWare networks.

O

octet A group of eight bits.

OCx (Optical Carrier) A system of levels of transmission speeds over a fiber optic network such as SONET. OC-1 offers 51.5 Mbps, and each level is a multiple of this value. For example, OC-12 is 622 Mbps.

Open Systems Interconnection *See* OSI.

Optical Carrier *See* OCx.

OSI (Open Systems Interconnection) The abstract model that defines a system of seven layered protocols that work together to enable full internetwork communication.

P

Packet Internet Groper *See* ping.

PDC (primary domain controller) On a Windows NT network, the PDC holds the main repository of security information for the domain.

peer to peer A system of networking whereby any network node can be a server or a client to the other nodes.

physical layer Layer 1 of the OSI reference model, which defines the media, signaling, and connectors to the network. Hubs and repeaters work at the physical layer.

ping (Packet Internet Groper) A utility in the TCP/IP stack that uses an echo and response to test connectivity to remote systems.

Point-to-Point Protocol *See* PPP.

Point-to-Point Tunneling Protocol *See* PPTP.

port number The value assigned to data to identify which application that data is intended for.

PPP (Point-to-Point Protocol) A protocol used for dial-up and VPN connections, commonly used for the Internet.

PPTP (Point-to-Point Tunneling Protocol) A tunneling protocol used in a VPN that encapsulates data in IP packets.

presentation layer Layer 6 of the OSI model, which defines the way data is presented to the application layer, including how data is compressed, expanded, encrypted, or decrypted and which character set is used.

PRI (Primary Rate Interface) Used in ISDN, PRI offers 23 B channels and 1 D channel, all at the rate of 64 Kbps, to provide a total of 1.544 Mbps.

primary domain controller *See* PDC.

Primary Rate Interface *See* PRI.

protocol The set of rules or standards governing the transmission of data on a network.

R

radio frequency interference *See* RFI.

RAID (redundant array of independent disks) A fault-tolerant way to provide multiple disks in configurations intended to allow redundancy. RAID 0 is disk striping without redundancy. RAID 1 is disk mirroring. RAID 5 is three or more disks striped with parity.

redundant array of independent disks *See* RAID.

repeater The device that receives a signal from one port, copies it, and regenerates the signal to send out its other port.

remote node A computer that connects to a private network through dial-up or VPN connection.

RFI (radio frequency interference) The noise created by radio waves caused by wireless devices such as wireless telephones that affect the signal in a wireless network.

ring topology The physical topology of computers daisy-chained in a ring configuration. A logical ring topology will transmit data from station to station in a logical ring; it can do so on a physical star topology as well as a physical ring.

router A network device that moves data between networks based on the logical addresses in the network layer of the data packets.

S

SAM (Security Account Manager) The database containing user, group, computer, and security information in a Windows NT domain.

SAN (Storage Area Network) A separate network of data storage units that sits outside the LAN and provides data access through servers that connect both to the SAN and the LAN.

Security Account Manager *See* SAM.

session layer The OSI model layer that is responsible for the initiation, maintenance, and termination of sessions between applications on separate computers.

SONET (Synchronous Optical Network) A physical-layer network technology designed to carry large data transmissions over fiber optic cabling.

star topology A physical topology consisting of a central hub with individual cables extending to network devices.

Storage Area Network *See* SAN.

subnet mask A 32-bit number that is used to screen a portion of the IP address to differentiate the part considered the logical network address from the part that is the logical node address.

Synchronous Optical Network *See* SONET.

T

T1 A digital leased line offered by telephone carriers used for WANs providing 1.544 Mbps.

TCP (Transmission Control Protocol) A connection-oriented transport-layer protocol used in the TCP/IP stack.

Telnet A protocol in the TCP/IP stack that provides character-based terminal emulation.

ThickNet Copper coaxial cabling with a thick outer shielding, also known as either RG-8 or RG-10

ThinNet Copper coaxial cabling with a flexible outer shielding, also known as RG-58

token ring A physical and data-link layer specification that provides for data transmission in a logical ring using a token-passing media access method.

Transmission Control Protocol *See* TCP.

transport layer Layer 4 of the OSI model, which receives information from upper layers, segments them into a data stream, and provides end-to-end transport service with either guaranteed or non-guaranteed data delivery.

U

UDP (User Datagram Protocol) A transport-layer protocol in the TCP/IP stack that provides connectionless data delivery.

unshielded twisted pair *See* UTP.

User Datagram Protocol *See* UDP.

UTP (unshielded twisted pair) A type of cabling with multiple internal wires twisted in pairs.

V

virtual local area network *See* VLAN.

virtual private network *See* VPN.

VLAN (virtual local area network) A logical grouping of stations by MAC address or IP address to provide segmentation of traffic on a switch.

VPN (virtual private network) A connection that tunnels through a public network to securely transmit data to a private network.

W

WAN (wide area network) A network that spans a large geographic area.

WAP (wireless access point) A device that provides wireless connectivity to wireless devices within a certain range.

wide area network *See* WAN.

Windows Internet Name Service *See* WINS.

WINS (Windows Internet Name Service) A Windows network service that resolves NetBIOS names to IP addresses.

wireless access point *See* WAP.

Index

Numerics

A

B

C

D

Q-R

S

T

U

V

W

X-Z

About the CD

The companion CD contains a fully searchable electronic version of this book. It also contains an electronic assessment, which allows you to take a timed practice test that prepares you for the actual Network+ Certification exam. Explanations of the answers to each question on the assessment can be found in a separate file on the CD.

Follow these instructions to use the companion CD:

1. Insert the companion CD into your CD drive.
2. If a license agreement does not appear automatically, double-click StartCD.exe in the root folder of the CD. Accept the license agreement to display the starting menu screen.

   The menu provides you with links to all the resources available on the CD.

System Requirements

To install and run the contents of this CD, your system must meet the following minimum requirements:

- Operating system:
  - Microsoft Windows 95 or Microsoft Windows NT 4 with Service Pack 6a or later
  - Microsoft Windows 98, Microsoft Windows Me or Microsoft Windows 2000
  - Microsoft Windows XP or Microsoft Windows Server 2003
- Pentium class with 166 megahertz (MHz) or higher processor
- Memory required:
  - Microsoft Windows 95: 12 MB RAM
  - Microsoft Windows 98, Me, and NT 4 SP6a: 16 MB RAM
  - Microsoft Windows 2000, Microsoft Windows XP and Microsoft Windows Server 2003: 64 MB RAM

- Microsoft Internet Explorer 5.01 or later. The electronic assessment requires a full installation of Internet Explorer 5.01 or later. If you do not have Internet Explorer 5.01 or later, you can install Internet Explorer 6 from the Companion CD.
- Hard disk space required:
  - To install the eBook: 10 MB
  - To install Internet Explorer 6 SP1 from this CD: 32 MB
  - To install the electronic assessment: 17 MB
- A double-speed CD drive or better
- 800x600 with high color (16-bit) display settings
- Microsoft Mouse or compatible pointing device

Melissa Craft

Melissa Craft (CCNA, MCNE, MCSE, Network+, CNE-3, CNE-4, CNE-GW, CNE-5, CCA) is the vice president and CIO for Dane Holdings, Inc., a financial services corporation located in Phoenix, Arizona, where she manages the Web development, LAN, and WAN for the company. During her career, Melissa has developed enterprise-wide technology solutions and methodologies focused on client organizations. These solutions touch every part of a system's life cycle, from assessing the need, determining the return on investment, network design, testing, and implementation to operational management and strategic planning.

In 1997, Melissa began writing magazine articles on networking and the information technology industry. In 1998, Syngress Publishing hired Melissa to contribute to an MCSE certification guide. Since then, she has continued to write about various technology and certification subjects.

Melissa holds a bachelor's degree from the University of Michigan and is a member of the IEEE, the Society of Women Engineers, and American Mensa, Ltd. Melissa lives in Glendale, Arizona, with her family, Dan, Justine, and Taylor.

The manuscript for this book was prepared and submitted to Microsoft Press in electronic form. Pages were composed by Microsoft Press using Adobe FrameMaker+SGML for Windows, with text in Garamond and display type in ITC Franklin Gothic Condensed. Composed pages were delivered to the printer as electronic pre-press files.

Cover designer:	Tim Girvin Design
Interior Graphic Designer:	James D. Kramer
Principal Compositor:	Dan Latimer
Interior Graphic Artist:	Michael Kloepfer
Principal Proofreader:	nSight, Inc.
Indexer:	Pamona Corporation

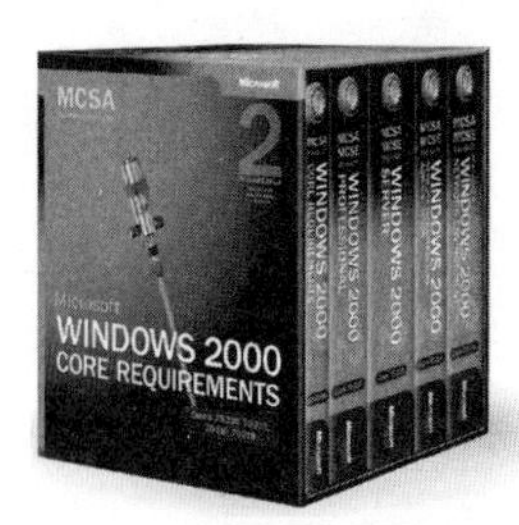

MCSA
WINDOWS 2000
CORE REQUIREMENTS

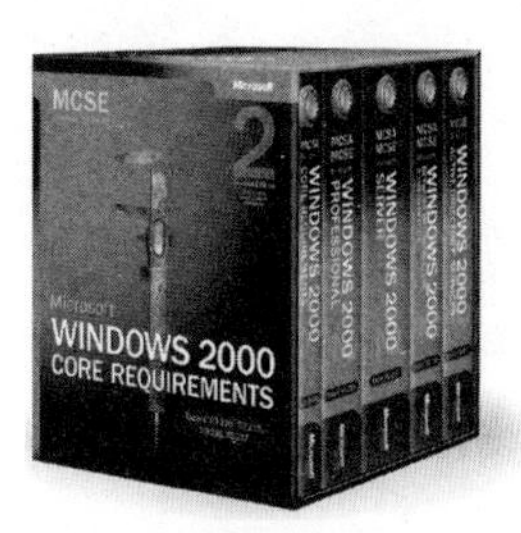

MCSE
WINDOWS 2000
CORE REQUIREMENTS

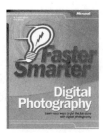

Proof of Purchase

Use this page as proof of purchase if participating in a promotion or rebate offer on this title. Proof of purchase must be used in conjunction with other proof(s) of payment such as your dated sales receipt—see offer details.

Faster Smarter Network+ Certification

0-7356-1932-8

____

CUSTOMER NAME

Microsoft Press, PO Box 97017, Redmond, WA 98073-9830

MICROSOFT LICENSE AGREEMENT

Book Companion CD

IMPORTANT—READ CAREFULLY: This Microsoft End-User License Agreement ("EULA") is a legal agreement between you (either an individual or an entity) and Microsoft Corporation for the Microsoft product identified above, which includes computer software and may include associated media, printed materials, and "online" or electronic documentation ("SOFTWARE PRODUCT"). Any component included within the SOFTWARE PRODUCT that is accompanied by a separate End-User License Agreement shall be governed by such agreement and not the terms set forth below. By installing, copying, or otherwise using the SOFTWARE PRODUCT, you agree to be bound by the terms of this EULA. If you do not agree to the terms of this EULA, you are not authorized to install, copy, or otherwise use the SOFTWARE PRODUCT; you may, however, return the SOFTWARE PRODUCT, along with all printed materials and other items that form a part of the Microsoft product that includes the SOFTWARE PRODUCT, to the place you obtained them for a full refund.

SOFTWARE PRODUCT LICENSE

The SOFTWARE PRODUCT is protected by United States copyright laws and international copyright treaties, as well as other intellectual property laws and treaties. The SOFTWARE PRODUCT is licensed, not sold.

1. **GRANT OF LICENSE.** This EULA grants you the following rights:

   a. **Software Product.** You may install and use one copy of the SOFTWARE PRODUCT on a single computer. The primary user of the computer on which the SOFTWARE PRODUCT is installed may make a second copy for his or her exclusive use on a portable computer.

   b. **Storage/Network Use.** You may also store or install a copy of the SOFTWARE PRODUCT on a storage device, such as a network server, used only to install or run the SOFTWARE PRODUCT on your other computers over an internal network; however, you must acquire and dedicate a license for each separate computer on which the SOFTWARE PRODUCT is installed or run from the storage device. A license for the SOFTWARE PRODUCT may not be shared or used concurrently on different computers.

   c. **License Pak.** If you have acquired this EULA in a Microsoft License Pak, you may make the number of additional copies of the computer software portion of the SOFTWARE PRODUCT authorized on the printed copy of this EULA, and you may use each copy in the manner specified above. You are also entitled to make a corresponding number of secondary copies for portable computer use as specified above.

   d. **Sample Code.** Solely with respect to portions, if any, of the SOFTWARE PRODUCT that are identified within the SOFTWARE PRODUCT as sample code (the "SAMPLE CODE"):

      i. **Use and Modification.** Microsoft grants you the right to use and modify the source code version of the SAMPLE CODE, *provided* you comply with subsection (d)(iii) below. You may not distribute the SAMPLE CODE, or any modified version of the SAMPLE CODE, in source code form.

      ii. **Redistributable Files.** Provided you comply with subsection (d)(iii) below, Microsoft grants you a nonexclusive, royalty-free right to reproduce and distribute the object code version of the SAMPLE CODE and of any modified SAMPLE CODE, other than SAMPLE CODE, or any modified version thereof, designated as not redistributable in the Readme file that forms a part of the SOFTWARE PRODUCT (the "Non-Redistributable Sample Code"). All SAMPLE CODE other than the Non-Redistributable Sample Code is collectively referred to as the "REDISTRIBUTABLES."

      iii. **Redistribution Requirements.** If you redistribute the REDISTRIBUTABLES, you agree to: (i) distribute the REDISTRIBUTABLES in object code form only in conjunction with and as a part of your software application product; (ii) not use Microsoft's name, logo, or trademarks to market your software application product; (iii) include a valid copyright notice on your software application product; (iv) indemnify, hold harmless, and defend Microsoft from and against any claims or lawsuits, including attorney's fees, that arise or result from the use or distribution of your software application product; and (v) not permit further distribution of the REDISTRIBUTABLES by your end user. Contact Microsoft for the applicable royalties due and other licensing terms for all other uses and/or distribution of the REDISTRIBUTABLES.

2. **DESCRIPTION OF OTHER RIGHTS AND LIMITATIONS.**

   - **Limitations on Reverse Engineering, Decompilation, and Disassembly.** You may not reverse engineer, decompile, or disassemble the SOFTWARE PRODUCT, except and only to the extent that such activity is expressly permitted by applicable law notwithstanding this limitation.

   - **Separation of Components.** The SOFTWARE PRODUCT is licensed as a single product. Its component parts may not be separated for use on more than one computer.

   - **Rental.** You may not rent, lease, or lend the SOFTWARE PRODUCT.

   - **Support Services.** Microsoft may, but is not obligated to, provide you with support services related to the SOFTWARE PRODUCT ("Support Services"). Use of Support Services is governed by the Microsoft policies and programs described in the user manual, in "online" documentation, and/or in other Microsoft-provided materials. Any supplemental software code provided to you as part of the Support Services shall be considered part of the SOFTWARE PRODUCT and subject to the terms and conditions of this EULA. With respect to technical information you provide to Microsoft as part of the Support Services, Microsoft may use such information for its business purposes, including for product support and development. Microsoft will not utilize such technical information in a form that personally identifies you.

- **Software Transfer.** You may permanently transfer all of your rights under this EULA, provided you retain no copies, you transfer all of the SOFTWARE PRODUCT (including all component parts, the media and printed materials, any upgrades, this EULA, and, if applicable, the Certificate of Authenticity), **and** the recipient agrees to the terms of this EULA.
- **Termination.** Without prejudice to any other rights, Microsoft may terminate this EULA if you fail to comply with the terms and conditions of this EULA. In such event, you must destroy all copies of the SOFTWARE PRODUCT and all of its component parts.

3. **COPYRIGHT.** All title and copyrights in and to the SOFTWARE PRODUCT (including but not limited to any images, photographs, animations, video, audio, music, text, SAMPLE CODE, REDISTRIBUTABLES, and "applets" incorporated into the SOFTWARE PRODUCT) and any copies of the SOFTWARE PRODUCT are owned by Microsoft or its suppliers. The SOFTWARE PRODUCT is protected by copyright laws and international treaty provisions. Therefore, you must treat the SOFTWARE PRODUCT like any other copyrighted material **except** that you may install the SOFTWARE PRODUCT on a single computer provided you keep the original solely for backup or archival purposes. You may not copy the printed materials accompanying the SOFTWARE PRODUCT.

4. **U.S. GOVERNMENT RESTRICTED RIGHTS.** The SOFTWARE PRODUCT and documentation are provided with RESTRICTED RIGHTS. Use, duplication, or disclosure by the Government is subject to restrictions as set forth in subparagraph (c)(1)(ii) of the Rights in Technical Data and Computer Software clause at DFARS 252.227-7013 or subparagraphs (c)(1) and (2) of the Commercial Computer Software—Restricted Rights at 48 CFR 52.227-19, as applicable. Manufacturer is Microsoft Corporation/One Microsoft Way/Redmond, WA 98052-6399.

5. **EXPORT RESTRICTIONS.** You agree that you will not export or re-export the SOFTWARE PRODUCT, any part thereof, or any process or service that is the direct product of the SOFTWARE PRODUCT (the foregoing collectively referred to as the "Restricted Components"), to any country, person, entity, or end user subject to U.S. export restrictions. You specifically agree not to export or re-export any of the Restricted Components (i) to any country to which the U.S. has embargoed or restricted the export of goods or services, which currently include, but are not necessarily limited to, Cuba, Iran, Iraq, Libya, North Korea, Sudan, and Syria, or to any national of any such country, wherever located, who intends to transmit or transport the Restricted Components back to such country; (ii) to any end user who you know or have reason to know will utilize the Restricted Components in the design, development, or production of nuclear, chemical, or biological weapons; or (iii) to any end user who has been prohibited from participating in U.S. export transactions by any federal agency of the U.S. government. You warrant and represent that neither the BXA nor any other U.S. federal agency has suspended, revoked, or denied your export privileges.

DISCLAIMER OF WARRANTY

NO WARRANTIES OR CONDITIONS. MICROSOFT EXPRESSLY DISCLAIMS ANY WARRANTY OR CONDITION FOR THE SOFTWARE PRODUCT. THE SOFTWARE PRODUCT AND ANY RELATED DOCUMENTATION ARE PROVIDED "AS IS" WITHOUT WARRANTY OR CONDITION OF ANY KIND, EITHER EXPRESS OR IMPLIED, INCLUDING, WITHOUT LIMITATION, THE IMPLIED WARRANTIES OF MERCHANTABILITY, FITNESS FOR A PARTICULAR PURPOSE, OR NONINFRINGEMENT. THE ENTIRE RISK ARISING OUT OF USE OR PERFORMANCE OF THE SOFTWARE PRODUCT REMAINS WITH YOU.

LIMITATION OF LIABILITY. TO THE MAXIMUM EXTENT PERMITTED BY APPLICABLE LAW, IN NO EVENT SHALL MICROSOFT OR ITS SUPPLIERS BE LIABLE FOR ANY SPECIAL, INCIDENTAL, INDIRECT, OR CONSEQUENTIAL DAMAGES WHATSOEVER (INCLUDING, WITHOUT LIMITATION, DAMAGES FOR LOSS OF BUSINESS PROFITS, BUSINESS INTERRUPTION, LOSS OF BUSINESS INFORMATION, OR ANY OTHER PECUNIARY LOSS) ARISING OUT OF THE USE OF OR INABILITY TO USE THE SOFTWARE PRODUCT OR THE PROVISION OF OR FAILURE TO PROVIDE SUPPORT SERVICES, EVEN IF MICROSOFT HAS BEEN ADVISED OF THE POSSIBILITY OF SUCH DAMAGES. IN ANY CASE, MICROSOFT'S ENTIRE LIABILITY UNDER ANY PROVISION OF THIS EULA SHALL BE LIMITED TO THE GREATER OF THE AMOUNT ACTUALLY PAID BY YOU FOR THE SOFTWARE PRODUCT OR US$5.00; PROVIDED, HOWEVER, IF YOU HAVE ENTERED INTO A MICROSOFT SUPPORT SERVICES AGREEMENT, MICROSOFT'S ENTIRE LIABILITY REGARDING SUPPORT SERVICES SHALL BE GOVERNED BY THE TERMS OF THAT AGREEMENT. BECAUSE SOME STATES AND JURISDICTIONS DO NOT ALLOW THE EXCLUSION OR LIMITATION OF LIABILITY, THE ABOVE LIMITATION MAY NOT APPLY TO YOU.

MISCELLANEOUS

This EULA is governed by the laws of the State of Washington USA, except and only to the extent that applicable law mandates governing law of a different jurisdiction.

Should you have any questions concerning this EULA, or if you desire to contact Microsoft for any reason, please contact the Microsoft subsidiary serving your country, or write: Microsoft Sales Information Center/One Microsoft Way/Redmond, WA 98052-6399.

PN 097-0002296